Fundamentals of Numerical Weather Prediction

Numerical models have now become essential tools in environmental science, particularly in weather forecasting and climate prediction. This book provides a comprehensive overview of the techniques used in these fields, with emphasis on the design of the most recent numerical models of the atmosphere. It presents a short history of numerical weather prediction and its evolution, before providing step-by-step descriptions of the various model equations and how to solve them numerically. It outlines the main elements of a meteorological forecast suite, and the theory is illustrated throughout with practical examples of operational models and parameterizations of the main physical processes.

This book is founded on the author's many years of experience, working as a scientist at Météo-France and teaching university-level courses. It is a practical and accessible textbook for graduate courses and is a handy resource for researchers and professionals in atmospheric physics, meteorology, and climatology, as well as the related disciplines of fluid dynamics, hydrology, and oceanography.

Jean Coiffier is now retired from Météo-France, where he was Ingénieur en Chef des Ponts et Chaussées, and is a member of the Société Météorologique de France. His involvement in meteorological science began in 1968 at the new Algerian Meteorological Service, implementing elements of a modest meteorological forecast suite on a small computer, before joining the Direction de la Météorologie Nationale (later Météo-France) where he took part in the development and implementation of operational models. In 1989, he became the head of the General Forecast Office. He worked there until his retirement, also giving regular lectures on numerical weather prediction to students of the École Nationale de la Météorologie and training courses to professional forecasters. He also played an active role in realizing Computer Aided Learning modules devoted to numerical modelling and forecasting methods.

Fundamentals of Numerical Weather Prediction

JEAN COIFFIER

formerly at Météo-France

Translated by

CHRISTOPHER SUTCLIFFE

CAMBRIDGE
UNIVERSITY PRESS

CAMBRIDGE
UNIVERSITY PRESS

University Printing House, Cambridge CB2 8BS, United Kingdom

One Liberty Plaza, 20th Floor, New York, NY 10006, USA

477 Williamstown Road, Port Melbourne, VIC 3207, Australia

314-321, 3rd Floor, Plot 3, Splendor Forum, Jasola District Centre, New Delhi - 110025, India

103 Penang Road, #05-06/07, Visioncrest Commercial, Singapore 238467

Cambridge University Press is part of the University of Cambridge.

It furthers the University's mission by disseminating knowledge in the pursuit of
education, learning and research at the highest international levels of excellence.

www.cambridge.org
Information on this title: www.cambridge.org/9781107001039

First published by Météo-France, Paris as *Les bases de la prévision numérique du temps,* 2009
This edition in English 2011

Ouvrage publié avec le concours du Centre national du livre – ministère français chargé de la Culture
This edition published with the support of the Centre national du livre – French Ministry of Culture

A catalogue record for this publication is available from the British Library

Library of Congress Cataloging in Publication data
Coiffier, Jean.
[Bases de la prévision numérique du temp. English]
Fundamentals of numerical weather prediction / Jean Coiffier ; translated by Christopher Sutcliffe.
p   cm.
Includes bibliographical references and index.
ISBN 978-1-107-00103-9
1. Numerical weather forecasting.   2. Weather forecasting – Mathematical models.   I. Title.
QC996.C65   2011
551.63´4–dc23   2011026294

ISBN 978-1-107-00103-9 Hardback

Contents

Foreword to the French Edition

It is a pleasure to present this book by Jean Coiffier. After many years of teaching at the *École Nationale de la Météorologie*, he wanted to give a lasting form to the work at which he had laboured so long. The result is what is probably the first book (at any rate it is the first in French) to present the fundamentals and the current methods of numerical weather prediction in a comprehensive and consistent way.

Numerous books, some of them excellent, deal with dynamic meteorology, the science of the moving atmosphere. But although numerical modelling has become a fundamental tool not just for weather forecasting but also for dynamic meteorology itself as well as for oceanography and climate studies, there is no one book that presents the methods and techniques in a systematic manner. New users of numerical models of the atmosphere, whether students on placement or experienced scientists making career changes, find it hard to make their way into the world of numerical modelling. It is not that documentation is lacking; most models now come with detailed guides to help users take their first steps; and copious academic literature including some outstanding papers is also readily available.

What is lacking is an overview of the techniques and methods; and that is what *Fundamentals of Numerical Weather Prediction* provides. Without requiring from readers more knowledge than the standard science graduate should have acquired, Jean Coiffier guides them with pragmatism and pedagogy. The main points are explained in simple but, at the same time, clear and rigorous terms. A few years ago, I proposed to a group of students at the *École Polytechnique* to develop a general circulation model of the atmosphere from scratch as a shared coursework project. It was Jean Coiffier's lecture notes (those on which this book is based) that appeared to me the most suitable document to take them straight to the crux of what they needed to learn.

Numerical weather prediction is now more than 50 years old. One might wonder what its pioneers would make of what it has become. Perhaps they would be surprised at how the definition of initial conditions and data assimilation have grown steadily more important over the years. Jean Coiffier deals with all these aspects of numerical modelling, introducing the essential points of assimilation methods (although so very different from what he describes elsewhere) with the same precision and the same rigour as he presents the discretization algorithms.

Jean Coiffier is perhaps a little too modest. He confines the title of his book to 'numerical weather prediction,' but the methods set out in it are the same as are used for numerical modelling of the climate, which is now so crucial for forecasting the changes in store for us and for preventing or at least mitigating their harmful effects.

This book is aimed at students and it will be very helpful to them. But I am sure that a number of engineers and experienced scientists too will be using it, for they can be sure to find in it clear answers to precise questions. I am confident that many will join me in thanking Jean Coiffier for his *Fundamentals of Numerical Weather Prediction*.

Olivier Talagrand – 25 September 2007

Foreword to the English Edition

I am delighted to have been given the opportunity to write a foreword to this excellent textbook by Jean Coiffier on the mathematical and physical basis for NWP (Numerical Weather Prediction). It is an English translation of one originally published in French by Météo-France entitled '*Les bases de la prévision numérique du temps*'. To reflect the increasing use and importance of nonhydrostatic models in NWP, both for research and for operational forecasting, a valuable new appendix has been added in the English version. This gives an overview of two currently used nonhydrostatic models, one developed in Europe and the other in the United States: whilst both models are based on the same underlying continuous equation set, by way of contrast the first has an implicit time discretization, whereas the second has an (almost) explicit one with a smaller time step.

I fully endorse Olivier Talagrand's assessment, reproduced herein after translation, of the virtues of the original French version of this book. Assuming only a basic knowledge of physics and mathematics, the student is provided with an elegantly written synthesis of the essentials of NWP. Particularly noteworthy is the logical progression from simple equation sets, and their properties and discretization, to increasingly complex ones.

Whilst this book has been primarily written with the student in mind, it is also a valuable reference book for the experienced practitioner wishing to refresh his memory on specific aspects of the subject and to locate associated journal papers from the extensive references. To date, only French-speaking readers have been able to benefit from Jean's well-balanced, clear, and rigorous exposition of the subject. With this in mind, he is to be congratulated for finding further time during his retirement years to collaborate with Cambridge University Press on an English translation of his book, thereby making it available to a much wider audience.

It remains only for me to highly recommend this book to anyone, student or experienced practitioner, who is interested in the scientific foundation of weather prediction and climate simulation.

Andrew Staniforth – 15 October 2010

Preface

Fundamentals of Numerical Weather Prediction is intended to introduce students to current techniques for developing numerical weather prediction models. It is based on lecture notes for the course I taught on numerical weather prediction at the *École Nationale de la Météorologie* in the 1990s.

Numerical weather prediction consists of automatically performing meteorological forecasts and involves implementing a series of clearly identified processes: data collection and control, determination of the initial state of the atmosphere (analysis), computation of the final state at a given range (forecast), computation of the characteristic weather parameters at the local scale, tailoring and dissemination of results. This book does not purport to describe exhaustively all the techniques used to implement all of these processes in practice but focuses instead on the forecasting process proper. This consists in determining, with the help of numerical computation techniques, the solutions of a system of equations describing the behaviour of the atmosphere. The choice of an appropriate system of equations and of the series of numerical calculations to be performed to determine approximate solutions for this system defines what is commonly called a numerical prediction model. This basic tool is used both for weather forecasting and for climate simulation.

This book is for students looking to take their first steps in the techniques of numerical prediction and does not require any particular mathematical knowledge beyond what is expected of science graduates. However, it does assume knowledge of the fundamentals of dynamic meteorology with regard to developing and justifying the systems of equations and of the main methods of numerical analysis for solving partial differential equations.

Chapter 1 provides an overview of the history of numerical weather prediction from its beginnings in the 1950s until the late 1990s, highlighting the major advances and the evolution of computing tools.

Chapter 2 introduces the systems of equations most widely used for simulating atmospheric motion, and especially the 'primitive equations' and the way they are formulated in various coordinate systems.

Chapters 3 and 4 explain the principal techniques for representing meteorological fields for numerical weather prediction: the finite difference method and the spectral method.

Chapter 5 examines the effects of the various finite difference schemes in space and time; the linear shallow water model is used for comparing exact solutions with numerical ones.

Chapter 6 describes a few barotropic models using the numerical techniques studied before with regard to spatial discretization and implementing the various time integration schemes.

Chapter 7 explains the formulation of the primitive equations used in the baroclinic models and specifies their properties with regard to global invariants; a general formulation encompassing the primitive (hydrostatic) equations and the Euler (nonhydrostatic) equations is also presented.

Chapter 8 describes in detail the vertical discretization of the equations and the implementation of baroclinic models using explicit and semi-implicit time integration algorithms, while the fields are represented horizontally using the finite difference technique or the spectral method.

Chapter 9 deals with the parameterization of physical processes. Instead of explaining in detail all the parameterization schemes used in forecasting models (for which an abundant literature is continuously updated in the specialized journals), I have opted to describe a few examples emphasizing how the model dynamics and model physics hang together. Plentiful bibliographic references to other parameterization schemes are also given.

Chapter 10 is an overview of the processes making up an operational forecasting suite, with special emphasis on data assimilation, which has grown ever more important in recent years. The chapter ends with a look at future prospects for NWP.

Appendix A was written especially for the English edition of the book. It describes two types of models based on the nonhydrostatic equations that have been increasingly utilized for operational mesoscale forecasting in recent years.

Last, a copious further reading list completes the text and should enable readers to find more details about the various NWP techniques explained in the book.

It is no accident that many of the examples of NWP methods and techniques presented refer to models developed and operated by Météo-France under various acronyms (Améthyste, Sisyphe, Émeraude, Péridot, Arpège/IFS Aladin, and Arome), since I spent most of my working life with the French Meteorological Service. And it was at Météo-France that I was fortunate enough to work with Jean Lepas, Daniel Rousseau, and Jean-François Geleyn, who constantly helped me to acquire the essential knowledge that has gone into this book.

My hope is, then, that this book will enable students to acquire the basic knowledge and essential techniques of NWP so that they can quickly and effectively become active members of teams engaged in atmospheric modelling.

Jean Coiffier

Acknowledgments

I would like to thank here all of those colleagues who were patient enough to work through the chapters of the French version of this book in its various stages of development and so helped to make the whole more readable and intelligible: Frédéric Atger, Éric Bazile, Pierre Bénard, Yves Bouteloup, François Bouttier, François Bouyssel, Jean-Marie Carrière, Gérard De Moor, Michel Déqué, Jean-François Geleyn, Gwenaelle Hello, Jean-Pierre Javelle, Régis Juvanon du Vachat, Sylvie Malardel, Pascal Marquet, Jean Nicolau, Jean Pailleux, Jean-Marcel Piriou, Michel Rochas, Daniel Rousseau, Yann Seity, Joël Stein, and Karim Yessad.

My special thanks to Gérard De Moor and Jacques Siméon, who constantly encouraged me to continue writing this book from my lecture material, and to Claude Sinolecka, who painstakingly read through the entire French manuscript to bring it up to scratch for publication.

In the time that elapsed between completing the French version and writing the English version of the book, I have been able to add an appendix describing two types of models based on nonhydrostatic equations. Pierre Bénard at Météo-France and Stan Benjamin and Jimy Dudhia at the National Center for Atmospheric Research have been kind enough to help me to finalize the parts on the AROME model and WRF/ARW model respectively. The figures depicting the evolution of skill in operational numerical prediction have also been updated to take account of the latest data, thanks to help from Jeff McQueen, Geoff Di Mego, Michelle Mainelli, and Denis Staley at NCEP, Dominique Marbouty and Robert Hine at ECMWF, and Bruno Lacroix and Marc Tardy at Météo-France. I owe special thanks to Jean Pailleux and Peter Lynch for their critical review of the English manuscript, and to Jean-François Geleyn, who has painstakingly amended, improved, and enhanced the chapter on physical parameterizations.

The English version of this book would certainly never have seen the light of day without the efforts made by my former colleagues at Météo-France: Anne Guillaume, who convinced me to buckle down to the job and who helped me to present the project to the publishers; and Michel Hontarrède, who managed to solve sundry administrative headaches. I am grateful to Cambridge University Press and in particular to Susan Francis for trusting me to see the project through, and to Christopher Hudson, Laura Clark, and Christopher Miller, who were always there to help me in preparing this book.

I have nothing but praise for the way Christopher Sutcliffe has cooperated with me on the English translation of this book, and I am most grateful to him for the time spent discussing the relevance of the scientific terms used.

The splendid picture of the Earth captured in its grid on the cover is the meticulous labour of Pascal Lamboley on a Meteosat image provided by the *Centre de Météorologie Spatiale de Météo-France*, thanks to the good offices of Patrick Donguy.

My warm thanks to my friends Olivier Talagrand, Director of Research at the CNRS, and Andrew Staniforth, Met Office scientist, who agreed to write forewords for the French and English versions respectively.

Finally, I am especially grateful to my wife, Florence, for tolerating a husband who has spent most of his time riveted to his computer throughout the making of this book.

Partial list of symbols

Latin letters

a	Mean radius of the Earth; length.
b	Coefficient; length.
c	Wave velocity.
c_R	Rossby wave velocity.
d	Distance.
e	Depth inside the soil.
e	mathematical constant.
e_N	2.71828 is the numerical value of the mathematical constant e. Residue of an expansion containing N terms.
e_T	Turbulent kinetic energy; soil thermal emissivity.
f	Coriolis parameter: $2\Omega \sin \varphi$; function.
g	Gravity acceleration; function.
g^*	Newtonian acceleration.
h	Standard deviation of subgrid-scale orography; weighting coefficient of digital filter; rain/snow discrimination function.
i	$\sqrt{-1}$
i, j	Horizontal location indices.
k	Wave number; vertical level index; time step index.
$\mathbf{k}$	Unit vector in the vertical.
l	Obstacle width; length.
ℓ	Prandtl length.
m	Map scale factor; zonal wavenumber.
$\tilde{m}$	Map scale factor in the case of variable resolution spectral treatment.
n	Global wavenumber.
p	Pressure; vertical coordinate.
p_s	Surface pressure.
q	Specific humidity of the air; concentration.
q_{sat}	Saturation specific humidity.
q_S	Surface specific humidity.
q_c	Specific concentration of condensed water inside the cloud.
q_d	Specific concentration of dry air.
q_v	Specific concentration of water vapour.
q_l	Specific concentration of suspended liquid water.
q_s	Specific concentration of suspended solid water.
q_r	Specific concentration of precipitating liquid water.

q_i	Specific concentration of precipitating ice.
$\mathbf{r}$	Radius vector.
r	Radial distance.
$\mathbf{r}_i$	Proportion of ice.
r_f	Proportion of snow within a layer.
s	General vertical coordinate; dry static energy.
s_s	Surface dry static energy.
$\dot{s}$	Vertical velocity: ds/dt.
t	Time.
u,v	Horizontal wind components.
u^g	Quantity of gas radiation passes through.
u^*	Friction velocity.
s^*	Scaling dry static energy in the surface boundary layer.
q^*	Scaling specific humidity in the surface boundary layer.
w	Vertical velocity in z-system; weighting coefficient.
w_c	Vertical velocity within the cloud.
$w^g_{\Delta v}$	Equivalent line bandwith relating to the gas g and to the spectral interval Δv.
w_k	Gauss weight at latitude μ_k.
x,y	Horizontal Cartesian coordinates.
z	Height.
z_0	Roughness length.
A	Generic variable identifying a 2-dimensional field $A(\lambda, \mu)$ or $A(x, y)$; coefficient.
A_n^m	Coefficient of series expansion of A in terms of surface spherical harmonics.
$A_m(\mu)$	Fourier coefficient at latitude μ.
A_T	Terrestrial albedo for solar radiation.
B	Planck function; buoyancy.
C	Constant; phase change coefficient for water vapour.
C_p	Specific heat at constant pressure for air.
C_v	Specific heat at constant volume for air.
C_{p_d}	Specific heat at constant pressure for dry air.
C_{p_v}	Specific heat at constant pressure for water vapour.
C_l	Specific heat at constant pressure for liquid water.
C_i	Specific heat at constant pressure for ice.
$\hat{C}$	Specific heat for the comprehensive set of non-precipitating phases.
C_{Sol}	Soil characteristic constant.
C_S	Soil specific heat.
C_D	Drag coefficient.
C_H, C_E	Transfer coefficients for dry static energy and specific humidity.
C_N	Net condensation rate.
C_{BCC}	Buoyant convective condensation rate.
C_{UCC}	Non-buoyant convective condensation rate.

D	Discretized divergence of the horizontal wind; raindrop diameter.
$\mathcal{D}$	Thermal diffusion coefficient.
$\tilde{D}_k$	Numerical divergence of the momentum at level k.
E	Surface evaporation flux.
E_c	Kinetic energy.
E_p	Potential energy.
F	Inverse Froude number: NH/U.
$\mathbf{F}$	Friction force.
$F_\downarrow$	Downward radiation flux.
$F_\uparrow$	Upward radiation flux.
$F_\downarrow^*$	Modified downward flux.
$F_\uparrow^*$	Modified upward flux.
G	Radiation flux absorbed by the soil.
H	Effective obstacle height.
H_L	Latent heat flux.
H_s	Sensible heat flux.
Hu	Relative humidity.
$H_n^m(x, y)$	Product of complex exponential functions with wavenumbers n and m.
I_0	Solar constant.
J	Turbulent flux; cost function.
J, K, L, M	Number of grid points.
K_{abs}^c	Absorption coefficient for the element c.
K	Kinetic energy.
K_s	Tuning coefficient.
K_M	Exchange coefficient for momentum.
K_H, K_E	Exchange coefficient for dry static energy and specific humidity.
K_{DH}	Horizontal diffusion coefficient.
K_R	Relaxation factor.
$\mathbf{K}$	Wave vector.
L	Wavelength.
L_M	Monin-Obukhov length.
L_R	Rossby radius of deformation.
$L_l(T_0)$	Latent heat of vaporization of water at $T_0 = 0$ K.
$L_i(T_0)$	Latent heat of melting of ice at $T_0 = 0$ K.
M	Source of water vapour by mass unit and time unit; mass of water; maximum zonal wavenumber.
M_i	Mass of snow.
M_c	Mass flux.
N	Cloud cover; cost function; Brunt-Vaïsälä frequency; number of levels in the vertical; number of time steps; maximum global wavenumber; number of raindrops.
P	Term introduced to express the linearized pressure force in the semi-implicit algorithm; upper air precipitation flux.

P_l	Rain flux.
P_i	Snow flux.
P_L	Surface rain flux.
P_I	Surface snow flux.
P_T	Total precipitation flux at the surface: $P_L + P_I$
P_S	Precipitation flux at the surface.
$P_n^m(\mu)$	Associated Legendre function.
Q	Heat source per mass unit and time unit; wave velocity reduction factor.
Q_1	Heat source per unit time.
Q_2	Energy loss per unit time corresponding to the water vapour deficit.
Q_R	Heat source due to radiation.
Q_1^c	Static energy tendency: $Q_1 - Q_R$ due to convection within the mesh.
Q_2^c	Specific humidity tendency: $-Q_2/L$ due to convection within the mesh.
Q_3^c	Momentum tendency due to convection within the mesh.
R	Specific ideal-gas constant relative to the air; influence radius; wave amplitude reduction factor.
Ri	Richardson number.
R_d, R_v	Ideal-gas constants relative to dry air, to water vapour.
R_T	Terrestrial radiation.
R_A	Atmospheric radiation.
R_G	Global radiation.
S	Solar radiation flux.
S_0	Solar radiation flux at the top of the atmosphere.
T	Thermodynamic temperature; truncation.
T_S	Surface temperature.
T_{00}	Temperature of the water triple point.
U, V	Reduced components of the horizontal wind; velocities.
V	Volume.
$\mathbf{V}$	Horizontal wind (general case).
$\mathbf{V}_H$	2-dimensional horizontal wind.
$\mathbf{V}_3$	3-dimensional wind velocity.
Veg	Vegetation proportion.
w	Velocity of falling raindrops.
w_p	Relative deep layer water content.
w_s	Relative surface water content.
$Y_n^m(\lambda, \mu)$	Surface spherical harmonic.
Z	Natural logarithm of surface pressure; height; length.
Z_s	Height of topography.

Gothic letters

$\mathfrak{M}$	Total mass.
$\mathfrak{E}$	Total energy.
$\mathfrak{J}$	Total angular momentum.
$\mathfrak{K}$	Total kinetic energy.

$\mathfrak{Z}$	Total potential absolute vorticity.
$\mathfrak{H}$	Total absolute potential enstrophy.
$\mathfrak{Q}$	Total internal energy.

Greek letters

α	Angle; coefficient; length.
β	Rossby parameter: $\partial f/\partial y$; fractional part of backscattered radiation; coefficient; length.
γ	Length; coefficient; binary variable.
δ	Discretized divergence; optical path; length; binary indicator; amplitude of an orographic wave; root mean square (RMS) error.
δp_k	Thickness of layer k in p-system.
$\delta \sigma_k$	Thickness of layer k in sigma-system.
δ_x	Mesh size along x-axis.
δ_y	Mesh size along y-axis.
ε	Entrainment rate of water vapour inside the cloud; coefficient; infinitesimal quantity.
ε_T	Truncation error.
ζ	Vorticity.
η	Divergence.
ζ'	Reduced value of vorticity.
η'	Reduced value of divergence.
θ	Co-latitude; potential temperature.
κ	Ratio R/C_p; Von Karman's constant.
λ	Longitude; wavelength; asymptotic mixing length; coefficient; variable.
λ_S	Soil thermal conductivity factor.
μ	Sine of latitude.
μ_k	Gaussian latitude.
μ_0	Cosine of the zenith angle.
ν, ν'	Numerical coefficients.
ξ	Absolute vorticity: $\zeta + f$.
$\xi*$	Absolute potential vorticity ξ/Φ.
π	value of π is 3.14159..., mathematical constant.
π	Mass-type vertical coordinate or hydrostatic pressure when the pressure at the top of the working domain vanishes.
$\dot{\pi}$	Vertical velocity: $d\pi/dt$.
ρ	Air density.
ρ_w	Water vapour density.
ρ_r	Rain density.
ρ_s	Snow density.
σ	Vertical coordinate sigma; Stefan-Boltzman constant; partition coefficient; frequency; standard deviation.
$\dot{\sigma}$	Vertical velocity: $d\sigma/dt$.
$\dot{\varsigma}_{\tilde{k}}$	Vertical velocity term $\dot{s}\,\partial p/\partial s$.

τ	Radiation transmission factor; e-folding time.
$\boldsymbol{\tau}$	Momentum flux.
φ	Latitude.
χ	Velocity potential function.
ψ	Stream function; generic function.
ω	Vertical velocity: dp/dt; frequency (angular frequency).
ω^*	Mass flux.
ϖ	Simple scattering albedo.
Γ	Gamma function; transfer coefficient of momentum in the vertical.
Δt	Time step.
Δx	Mesh size along x-axis.
Δy	Mesh size along y-axis.
$\Delta \delta$	Optical depth.
Δv	Absorption bandwidth.
Θ	Angle.
Φ	Geopotential.
Φ_s	Surface geopotential.
Φ_1, Φ_2	Universal functions for the surface boundary layer.
Ψ	Angle.
Ω	Angular velocity of the Earth; angular velocity.
$\boldsymbol{\Omega}$	Angular velocity vector of the Earth.

Generalized vectors, matrices, and operators

$\mathbf{A}$	Linearized energy conversion term matrix; analysis error covariance matrix.
$\mathbf{B}$	Linearized hydrostatic relation matrix; background field error covariance matrix.
$\tilde{\mathbf{D}}$	Column vector of momentum.
$\mathbf{H}$	Linearized observation operator matrix.
$\mathbf{E}$	Error vector.
$\mathbf{H}_r$	Linearized observation operator matrix for the incremental method.
$\mathbf{K}$	Optimal interpolation weight matrix.
$\mathbf{M}$	Linearized barotropic system matrix; vertical structure matrix of the baroclinic linearized system for the semi-implicit algorithm.
$\mathbf{P}$	Column vector of linearized pressure force term.
$\mathbf{P}_f$	Background field error covariance matrix.
$\mathbf{P}_a$	Analysis error covariance matrix.
$\mathbf{Q}$	Matrix of the eigenvectors of the operator defined by the matrix $\mathbf{M}$.
$\mathbf{Q}(t_k)$	Model error at time step t_k.
$\mathbf{R}$	Observation error covariance matrix.
$\mathbf{R}^*$	Column vector describing the linearization profile RT^*.
$^t\mathbf{S}_k$	Row vector of the layer depths in the vertical up to level k followed by zeroes.
$\mathbf{T}$	Column vector of temperature.
$\mathbf{U}, \mathbf{V}$	Column vector of reduced wind components U and V.
$\mathbf{X}$	Generic vector; state vector.

$\mathbf{X}_a$ Analysis state vector.

$\mathbf{X}_b$ Background field state vector.

$\mathbf{X}_t$ State vector of the actual atmosphere at time t.

$\mathbf{X}_F^*$ Resulting state vector after application of the digital filter.

$\hat{\mathbf{Z}}$ Eigenmode of the linearized model $\mathcal{L}$.

$\delta\mathbf{X}'$ State vector deviation from an approximate value.

ε Error vector.

$\bar{\varepsilon}$ Mean error vector.

ζ Column vector of vorticity.

η Column vector of divergence.

$\dot{\varsigma}$ Column vector of vertical velocity.

$\mathbf{\Phi}$ Column vector of geopotential.

ψ Column vector of streamfunction.

χ Column vector of velocity potential.

Λ Diagonal matrix of the implicit system obtained with the baroclinic model.

S Nonlinear general operator.

$\mathcal{L}$ Linear operator.

$\mathcal{N}$ Nonlinear operator.

$\mathcal{F}$ Nonlinear operator acting on the state vector at time t.

$\nabla_x\mathcal{F}$ Tangent linear operator of $\mathcal{F}$.

$\mathcal{H}$ Observation operator.

$\mathcal{M}_k$ Model operator acting between times t_k and t_{k+1}.

Various mathematical notations

$\cdot$ Scalar or dot product.

$\times$ Vector product.

$|\ |$ Absolute value.

$\|\ \|$ Norm of a vector.

$<,>$ Scalar product.

∇ Gradient of a scalar function.

$\nabla\cdot$ Divergence.

$\nabla\times$ Curl.

∇^2 Laplacian.

∇'^2 Reduced Laplacian.

$J(\ ,\ )$ Jacobian.

$\nabla_\mathbf{x}$ Gradient of a scalar function of a vector $\mathbf{X}$ with respect to the components of this vector.

$\partial/\partial l$ Partial derivative of a multivariable function with respect to the variable l.

d/dt Total derivative.

$\bar{A}^x$ Mean value: $[A(x + \Delta x/2) + A(x + \Delta x/2)]/2$.

A_x Finite difference: $[A(x + \Delta x/2) - A(x + \Delta x/2)]/\Delta x$.

1   Half a century of numerical weather prediction

1.1 Introduction

Numerical weather prediction (NWP) is a very young discipline that developed essentially in the second half of the twentieth century with the continual benefit of advances in computing. The techniques implemented are used to solve equations describing the behaviour of the atmosphere, that is, to numerically compute future values of the atmosphere's characteristic parameters from initial values that are known from meteorological observations.

The equations used are the general equations of fluid mechanics that were already well established by the early twentieth century and to which certain simplifications are applied. Those simplifications are justified by the orders of magnitude of the various terms in the specific instance of the Earth's atmosphere and by the scales to be described. Computers are essential for solving these systems of nonlinear equations, which, in the general case, cannot be solved analytically.

A *numerical model* of the atmosphere is constructed in two separate stages: first, a system of equations is established to govern the continuous behaviour of the atmosphere; then, by the process of *discretization*, the equations relating to continuous variables are replaced by equations relating to discrete variables, the solutions to which are obtained by an appropriate algorithm. The results of a numerical prediction (that is, the solutions of discretized equations of dynamic meteorology) depend therefore on the discretization process employed.

Implementing this algorithm requires a sufficiently powerful computing tool. This is why advances in numerical weather prediction have followed in the wake of the fantastic development of electronic computers since they came into being at the end of the Second World War.

And last, it should be emphasized that weather forecasting achieved by forecasters using numerical models owes its success to the implementation of the global weather observing system that relies on both conventional and satellite measurements and provides an admittedly imperfect but nonetheless effective description of the atmosphere at a given initial time.

1.2 The early days

The history of numerical weather prediction features a number of stages that proved decisive in the development of the discipline.

Back in 1904, the Norwegian Vilhelm Bjerknes recognized that weather forecasting is fundamentally a deterministic *initial-value* problem in the mathematical sense (Bjerknes, 1904):

> *If it is true, as every scientist believes, that subsequent atmospheric states develop from the preceding ones according to physical law, then it is apparent that the necessary and sufficient conditions for the rational solution of forecasting problems are the following:*
>
> – *A sufficiently accurate knowledge of the state of the atmosphere at the initial time.*
> – *A sufficiently accurate knowledge of the laws according to which one state of the atmosphere develops from another.*

However, he realized that the difficulty lay in the need to solve a system of nonlinear partial differential equations for which there were no analytical solutions, in the general case.

Between 1916 and 1922, the Englishman Lewis Fry Richardson tried to solve weather forecast equations by numerical methods. He even made a 6-hour forecast by hand, although it proved quite unrealistic. Undaunted, though, he sought out the causes of his failure. His work was published in 1922 in his famous and truly visionary *Weather Prediction by Numerical Process* (Richardson, 1922). Noting that '$32 \times 2000 = 64,000$ *computers would be needed to race the weather for the whole globe*', Richardson let his imagination roam and dreamed of a weather-forecasting factory, with a myriad of people making synchronized computations under the control of a supervisor tasked with the orchestration of operations (Figure 1.1).

In 1928, the German mathematicians Courant, Friedrichs, and Lewy systematically studied how to solve partial derivative equations by using finite differences and specified the constraints to comply with when performing discretization (Courant et al., 1928).

In 1939, the Swede Carl-Gustav Rossby showed that the *absolute vorticity conservation equation* provided a correct interpretation of the observed displacement of atmospheric centres of action (Rossby, 1939).

In 1946, the first electronic computer, the ENIAC (Electronic Numerical Integrator and Computer) was installed at Pennsylvania University, in Philadelphia, while the Hungarian-born U.S. mathematician John von Neumann was also working on building improved machines at the Institute for Advanced Studies in Princeton.

In 1948, the American Jule Charney proposed a simplification of the general system of equations, known as the *quasi-geostrophic approximation*, and found, as a specific instance, the equation studied by Rossby (Charney, 1948).

Figure 1.1   Richardson's 'dream'. (Artist's impression by Alf Lannerbaeck, published by the Swedish newspaper *Dagens Nyheter*, 22 September 1984)

Finally, in 1950 Jule Charney, the Norwegian Ragnar Fjörtoft, and John von Neumann made the first numerical weather prediction (Charney et al., 1950): they used the absolute vorticity conservation equation for this experiment and did the computing on the ENIAC at Aberdeen (Maryland). The results obtained for the forecast of geopotential height of the 500 hPa isobaric surface, characteristic of the middle atmosphere, were most encouraging, and the experiment marked the starting point of modern numerical prediction (Platzman, 1979). In answer to Charney, who had sent the paper describing the experiment to him, Richardson wrote in 1952: '*Allow me to congratulate you and your collaborators on the remarkable progress which has been made in Princeton; and on the prospects of further improvement which you indicate to establish a science of meteorology, with the aim of predicting future states of the atmosphere from the present state*' (Ashford, 1985).

1.3   Half a century of continual progress

The success of Charney, Fjörtoft, and von Neumann's experiment was to lead from the mid 1950s onwards to the development for operational purposes of a large number of increasingly complex prediction models of ever greater spatial resolution, allowing ever smaller scales to be covered.

1.3.1 The need to be fast and accurate

Richardson had fully understood that numerical weather prediction was a race between the computing process and the actual evolution of the atmosphere. The speed of computation depends on the various characteristics of the prediction model and on the speed of the computer used, in a form we shall examine in detail.

Suppose that the equations are discretized by dividing space into boxes defined by a horizontal grid and a number of vertical levels. Within each box, the atmosphere is assumed to be homogeneous and so it suffices to know the values of the various atmospheric quantities at some point within the box. The time required to make a prediction for a given time range can then be calculated by taking account of the various factors involved:

- The total number N_v of variables to be processed. The state of the atmosphere being described by a limited number of quantities (the two components of horizontal wind, temperature, specific humidity, and surface pressure), the number of variables is equal to the product of that number of parameters by the total number of points processed, which varies with the size of the geographical domain and the spatial resolution adopted horizontally and vertically.
- The number of calculations N_c to be made per variable for a time step Δt. This number of elementary arithmetical operations depends on the complexity of the model, with greater allowance for interactions among variables being reflected by an increased number of calculations.
- The number of time steps N_t needed to reach a given time range H, namely, $N_t = H/\Delta t$. This time step Δt depends on the spatial resolution characterized by the mesh size Δx of the grid, for it must satisfy the Courant, Friedrichs, and Lewy (CFL) condition that is expressed as:

$$U\Delta t/\Delta x < C,$$

where U is the speed of propagation of the fastest waves described by the equations and C is a dimensionless number dependent upon the problem geometry and the chosen discretization. While some algorithms, to be discussed later, can take us beyond this limit, it is nonetheless true that the time step Δt must be reduced concomitantly with the mesh size Δx to process the space and time scales of mesoscale atmospheric phenomena with similar accuracy, as shown by examination of Table 1.1.
- The computer's calculating speed R. This is expressed as the number of elementary floating point operations per second, or *flops*, whether done by one computer or several computers in parallel.

The time T required to make a prediction for a given time range H is given by the ratio:

$$T = N_v N_c N_t / R.$$

Table 1.1 The different meteorological phenomena with their respective time and space scales. (After Orlanski (1975). *Bull. Am. Meteor. Soc.*, **56**, 528 © Amer. Met. Soc.)

CLASSIFICATION OF SCALES	Time	Climatological scale		Planetary and synoptic scales	Meso-scale	Micro-scale	
Length	T / L	1 month		1 day	1 hour	1 minute	1 sec.
Macro α scale (10 000 km)		Standing waves	Ultra-long waves	Tidal waves			
Macro β scale (2 000 km)				Baroclinic waves			
Meso α scale (200 km)				Fronts and Hurricanes			
Meso β scale (20 km)					Nocturnal low level jet / Squall lines / Inertial waves / Cloud clusters / Mtn and lake disturbances		
Meso γ scale (2 km)						Thunderstorms / Internal gravity waves / Clear air turbulence / Urban effects	
Micro α scale (200 m)						Tornadoes / Deep convection / Short gravity waves	
Micro β scale (20 m)							Dust devils / Thermals / Wakes
Micro γ scale							Plumes / Roughness / Turbulence

To take the example of the ARPEGE operational model used by Météo-France in 1998, the number of variables to be processed was $N_v \approx 23.10^6$ (600×300 horizontal points, 31 levels, 4 three-dimensional variables, and 1 two-dimensional variable), the number of calculations to be made for one variable was $N_c \approx 7.10^3$, and the number of time steps for a 24-hour forecast was $N_t = 96$ (15-minute time steps). The calculations were made on a FUJI VPP700 multiprocessor computer credited with a computational speed of up to 20 gigaflops (20 billion floating point operations per second) and the time required for a 24-hour forecast was just under a quarter of an hour.

As the time T is imposed by operational constraints, any increase in the computer's speed means the model's resolution may be augmented (horizontal grid spacing and number of vertical levels) as may the number of calculations made for each of the variables. This evolution towards greater resolution and increased complexity has been the rule in recent decades; it has also been facilitated by the development of new algorithms allowing longer time steps.

1.3.2 The use of filtered equations

The earliest models used operationally relied on the quasi-geostrophic approximation that imposes a diagnostic (that is, time-independent) relation between the pressure field and the wind field, which reduces the number of degrees of freedom of the model. This approximation also has the effect of conserving only slow waves, known as *Rossby waves*, as solutions and eliminating rapidly propagating *inertia-gravity waves*; it thus allows us to use a comparatively large time step compatible with the CFL condition. Because of the filtering effect so obtained, the simplified equations are known as *filtered equations*. Such a three-level model (Charney, 1954) was put into service for operational forecasting in May 1955 by the U.S. Weather Bureau. However, it was not until it was improved by Cressman (1963) that forecasters could really use the tool (Shuman, 1989). In the 1960s and until the mid 1970s models with filtered equations were widely used by the leading meteorological services (Bushby, 1987; Pône, 1993; Cressman, 1996; Rochas and Javelle, 1993). Enhanced computer performances were then used to extend the domain and increase the horizontal resolution and the number of vertical levels (thereby increasing the number of variables N_v) so as to better describe the dynamics of the atmosphere.

1.3.3 Back to the primitive equations and initialization

The growing calculating speed of computers meant it was then possible to return to the equations for the evolution of a fluid in hydrostatic equilibrium used earlier by Richardson, which were from then on termed the *primitive equations*. They admit rapidly propagating inertia-gravity waves as solutions, and for compliance with the CFL condition require the choice of a time step some six times smaller than with filtered equations, thereby increasing the number of time steps N_t. Work on the primitive equations begun by Eliassen (1956) led to successful tests in the United States (Smagorinsky, 1958) and in Germany (Hinkelmann, 1959). In the United States, the primitive equation model with six vertical levels developed by Shuman and using a 381 km mesh on an octagonal domain covering most of the northern hemisphere (Shuman and Hovermale, 1968) began its operational career on 6 June 1966, thus opening up the path to generalized use of this type of model for many meteorological services.

Primitive equation models are relatively easy to implement but require that the *initialization* problem be solved. Pressure and wind fields coupled through evolution equations must respect a certain balance at the initial time; otherwise they will give rise to substantial oscillations owing to the propagation of gravity waves of unrealistic amplitudes (Hinkelmann, 1951). The difficulty in obtaining a balanced initial state from pressure and wind observations was what brought about Richardson's unrealistic result in the first attempt at numerical prediction (Lynch, 1994).

Static initialization methods whereby the wind field is deduced from the pressure field using a linear or nonlinear equation proved comparatively ineffective; moreover, wind observations were not really used then for defining the initial state. It was in the late 1970s that an elegant solution to the problem of initialization of global fields was found, independently by Baer and Tribbia (1977) and by Machenhauer (1977). The idea was to decompose the initial state of the atmosphere into normal modes (that is, into solutions of a linearized version of the model) and then to correct the inertia-gravity modes in the initial state so as to make them stationary when the model evolved. This technique of *nonlinear normal mode initialization* meant primitive equation models could be used effectively to take full advantage of initial pressure and wind data.

1.3.4 Global processing and the spectral method

Elementary reasoning based on the speed at which perturbations move shows that the working area has to be extended and so the number of points N_v increased when one wishes to make predictions over longer time ranges (Figure 1.2).

The models for limited geographical areas were replaced by hemispheric models, and then finally by global models allowing interactions between the two hemispheres to be handled properly. This meant grids had to be defined on a sphere and the problem of instability resulting from smaller mesh size close to the poles had to be solved.

Alongside grid point models using the finite difference method for computing partial derivatives, the use of spectral models has developed. In these, fields defined on the sphere are represented by series expansion in terms of basis functions: the surface

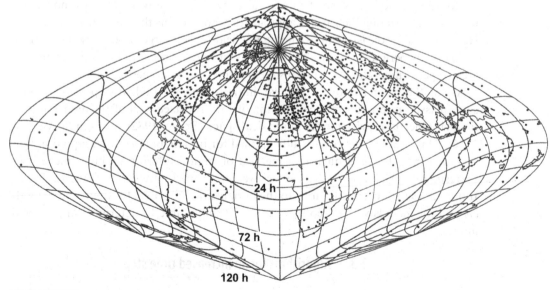

Figure 1.2   Worldwide distribution of radiosounding stations and indication of regions for which observations are required for making 1-, 3-, and 5-day forecasts over the central area Z. (ECMWF image)

spherical harmonics. This method allows better evaluation of wave speeds than by the finite difference method. It had long been reserved for models with just a few degrees of freedom, because of the high cost of direct computation of the expansion coefficients for nonlinear terms. With the advent of the fast Fourier transform (Cooley and Tukey, 1965), it proved far more advantageous to use the *transform method*, consisting in calculating the nonlinear terms at the nodes of an intermediate grid (Orszag, 1970; Eliassen et al., 1970). This technique made the spectral method highly competitive (Bourke, 1972), and in the 1980s it superseded the grid point method almost everywhere for producing global models.

1.3.5 Limited area models

In addition to extending the area, which was necessary to extend the time range of forecasts, it turned out to be advantageous for short-range forecasting (1–2 days) to continue working on a restricted area using a grid with a fine enough mesh to simulate small-scale motion correctly and reproduce features caused by topography. Thus, limited area models (LAM) were developed enabling short-range, small-scale predictions to be made (Rousseau et al., 1995). Mathematical analysis shows that field values have to be specified on the area boundary for each time step. The field values may be obtained by interpolating fields from a larger scale model. However, a dissipation term should be introduced into the limited area model to damp perturbations that are engendered by forcing fields at the boundary and that propagate towards the interior of the domain (Davies, 1976). This leads to the *nested models* that are the basis of operational prediction systems in most meteorological services.

As the spectral method and normal mode nonlinear initialization had proved effective for global models, it was tempting to apply the same techniques to models for limited areas. Among the various approaches proposed was that of Machenhauer and Haugen (1987), consisting in extending the fields over a larger domain so as to make them doubly periodic; this artefact allows the spectral method to be used on a limited area by performing the series expansion in terms of trigonometric functions.

As for the normal mode nonlinear initialization method, it is possible, under certain assumptions about the definition of the linearized part of the model, to stationarize inertia-gravity modes in physical space (Brière, 1982; Juvanon du Vachat, 1986). This process was successfully applied for initializing primitive equation limited area models. Subsequently a method of digital filtering of high frequencies corresponding to inertia-gravity waves was proposed by Lynch and Huang (1992). It provides satisfactory solutions to the initialization problem for limited area models and for models whose geometry makes it impossible to determine any normal modes.

1.3.6 Algorithms for an increased time step

The use of explicit time integration schemes with primitive equations requires the use of time steps six times smaller than those of the filtered models just to satisfy the

CFL condition. Robert (1969) came up with a fresh approach, proposing to process the terms responsible for gravity wave propagation implicitly. This *semi-implicit time integration* algorithm yields a new, much less restrictive CFL condition as it involves only the maximum speed of the synoptic wind and no longer the speeds of the fastest waves. This possibility of increasing the time step has its downside, as a system of linear equations then needs to be solved. Despite this, the semi-implicit algorithm maintains a clear lead, allowing the run time for grid point models to be divided fourfold and by even more for spectral models. This explains why it has been so popular and become so widespread since the 1970s.

Lagrangian processing of advection was initially used by Fjörtoft (1952) to solve a simple model graphically. The method then inspired Lepas (1963) in constructing a numerical prediction model. Krishnamurti (1962) and Sawyer (1963) also proposed using the technique to improve the accuracy of numerical advection schemes. However, the credit goes again to Robert (1981) for showing that the method used in conjunction with semi-implicit processing could free us from the CFL condition. Time discretization is performed on the total derivative (or *Lagrangian derivative*) and forces us to interpolate the model variables at the starting point of particles arriving at the grid points during their movement in one time step. We thus obtain the *semi-Lagrangian semi-implicit scheme* algorithm allowing us to further increase the time step Δt (and so reduce the number of time steps N_t) within the limits compatible with the required accuracy for representing the relevant time scales.

It should also be emphasized that this highly effective algorithm has also made the use of variable grid models (that is, with increasing resolution over a chosen area) into a competitive solution for the nested model system once the time step is no longer dependent on the smallest grid dimension in the working domain (Courtier and Geleyn, 1988; Côté et al., 1993).

1.3.7 The move to nonhydrostatic equations

The semi-Lagrangian semi-implicit scheme opened up new horizons for nonhydrostatic models that are essential for correctly handling spatial scales of the order of 1 km. Their operational implementation had until then come up against the problem of the very small time step arising from the need to comply with the CFL condition relative to the propagation of sound waves (also known as acoustic waves). However, Lagrangian processing of advection combined with implicit processing of the terms responsible for the propagation of gravity waves and sound waves leads to an unconditionally stable algorithm so that it is now possible to envisage using nonhydrostatic models (Tanguay et al., 1990; Laprise, 1992; Bubnova et al., 1995) to simulate atmospheric motion from the planetary scale to mesoscale (Table 1.1).

1.3.8 Physical processes

It soon proved necessary to evaluate the source and sink terms of momentum, heat, and water vapour resulting from more or less complex physical processes that have to be introduced into equations to reproduce the evolution of the atmosphere realistically. Given that the scales to be taken into account to accurately simulate the relevant physical processes are generally smaller than the scales described by the model variables (these are sometimes referred to as *subgrid* scales), these processes have to be parameterized: their average effect on the model variables alone is sought. These additional computations form the model *physics* and are grafted onto the numerical processing of equations, which is the model *dynamics*.

After allowing in a simple way for the effects of friction to avoid depressions deepening excessively, a real improvement was made in describing the atmospheric water cycle and its associated energy exchanges. The addition of an extra equation describing the transport of water vapour is required to have the means for handling the effects of changes in water phases and for calculating precipitation (Smagorinsky, 1962).

Proper description of the turbulent transfer mechanisms between the soil and the atmosphere, not just for momentum but also for sensible heat and water vapour, implies calculating turbulent fluxes near the surface (Businger et al., 1971; Deardorff, 1972; Louis, 1979). This calculation involves additional variables, apart from dynamic model variables calculated for the lowermost level, such as surface temperature and moisture as well as data characterizing the soil such as roughness length or plant cover (Deardorff, 1977).

It is obvious that the evolution of surface variables is directly related to energy inputs from radiation flux, which in turn is highly dependent upon the time of day and cloud cover. This is why it is essential to calculate the effects of interaction between radiation and the various constituents of the atmosphere, especially the water present in its various phases. The effects of absorption, scattering, and re-emission of radiation, which differ particularly depending on whether the atmosphere is clear or cloudy, must be computed (Rodgers and Walshaw, 1966; Katayama, 1974).

Because of the hydrostatic hypothesis, the primitive equations cannot deal explicitly with convective motion resulting from local vertical instability of the atmosphere. The *convective adjustment* methods (Manabe and Strickler, 1964), which were designed to correct the vertical profiles leading to unstable solutions for the model, have been superseded by more elaborate methods that account for the effects of interaction between convective clouds and their environment (Kuo, 1965, 1974; Arakawa and Schubert, 1974; Bougeault, 1985).

The comparatively recent inclusion of energy dissipation of the vertically propagating mountain waves has also improved the prediction of the intensity of jet streams above mountainous regions (Palmer et al., 1986).

1.3.9 Objective analysis and data assimilation

Alongside the improvement made to the forecasting models, very important theoretical and practical work has also been done in precisely determining a given state of the atmosphere allowing for the various observations available (Daley, 1980). This operation is referred to as *objective analysis* when it is intended to define a state of the atmosphere at a given time and as *data assimilation* when it is repeated to provide successive states of the atmosphere over a given period of time.

Objective analysis was first performed using geometric interpolation methods (Gilchrist and Cressman, 1954), then with successive correction methods of a *background field* provided by a forecast model (Bergthorsson and Döös, 1955). Taking into account the statistical properties of fields of meteorological variables (which is the basis of *optimal interpolation* methods) was an important step in making allowance for the specific characteristics of the various observations available and for deriving benefit from the linkages among the fields to be analysed (Gandin, 1963; Lorenc, 1981). A very general variational formulation of this problem (which came down to minimizing a *cost function* and can be solved by optimal control methods) was proposed in the mid 1980s (Talagrand and Courtier, 1987).

The accuracy with which an initial state of the atmosphere is obtained with the help of data assimilation has a decisive impact on the skill of weather forecasts made with numerical prediction models. The huge investments required to launch meteorological satellites carrying modern remote sensing systems are fully justified by the tremendous improvements in meteorological data assimilation.

1.4 Developments in computing

One cannot speak of numerical weather prediction without evoking the computational tools made available to meteorologists. Richardson's dream (Figure 1.1) has now become reality with the spectacular development of electronics and it is highly symbolic that John von Neumann, who is considered to be one of the fathers of computing science, should have been involved in the first experiment in numerical weather prediction. Since that heroic age, the major meteorological services have constantly sought to have the best machines on the market so as to be able to perform the ever increasing number of calculations required for modern weather forecasting as quickly as possible.

1.4.1 Computing power accompanies progress

The use of numerical prediction models has meant that the quality of weather forecasts has improved constantly. The plot in Figure 1.3 shows a steady improvement of weather forecasts over North America from 1955 to 2009 through operational use of successive generations of numerical models. Forecast skill S shown in the figure is derived from the score $S1$ (proposed by Teweles and Wobus (1954) for quantifying forecast quality as a standardized value of mean quadratic error relative to the geopotential gradient) by using the formula: $S = 100(1 - S1/70)$.

Forecast Skill S

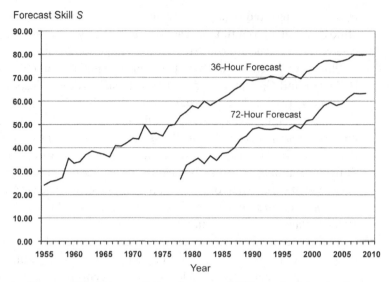

Figure 1.3   NCEP operational forecast skill for 36 hour and 72 hour forecasts at 500 hPa over North America. (Data courtesy NCEP)

It has also been shown, taking France as an example, that the skill of a 72-hour forecast in 2000 was greater than that of a 24-hour forecast in 1980 (Pailleux et al., 2000).

This overall improvement in forecasting skill has been achieved by all meteorological services that use numerical models for operational forecasting. It is the logical outcome of increased spatial resolution of models, more realistic allowance for physical processes, and more efficient computational algorithms. Of course, these improvements are reflected by a substantial increase in the number of calculations, which could not have been achieved without increasingly powerful computers (Figure 1.4).

1.4.2 From the ENIAC to scientific mainframes

The U.S. meteorological department was very quick to exploit the power of the early scientific computers manufactured by International Business Machines (IBM), and an IBM 701, capable of performing 3×10^3 operations per second, was installed at the *Weather Bureau* in 1955 for the needs of numerical prediction. The IBM 701 and IBM 704, based on electronic tube technology, were superseded around 1969 by the IBM 7090 and IBM 7094, which were far more powerful, by using semiconductor technology and ferrite memories, and were capable of executing up to 1×10^5 flops.

From the mid 1960s, machines proposed by Control Data Corporation (CDC) became favourites with the major scientific computing centres and especially with meteorological departments. In 1966, the CDC 6600 designed by Seymour Cray, one of CDC's founders, made it possible to attain 2.5×10^6 flops; it in turn was superseded in 1970 by the CDC 7600, capable of executing up to 7×10^6 flops.

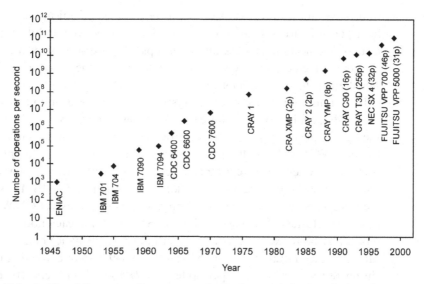

Growth in scientific mainframe computing power; sustained performance in floating point operations per second (flops).

1.4.3 Single and multiprocessor vector machines

In 1976, having left CDC to form Cray Research Inc., Seymour Cray (d. 1996) presented the first CRAY-1 supercomputer. This single processor machine, capable of sustained performance of 7×10^7 flops, used a *vector* processing unit, allowing it simultaneously to perform various stages of an arithmetical calculation on strings of operands (termed vectors) to output a result at each cycle of the internal clock. This type of machine rapidly made its mark as the standard for scientific computation because it was efficient and easy to program.

Cray Research Inc. went on to offer machines with a large central memory shared by a small number of vector processors working in parallel on the same calculation process. So the major meteorological departments in the 1980s used CRAY-XMP, CRAY-2, and CRAY-YMP multiprocessor machines, leading in 1991 to the CRAY YMP-C90, which was capable of executing up to 7×10^9 flops.

1.4.4 Massively parallel computers

The 1990s also saw the emergence of massively parallel machines. This arrangement was introduced because it was impossible to reduce the switching time of electronic circuits below a certain limit, but also because of the lower cost of manufacturing high-performance processors. The principle is to run many processors in parallel to perform a single task, which has been broken up beforehand. However, unlike the previous generation of machines, where memory was shared by the processors, this architecture uses distributed memory, with each processor having its own memory; data are transferred among processors or memories via a communication network. The really high performance systems now use a mixed *cluster* type architecture; an extremely fast communication network allows groups of processors (known as nodes) to share a common memory for exchanging the information they need to carry out the set of calculations distributed among all the processors.

In the range of massively parallel systems capable of executing 1×10^{10} flops, the T3D then T3E machines of Cray Research, Inc. (before it was bought by Silicon Graphics, Inc.) got some serious competition from Japanese machines such as NEC's SX-4 or Fujitsu's VPP-5000, credited with a power of 1×10^{11} flops. Recent installations of computing systems using cluster architecture for weather forecasting requirements can attain powers on the order of 10^{13} flops (10 teraflops). The IBM system installed at the European Centre for Medium-Range Weather Forecasts (ECMWF) in 2009 comprises two clusters of 286 nodes (P6–575 SMP servers), each node having 16 Power6-type scalar biprocessors, providing a sustained performance of 20 teraflops. In 2010, Météo-France installed a NEC SX-8R + 2 SX-9 system consisting of a 1×32-node cluster (8 vector processors per node) and 2×10-node clusters (16 vector processors per node), capable of attaining a sustained performance of 9 teraflops.

1.4.5 Software advances

The constant advance in hardware has meant modellers designing and writing numerical processing programmes have had to adapt to the technical characteristics of the machines available. The problem with the early scientific computers was that it was impossible to store all the data in the central memory; the programme had to be structured to carry out computations on slices of the atmosphere and to save the intermediate data on a drum or magnetic disc at each time step. The advent of vector computers meant programmers had to give precedence to *vectorizable* algorithms, making loop calculations on strings of operands, avoiding sequence breaks. With multiprocessor machines, the programmes had to be restructured and cut up into *independent* tasks that could be executed in parallel in a synchronized way on the available processors. The emergence of massively parallel machines and their distributed memory has again complicated the programmer's work: calculations must therefore be shared among independent modules operating on various processors while ensuring proper overall synchronization. To this end, standard interfaces have been developed to help programmers to optimize task parallelism and to secure memory access: the MPI (Message Passing Interface) is used to control inter-node parallelism and access to distributed memory, while the OpenMP interface can manage intra-node parallelism and access to shared memory.

2 Weather prediction equations

2.1 Introduction

The equations used to build the various types of model simulating the evolution of the atmosphere are obtained from the basic general equations by making a number of simplifications. These simplifications are justified by analysis of the order of magnitude of the various terms in the equations for the scales to be represented (see Table 1.1) and by the degree of simplification to be achieved so as to simulate the behaviour of the atmosphere.

From the equations describing the behaviour of a nonviscous fluid (also known as the *Euler equations*), the traditional approximation in meteorology consists in approximating the atmosphere to a thin layer and leads to a system of nonhydrostatic equations that allows for the proper handling of mesoscale atmospheric motion in particular. The hydrostatic approximation consists in neglecting vertical acceleration and leads to what are called the *primitive equations* (as opposed to the *filtered equations*, which involve an additional hypothesis of balance between mass and wind fields and that were used to build the first operational numerical models). Although they do not allow convective motion to be simulated explicitly, the primitive equations are widely used both for weather forecasting models and for general atmospheric circulation models.

Various characteristic phenomena of the atmosphere can be simulated by representing the atmosphere by a single layer of homogeneous fluid in the vertical; the appropriate system of equations for this kind of simplification is the *shallow water model*, which proves a very useful tool for testing the efficiency of numerical methods of solving these equations. Last, a further simplification, consisting in assuming that the horizontal wind is nondivergent, gives the *zero divergence model*, which was used for the first historical simulation by Charney, Fjörtoft, and von Neumann (1950).

After reviewing the various systems of equations and detailing the form of the primitive equation system in the pressure coordinates on the sphere, the form of these equations is examined for various mappings of the sphere onto a plane (polar stereographic, Mercator, and Lambert projections). Finally, an original way to work with a variable resolution on the sphere is presented by introducing a conformal mapping of the sphere onto itself.

2.2 The simplifications and the corresponding models

Starting from the equations in their general form, we need to detail the hypotheses that lead to the various systems of equations for the different models of the atmosphere.

2.2.1 The general form of the equations

Atmospheric models are built from the equation of motion, the mass conservation equation (or *continuity equation*), the energy conservation equation (or, to simplify, the *thermodynamic equation*), the water vapour conservation equation, and the equation of state. For unit mass, with a frame of reference attached to the Earth and having its origin located at the Earth's centre, the equations take the form:

$$\frac{d\mathbf{V}_3}{dt} = -2\mathbf{\Omega} \times \mathbf{V}_3 - \frac{1}{\rho}\nabla_3\, p - \nabla_3\,\Phi + \mathbf{F}, \text{ momentum equation,} \qquad (2.1)$$

$$\frac{dT}{dt} = \frac{R}{C_p}\frac{T}{p}\frac{dp}{dt} + \frac{Q}{C_p}, \text{ thermodynamic equation,} \qquad (2.2)$$

$$\frac{d\rho}{dt} = -\rho\,\nabla_3\,.\mathbf{V}_3, \text{ continuity equation,} \qquad (2.3)$$

$$\frac{dq}{dt} = M, \text{ water vapour equation,} \qquad (2.4)$$

$$p = \rho\,R\,T, \text{ equation of state.} \qquad (2.5)$$

The geopotential Φ is defined as the product of height z by acceleration due to gravity g, which combines only the effects of Newtonian gravity g^* and the centrifugal force (assuming that the Earth is isolated in space and so neglecting the effect of the other bodies of the solar system); it is expressed in J.kg^{-1} in SI units:

$$\Phi = g^*z - \frac{\Omega^2 r^2 \cos^2\varphi}{2} = gz,$$

where Ω is the angular velocity of rotation of the Earth, r the radial distance ($r = \|\,\mathbf{r}\,\|$, where $\mathbf{r}$ is the radius vector as measured from the Earth's centre), and $r\cos\varphi$, the distance to the Earth's rotation axis at the latitude φ.

In equations (2.1) to (2.5), $\mathbf{V}_3$ represents the three-dimensional wind velocity, $\mathbf{\Omega}$ the angular velocity vector of the Earth, ρ is air density, p pressure, T temperature, q specific humidity, R and C_p the perfect gas constant and specific heat at constant pressure for air. The symbol ∇_3 denotes the gradient operator, the symbol ∇_3. the divergence operator, and d/dt, the total derivative operator whose expression depends on the system of coordinates adopted. The term $-2\mathbf{\Omega} \times \mathbf{V}_3$ represents the Coriolis acceleration resulting from the choice of a frame of reference rotating with the Earth.

$\mathbf{F}$, Q and M represent the source and sink terms for momentum, heat, and specific humidity, respectively. Their somewhat complicated expression depends in particular on the scale of the atmospheric motion these models are intended to describe. For an adiabatic frictionless and water vapour conserving atmosphere, these quantities are equal to zero. These hypotheses are now maintained to explain the various forms of the equations and the way to solve them numerically. In this case, the water vapour equation reduces to an advection equation for a passive scalar (e.g. specific humidity) with no interaction with the other variables; it is discretized in a similar way to the advection term of temperature in the thermodynamic equation. This is why the water vapour equation will now be omitted from the system of equations, at least for the time being.

2.2.2 The traditional approximation and the nonhydrostatic equations

Several simplifications can be made taking into account the order of magnitude of the various terms in the equations for the terrestrial atmosphere when considered as a thin layer (Phillips, 1966).

- First, the ellipticity of the terrestrial geoid is ignored and acceleration due to gravity g is assumed constant: the quantity Φ/g represents the *geopotential height* and is expressed in *geopotential metres* (symbol gpm). Second, the radial distance r is replaced by the mean radius a of an assumed spherical Earth (*thin layer approximation*).
- The Earth's angular velocity vector $\mathbf{\Omega}$ is replaced by its local vertical component, which is written: $\mathbf{\Omega} = \Omega \mathbf{k} \sin \varphi = (f/2)\mathbf{k}$, where $\mathbf{k}$ indicates the radial unit vector (along the local vertical) and $f = 2\Omega \sin \varphi$ is the *Coriolis parameter*.
- Several metric terms (the radial components of the derivatives of the horizontal unit vectors and the derivative of the vector $\mathbf{k}$) are also ignored.

It is important to note that these last two simplifying assumptions are required to obtain a system that conserves angular momentum once r is replaced by a. This set of simplifications characterizes *the traditional approximation in meteorology*. The resulting simplified form of the equations can also be obtained directly by deducing the momentum equation from the *Lagrange equations* when replacing the initial metrics:

$$d\mathbf{r}^2 = r^2 \cos^2 \varphi \, d\lambda^2 + r^2 d\varphi^2 + dr^2 \text{ (where } \lambda \text{ is the longitude),}$$

by the metrics consistent with the thin layer approximation (Hinkelmann, 1969):

$$d\mathbf{r}^2 = a^2 \cos^2 \varphi \, d\lambda^2 + a^2 d\varphi^2 + dr^2.$$

Therefore, these equations verify the principle of conservation of angular momentum in a way that is consistent with the thin layer approximation (Phillips, 1966; Lorenz, 1967; Müller, 1989). The following equation is verified:

$$\frac{d}{dt}\left[a^2 \cos^2\varphi \left(\frac{d\lambda}{dt} + \Omega \right) \right] = -\frac{1}{\rho}\frac{\partial p}{\partial \lambda}, \qquad (2.6)$$

which expresses the fact that the total derivative of angular momentum (calculated when assuming that all the particles are situated at a distance a from the Earth's centre) is equal to the torque exerted by the longitudinal forces.

By writing $\mathbf{V}_3 = \mathbf{V} + w\mathbf{k}$ in the momentum equation (2.1) so as to separate the horizontal wind vector $\mathbf{V}$ (two-dimensional) and the vertical wind component w and now using ∇ and $\nabla.$ for the horizontal gradient and divergence, we obtain the system of nonhydrostatic equations (2.7) for an adiabatic frictionless atmosphere. These equations describe the evolution of five meteorological variables: the three components of wind velocity, plus temperature and pressure.

$$\left.\begin{aligned}
\frac{d\mathbf{V}}{dt} &= -f\,\mathbf{k}\times\mathbf{V} - \frac{RT}{p}\nabla p\,, \\[2mm]
\frac{dw}{dt} &= -\frac{RT}{p}\frac{\partial p}{\partial z} - g\,, \\[2mm]
\frac{dT}{dt} &= \frac{R}{C_p}\frac{T}{p}\frac{dp}{dt}\,, \\[2mm]
\frac{dp}{dt} &= -\frac{p}{1-\kappa}\left(\nabla.\,\mathbf{V} + \frac{dw}{dz} \right).
\end{aligned}\right\} \qquad (2.7)$$

The last equation is obtained by combining the continuity equation (2.3) and the thermodynamic equation (2.2) and by noting the constant R/C_p as κ.

This system of equations can be used to model atmospheric flows over a wide spectrum of spatial scales, from the planetary scale to the mesoscale. It allows us to simulate the propagation of Rossby waves, inertia-gravity waves, and even sound waves. Because of the thin layer approximation, it cannot be used to simulate geophysical fluids with large vertical extension (e.g. the gaseous planets).

2.2.3 The hydrostatic assumption and the primitive equations

If we are interested in the *synoptic scales* for which the vertical velocities are an order of magnitude smaller than the horizontal velocities, we can ignore the vertical acceleration *dw/dt* against the other terms of the vertical velocity equation. This equation then becomes a diagnostic relation (i.e. where the variable *t* does not appear) known as the *hydrostatic balance equation*:

$$\frac{\partial p}{\partial z} = -\rho g. \tag{2.8}$$

This assumption is justified for the terrestrial atmosphere if we are to look at phenomena whose horizontal scale exceeds 10 km or so.

By substituting the expression for density ρ, taken from the equation of state (2.5), in equation (2.8), we get another form of the hydrostatic relation which, when added to the other equations of system (2.7), gives the system (2.9) of *primitive equations* for an adiabatic frictionless atmosphere:

$$\left.\begin{aligned}
\frac{d\mathbf{V}}{dt} &= -f\,\mathbf{k} \times \mathbf{V} - \frac{RT}{p}\nabla p\,, \\
\frac{dT}{dt} &= \frac{R}{C_p}\frac{T}{p}\frac{dp}{dt}\,, \\
\frac{dp}{dt} &= -\frac{p}{1-\kappa}\left(\nabla.\mathbf{V} + \frac{dw}{dz}\right), \\
\frac{\partial p}{\partial z} &= -\frac{p}{RT}g.
\end{aligned}\right\} \tag{2.9}$$

This system of equations is relevant for simulating atmospheric motion whose horizontal space scale is greater than about 10 km, which excludes its use for the explicit modelling of convection. It allows us to take into account Rossby waves and inertia-gravity waves; but nevertheless, it eliminates sound waves because of the hydrostatic relation which has a *filtering* effect on them. The primitive equations are the basis of most numerical models used by meteorological services for weather forecasting (at any rate up to the late 1990s).

To address the different formulations of the primitive equations in various systems of horizontal coordinates, and for the sake of simplicity, the pressure vertical coordinate has been chosen, but the generalization to any other type of vertical coordinate is straightforward enough.

2.2.4 The primitive equations in the pressure coordinates

In the system using height z as the vertical coordinate, the horizontal derivatives are taken at a constant height z. In the system using pressure as the vertical coordinate, the horizontal derivatives are taken at constant pressure p (isobaric derivatives). We can switch from constant height derivatives to isobaric derivatives by the formula:

$$\left(\frac{\partial}{\partial l}\right)_z = \left(\frac{\partial}{\partial l}\right)_p - \frac{\partial}{\partial z}\left(\frac{\partial z}{\partial l}\right)_p, \tag{2.10}$$

where l is either of the two horizontal coordinates.

The use of the relations (2.10) shows that the geometrical gradient term $(RT/p)\nabla p$ in the equation for the horizontal wind $\mathbf{V}$ is replaced by the isobaric gradient $\nabla_p \Phi$. When developing the total derivative in the new system of coordinates, the equation of motion (2.1) becomes:

$$\frac{\partial \mathbf{V}}{\partial t} + \left(\mathbf{V} \cdot \nabla_p \right) \mathbf{V} + \omega \frac{\partial \mathbf{V}}{\partial p} = -f\, \mathbf{k} \times \mathbf{V} - \nabla_p \Phi,$$

and the thermodynamic equation:

$$\frac{\partial T}{\partial t} = -\mathbf{V} \cdot \nabla_p T - \omega \frac{\partial T}{\partial p} + \frac{R}{C_p} \frac{T}{p} \omega,$$

where $\omega = dp/dt$ is now the vertical velocity in the pressure coordinates.

By using the vector identity:

$$(\mathbf{V} \cdot \nabla_p)\mathbf{V} = \nabla_p \frac{\mathbf{V}^2}{2} + \zeta\, \mathbf{k} \times \mathbf{V},$$

where $\zeta = \mathbf{k} \cdot (\nabla_p \times \mathbf{V})$ is the vorticity (vertical component of curl) and $\mathbf{V}^2/2$ the kinetic energy per unit mass for the horizontal wind (noted K), we get the equation for the horizontal wind:

$$\frac{\partial \mathbf{V}}{\partial t} = -\omega \frac{\partial \mathbf{V}}{\partial p} - (f + \zeta)\, \mathbf{k} \times \mathbf{V} - \nabla_p (\Phi + K).$$

This formulation is said to be semi-invariant, because it involves vorticity ζ and kinetic energy K, two quantities that are independent of the system of coordinates.

The continuity equation (2.3) can be written:

$$\frac{d (\ln \rho)}{dt} = -\left(\nabla_z \cdot \mathbf{V} + \frac{\partial w}{\partial z} \right).$$

By taking into account equation (2.10) to develop the vertical derivative of vertical velocity $\partial w/\partial z = (\partial w/\partial p)(\partial p/\partial z)$, we get:

$$\frac{d(\ln \rho)}{dt} = -\left(\nabla_p \cdot \mathbf{V} - \frac{\partial \mathbf{V}}{\partial z} \nabla_p z + \frac{\partial w}{\partial p} \frac{\partial p}{\partial z} \right).$$

By replacing vertical velocity $w = dz/dt$ in this equation by its developed form:

$$w = \left(\frac{\partial z}{\partial t} \right)_p + \mathbf{V} \cdot \nabla_p z + \omega \frac{\partial z}{\partial p},$$

the continuity equation becomes:

$$\frac{d(\ln \rho)}{dt} = - \left(\nabla_p \cdot \mathbf{V} + \frac{\partial \omega}{\partial p} \right) - \frac{\partial p}{\partial z} \left(\frac{\partial}{\partial t} + \mathbf{V} \cdot \nabla_p + \omega \frac{\partial}{\partial p} \right) \left(\frac{\partial z}{\partial p} \right),$$

which can be re-written, after rearranging terms, as:

$$\frac{d}{dt} \left[\ln \left(\rho \frac{\partial z}{\partial p} \right) \right] = - \left(\nabla_p \cdot \mathbf{V} + \frac{\partial \omega}{\partial p} \right).$$

As the expression $\rho \partial z / \partial p$ is the constant $-1/g$, the continuity equation in pressure coordinates then reduces to:

$$\nabla_p \cdot \mathbf{V} + \frac{\partial \omega}{\partial p} = 0,$$

which is purely a diagnostic relation in which the time t does not appear.

We thus get the system (2.11) of primitive equations in pressure coordinates for an adiabatic and frictionless atmosphere:

$$\left. \begin{aligned}
\frac{\partial \mathbf{V}}{\partial t} &= -\omega \frac{\partial \mathbf{V}}{\partial p} - (f + \zeta) \, \mathbf{k} \times \mathbf{V} - \nabla_p (\Phi + K), \\
\frac{\partial T}{\partial t} &= -\mathbf{V} \cdot \nabla_p T - \omega \frac{\partial T}{\partial p} + \frac{R}{C_p} \frac{T}{p} \omega, \\
\nabla_p \cdot \mathbf{V} + \frac{\partial \omega}{\partial p} &= 0, \\
\frac{\partial p}{\partial z} &= -\frac{p}{RT} g.
\end{aligned} \right\} \qquad (2.11)$$

The detailed form of the primitive equations for the various systems of horizontal coordinates will be explained by using this form of the equations in the pressure coordinates.

2.2.5 The shallow water model equations

The system of equations can be further simplified by approximating the atmosphere to a fluid of limited depth for which the density ρ as well as the distribution of the horizontal velocity in the vertical are constant. With these assumptions, the state of the fluid at a given point is characterized by the horizontal wind vector $\mathbf{V}$ and the height of the free surface z, which is the level where the pressure p vanishes and is assumed to be zero (Figure 2.1).

In the equation of motion of system (2.11), the vertical advection term becomes zero, since the horizontal wind is constant in the vertical and remains so. The hydrostatic

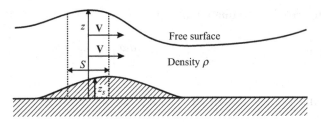

Figure 2.1   The shallow water model.

relation implies that $\partial(\nabla_p z)/\nabla p = 0$, since the density ρ is constant. Hence the term $\nabla_p \Phi$ is independent of the vertical and the pressure force can be evaluated as the gradient of the geopotential of the free surface $\Phi = gz$.

The continuity equation is obtained directly from the mass conservation principle for a column of fluid of constant density ρ, thickness $z - z_s$ (z_s being the altitude of the orography), and base area S.

$$\frac{d}{dt}\left[\rho g S(z - z_s)\right] = \rho g\left[S\frac{dz}{dt} + (z - z_s)\frac{dS}{dt}\right] = 0. \tag{2.12}$$

By introducing divergence:

$$\nabla \cdot \mathbf{V} = \frac{1}{S}\frac{dS}{dt},$$

we obtain the continuity equation (2.13):

$$\frac{\partial \Phi}{\partial t} = -\nabla \cdot \left[(\Phi - \Phi_s)\mathbf{V}\right], \tag{2.13}$$

where Φ_s is the geopotential at ground level.

Therefore the system of equations (2.14) corresponding to the shallow water model or homogeneous barotropic model is:

$$\left.\begin{array}{l}\dfrac{\partial \mathbf{V}}{\partial t} = -\left(f + \zeta\right)\mathbf{k} \times \mathbf{V} - \nabla\left(\Phi + K\right), \\[2mm] \dfrac{\partial \Phi}{\partial t} = -\nabla \cdot \left[(\Phi - \Phi_s)\mathbf{V}\right].\end{array}\right\} \tag{2.14}$$

This system of equations can also be obtained by integrating the equation of motion and the continuity equations of system (2.11) in the vertical, from the free surface where $p = 0$ to the ground surface where $p = \rho\,(\Phi - \Phi_s)$.

This system, often written setting $\Phi_s = 0$ (i.e. with no orography), describes the change in velocity of a column of fluid together with the change in the geopotential of its free surface (the level where the pressure vanishes). These are also known as the Saint-Venant equations after Barré de Saint-Venant (1871) who established them to study the motion of water in rivers. It is worth remembering that this model allows us to simulate both Rossby waves and inertia-gravity waves, just as the primitive equations do. That is why this model is often referred to as the primitive equation barotropic model. It is very widely used, especially for running tests, because it allows us to assess the properties of various numerical methods easily before using them in the general framework of the full primitive equations.

2.2.6 The zero divergence model equation

Returning to the equation for horizontal wind in system (2.14), by applying the operator $\mathbf{k}.\nabla \times$, we get the evolution equation for vorticity ζ:

$$\frac{\partial \zeta}{\partial t} = -\nabla \cdot \left[(\zeta + f)\mathbf{V}\right],$$

which is also written:

$$\frac{\partial \zeta}{\partial t} = -\mathbf{V} \cdot \nabla(\zeta + f) - (\zeta + f)\,\nabla \cdot \mathbf{V}. \tag{2.15}$$

This equation can be simplified by ignoring the divergence term of the horizontal wind:

$$\frac{\partial \zeta}{\partial t} = -\mathbf{V} \cdot \nabla(\zeta + f). \tag{2.16}$$

Because the Coriolis parameter f does not depend on time t, this equation becomes:

$$\frac{d}{dt}(\zeta + f) = 0. \tag{2.17}$$

This means that absolute vorticity $\zeta + f$ is conserved in purely rotational motion (i.e. with zero divergence).

Equation (2.16), like its Lagrangian formulation (2.17), is straightforward enough: as the wind divergence $\nabla \cdot \mathbf{V}$ is zero, the horizontal wind velocity $\mathbf{V}$ and its vorticity ζ can be expressed by a single variable, the stream function ψ.

This system describes Rossby waves only, because the zero divergence condition filters out inertia-gravity waves; this is why it is often called the barotropic filtered model. It was with the linear form of this equation that Rossby (1939) established the formula for computing the velocity of the waves that now carry his name; and it was with this equation too that Charney, Fjörtoft, and von Neumann (1950) successfully performed the first numerical weather prediction experiment. The model is still used as a nonlinear model for academic studies.

2.3 The equations in various systems of coordinates

2.3.1 Vector operators in curvilinear coordinates

Let's now consider a set of general orthogonal curvilinear coordinates x_1, x_2, x_3, that identify a point in geometrical space. The displacements in space of this point, ds_1, ds_2, ds_3, correspond to the infinitesimal variations of the coordinates dx_1, dx_2, dx_3, and are given by the relations:

$$ds_1 = h_1\, dx_1\,,\ \ ds_2 = h_2\, dx_2\,,\ \ ds_3 = h_3\, dx_3\,;$$

in which the scalar quantities h_1, h_2 and h_3 are locally defined and depend on the curvilinear coordinate system with which we have chosen to work.

The formulas for the gradient of geopotential Φ and for the divergence of the wind (with components u, v, and w, in the chosen curvilinear coordinate system) are:

$$\nabla_3 \Phi = \begin{pmatrix} \dfrac{1}{h_1}\dfrac{\partial \Phi}{\partial x_1} \\[2ex] \dfrac{1}{h_2}\dfrac{\partial \Phi}{\partial x_2} \\[2ex] \dfrac{1}{h_3}\dfrac{\partial \Phi}{\partial x_3} \end{pmatrix} \quad \text{and:}$$

$$\nabla_3 \cdot \mathbf{V}_3 = \frac{1}{h_1 h_2 h_3}\left[\frac{\partial}{\partial x_1}(h_2 h_3 u) + \frac{\partial}{\partial x_2}(h_1 h_3 v) + \frac{\partial}{\partial x_3}(h_1 h_2 w)\right].$$

The formula for the curl of the wind is then:

$$\nabla_3 \times \mathbf{V}_3 = \begin{pmatrix} \dfrac{1}{h_2 h_3}\left[\dfrac{\partial}{\partial x_2}(h_3 w) - \dfrac{\partial}{\partial x_3}(h_2 v)\right] \\[3ex] \dfrac{1}{h_1 h_3}\left[\dfrac{\partial}{\partial x_3}(h_1 u) - \dfrac{\partial}{\partial x_1}(h_3 w)\right] \\[3ex] \dfrac{1}{h_1 h_2}\left[\dfrac{\partial}{\partial x_1}(h_2 v) - \dfrac{\partial}{\partial x_2}(h_1 u)\right] \end{pmatrix}.$$

The third component of curl is the vorticity ζ.

One can just apply these classical formulas, which are established in any book covering vector analysis, to obtain the appropriate system of equations in any set of horizontal coordinates once the corresponding quantities h_1, h_2 and h_3 have been computed.

2.3.2 The equations in geographical coordinates

A point on a sphere is commonly located by its geographical coordinates: longitude λ and latitude φ (Figure 2.2).

The system of horizontal coordinates is then defined by:

$$x_1 = \lambda, \; x_2 = \varphi.$$

Along the vertical, the results already obtained in the pressure coordinates are used and, as there is no ambiguity, the notations ∇ and $\nabla \cdot$ now denote the gradient and the divergence on isobaric (constant pressure) surfaces.

To be consistent with the thin layer approximation ($r = a$), the formulas for the displacements corresponding to infinitesimal variations $d\lambda$ along a latitude circle and $d\varphi$ along a meridian are given by:

$$ds_1 = a \cos \varphi \, d\lambda, \quad ds_2 = a \, d\varphi,$$

while the pressure coordinate is used for the vertical:

$$ds_3 = dp.$$

These relations allow us to determine the quantities h_1, h_2, and h_3:

$$h_1 = a \cos \varphi, \, h_2 = a, \, h_3 = 1.$$

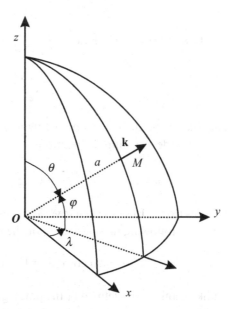

Figure 2.2   The system of natural geographical coordinates.

Hence, the primitive equations in geographical coordinates are:

$$
\left.
\begin{array}{l}
\dfrac{\partial u}{\partial t} = (\zeta + f)v - \omega \dfrac{\partial u}{\partial p} - \dfrac{1}{a \cos \varphi} \dfrac{\partial}{\partial \lambda}(K + \Phi)\,, \\[2ex]
\dfrac{\partial v}{\partial t} = -(\zeta + f)u - \omega \dfrac{\partial v}{\partial p} - \dfrac{1}{a} \dfrac{\partial}{\partial \varphi}(K + \Phi)\,, \\[2ex]
\dfrac{\partial T}{\partial t} = -\dfrac{u}{a \cos \varphi} \dfrac{\partial T}{\partial \lambda} - \dfrac{v}{a} \dfrac{\partial T}{\partial \varphi} - \omega \dfrac{\partial T}{\partial p} + \dfrac{R}{C_p} \dfrac{T}{p} \omega\,, \\[2ex]
\nabla \cdot \mathbf{V} + \dfrac{\partial \omega}{\partial p} = 0\,, \\[2ex]
\dfrac{\partial \Phi}{\partial \ln p} = -RT.
\end{array}
\right\}
\qquad (2.18)
$$

The quantities u and v represent the zonal and the meridional components of the wind $\mathbf{V}$ respectively.

Vorticity is given by:

$$
\zeta = \frac{1}{a^2 \cos \varphi}\left[\frac{\partial}{\partial \lambda}(av) - \frac{\partial}{\partial \varphi}(au \cos \varphi)\right],
$$

and divergence by:

$$
\nabla \cdot \mathbf{V} = \frac{1}{a^2 \cos \varphi}\left[\frac{\partial}{\partial \lambda}(au) + \frac{\partial}{\partial \varphi}(av \cos \varphi)\right].
$$

The quantity K represents the kinetic energy per unit mass and reads:

$$
K = \frac{u^2 + v^2}{2}.
$$

These equations can be further simplified by using the sine of the latitude, noted μ, instead of the latitude φ, giving the relations:

$$
\mu = \sin \varphi\,, \quad \sqrt{1 - \mu^2} = \cos \varphi\,, \quad d\mu = \cos \varphi \, d\varphi.
$$

Then, μ varies from -1 to $+1$ when φ varies from $-\pi/2$ to $+\pi/2$.
Introducing the following new variables for the horizontal wind:

$$
U = au \cos \varphi \text{ and } V = av \cos \varphi,
$$

which, unlike u and v, are defined at the poles, gives the simplified system (2.19):

$$\frac{\partial U}{\partial t} = (\zeta + f)V - \omega\frac{\partial U}{\partial p} - \frac{\partial}{\partial \lambda}(K + \Phi),$$

$$\frac{\partial V}{\partial t} = -(\zeta + f)U - \omega\frac{\partial V}{\partial p} - (1 - \mu^2)\frac{\partial}{\partial \mu}(K + \Phi),$$

$$\frac{\partial T}{\partial t} = -\frac{1}{a^2(1 - \mu^2)}\left[U\frac{\partial T}{\partial \lambda} + (1 - \mu^2)V\frac{\partial T}{\partial \mu}\right] - \omega\frac{\partial T}{\partial p} + \frac{R}{C_p}\frac{T}{p}\omega, \qquad (2.19)$$

$$\nabla.\mathbf{V} + \frac{\partial \omega}{\partial p} = 0,$$

$$\frac{\partial \Phi}{\partial \ln p} = -RT.$$

$$\text{with: } \zeta = \frac{1}{a^2(1 - \mu^2)}\left[\frac{\partial V}{\partial \lambda} - (1 - \mu^2)\frac{\partial U}{\partial \mu}\right],$$

$$\nabla \cdot \mathbf{V} = \frac{1}{a^2(1 - \mu^2)}\left[\frac{\partial U}{\partial \lambda} + (1 - \mu^2)\frac{\partial V}{\partial \mu}\right]$$

$$\text{and: } K = \frac{U^2 + V^2}{2a^2(1 - \mu^2)}.$$

2.3.3 Formulation of the equations for a conformal projection

A conformal projection (also called conformal mapping) is characterized by the following property: the ratio between any displacement on the map and its corresponding displacement on the sphere does not depend on the direction of the displacement; this ratio is called the map scale factor and is noted m. Such a projection preserves the angles, hence its popularity among sailors, who need to keep to a heading when steering a course.

On the projected map, the Cartesian coordinates x, y are used together with the relations:

$$x_1 = x, \qquad ds_1 = \frac{dx}{m},$$

$$x_2 = y, \qquad ds_2 = \frac{dy}{m},$$

defining a conformal projection of the sphere onto a plane.

From the formulas giving the gradient, divergence, and vorticity in curvilinear coordinates, and replacing h_1 and h_2 by $1/m$ and keeping $h_3 = 1$, we get the formulas for the primitive equation model (2.20):

$$\frac{\partial u}{\partial t} = (\zeta + f)v - \omega\frac{\partial u}{\partial p} - m\frac{\partial}{\partial x}(K + \Phi),$$

$$\frac{\partial v}{\partial t} = -(\zeta + f)u - \omega\frac{\partial v}{\partial p} - m\frac{\partial}{\partial y}(K + \Phi),$$

$$\frac{\partial T}{\partial t} = -m\left[u\frac{\partial T}{\partial x} + v\frac{\partial T}{\partial y}\right] - \omega\frac{\partial T}{\partial p} + \frac{R}{C_p}\frac{T}{p}\omega,$$

$$\nabla\cdot\mathbf{V} + \frac{\partial\omega}{\partial p} = 0,$$

$$\frac{\partial\Phi}{\partial\ln p} = -RT,$$

(2.20)

$$\text{with: } \zeta = m^2\left[\frac{\partial}{\partial x}\left(\frac{v}{m}\right) - \frac{\partial}{\partial y}\left(\frac{u}{m}\right)\right],$$

$$\nabla\cdot\mathbf{V} = m^2\left[\frac{\partial}{\partial x}\left(\frac{u}{m}\right) + \frac{\partial}{\partial y}\left(\frac{v}{m}\right)\right]$$

$$\text{and: } K = \frac{u^2 + v^2}{2}.$$

These equations can be further simplified by introducing the reduced wind components for the horizontal wind (Hollmann, 1959):

$$U = \frac{u}{m} \quad\text{and}\quad V = \frac{v}{m}.$$

They then reformulate into:

$$\frac{\partial U}{\partial t} = (\zeta + f)V - \omega\frac{\partial U}{\partial p} - \frac{\partial}{\partial x}(K + \Phi),$$

$$\frac{\partial V}{\partial t} = -(\zeta + f)U - \omega\frac{\partial V}{\partial p} - \frac{\partial}{\partial y}(K + \Phi),$$

$$\frac{\partial T}{\partial t} = m^2\left[U\frac{\partial T}{\partial x} + V\frac{\partial T}{\partial y}\right] - \omega\frac{\partial T}{\partial p} + \frac{R}{C_p}\frac{T}{p}\omega,$$

$$\nabla\cdot\mathbf{V} + \frac{\partial\omega}{\partial p} = 0,$$

$$\frac{\partial\Phi}{\partial\ln p} = -RT,$$

(2.21)

$$\text{with: } \zeta = m^2\left[\frac{\partial V}{\partial x} - \frac{\partial U}{\partial y}\right],$$

$$\nabla.\mathbf{V} = m^2 \left[\frac{\partial U}{\partial x} + \frac{\partial V}{\partial y} \right]$$

$$\text{and: } K = m^2 \left(\frac{U^2 + V^2}{2} \right).$$

2.4 Some typical conformal projections

2.4.1 Polar stereographic projection

This projection is widely used for mapping an entire hemisphere. The projection plane is orthogonal to the axis of the poles and intersects the Earth at the latitude φ_0 (Figure 2.3). A point M on the sphere is projected onto M', the intersection of the line SM, connecting the South Pole S to the point M, with the plane. On the projection, the parallels are represented by concentric circles around P', the projection of the North Pole P, while the meridians are represented by half-lines starting at P'. The scale depends on the latitude and takes the value 1 at the latitude φ_0 where the plane intersects the sphere.

The value of the map scale factor is calculated by examining a displacement along either a parallel or a meridian.

A point on the projection plane is identified either by its polar coordinates $\hat{\rho}$ (which in this case is the distance $P'M'$) and λ (its longitude), or by its Cartesian coordinates

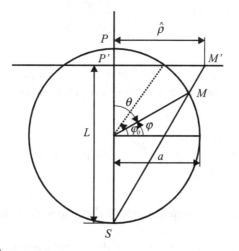 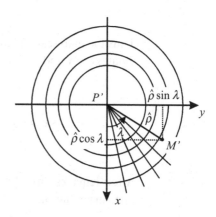

Figure 2.3   The polar stereographic projection.

x and y. By making the x-axis coincide with the origin meridian ($\lambda = 0$), the Cartesian coordinates are obtained from the polar coordinates by the relations:

$$x = \hat{\rho} \, \cos \lambda, \quad y = \hat{\rho} \sin \lambda,$$
$$\hat{\rho} = L \tan (\theta/2)$$

in which θ is the co-latitude, defined by $\theta = \pi/2 - \varphi$, and $L = a\,(1 + \sin \varphi_0)$ is the distance between the South Pole and the projection plane.

Based on the geometrical construction, the map scale factor is easily calculated from a displacement along a meridian and is:

$$m = -\frac{d\hat{\rho}}{a\,d\varphi} = \frac{-L\left(1 + \tan^2 \dfrac{\theta}{2}\right) d\theta}{-2\,a\,d\theta} \,.$$

The map scale factor calculated from a displacement along a parallel is:

$$m = \frac{\hat{\rho}\,d\lambda}{a \cos \varphi \, d\lambda} = \frac{L \, \tan \dfrac{\theta}{2}\left(1 + \tan^2 \dfrac{\theta}{2}\right)}{2a \tan \dfrac{\theta}{2}} \,.$$

It is verified that the projection is conformal since the expression of the map scale factor is the same in both cases. Taking into account the definition of the co-latitude θ, its expression simplifies as:

$$m = (1 + \sin \varphi_0) \left(\frac{1 + \tan^2 \dfrac{\theta}{2}}{2} \right) = \frac{1 + \sin \varphi_0}{1 + \sin \varphi} \,.$$

Thus the map scale factor m depends on the latitude φ alone. For a point whose coordinates are x and y on the projection plane, it is easy to compute $\sin \varphi$:

$$\sin \varphi = \cos \theta = \frac{1 - \tan^2 \dfrac{\theta}{2}}{1 + \tan^2 \dfrac{\theta}{2}} = \frac{L^2 - \hat{\rho}^2}{L^2 + \hat{\rho}^2}, \text{ with } \hat{\rho}^2 = x^2 + y^2.$$

A few remarks should be made on the practical use of the map scale factor when discretizing equations.

With a conformal projection, the discretization of the spatial derivatives by means of finite differences with a mesh Δx_0 always induces a term $m/\Delta x_0$:

$$m\frac{\partial\Phi}{\partial x}\approx\frac{m}{\Delta x_0}(\Phi_{i+1}-\Phi_{i-1});$$

by definition, this ratio is the inverse of the mesh size Δx_r on the sphere and we can set:

$$m'=\frac{m}{\Delta x_0}=\frac{1}{\Delta x_r};$$

the discretized equations can then be written by replacing m with m' while at the same time replacing the derivatives by simple differences. This working practice enables us to factorize multiplier m' and saves computing time.

Because of the expression of the map scale factor for this projection, after choosing a plane intersecting the sphere at the latitude φ_0, we have $\Delta x_r < \Delta x_0$ in the equatorial regions while $\Delta x_r > \Delta x_0$ in the polar regions. The actual mesh size varies with a ratio ranging from 1 to 2 between equatorial and polar regions. However, when using the Courant, Friedrichs, and Lewy criterion ($U\Delta t/\Delta x < 1$) to set the maximum time step, we need to take into account the minimum mesh size $\Delta x = \Delta x_{min}$ on the working domain; this intensifies the constraint on the maximum time step for relatively wide domains.

2.4.2 The Mercator projection

The polar stereographic projection is not well suited for mapping tropical and equatorial regions. Areas close to the equator can be better represented by using the Mercator projection. This consists in projecting the terrestrial sphere onto a cylinder of radius a' with its axis along the North–South Pole axis. The cylinder is then developed (Figure 2.4). A point M on the sphere is projected onto M' on the cylinder. The meridians then turn into equidistant parallel lines and the parallels become parallel line segments perpendicular to the meridians.

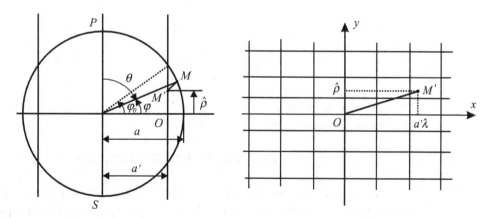

Figure 2.4   The Mercator projection.

Choosing the Cartesian axes in an appropriate way on the projection plane, with the equator as x-axis and with the y-axis set on the origin meridian, makes it easy to identify a point on the projection map from its coordinates as follows:

$$x = a'\lambda,$$

$$y = \hat{\rho},$$

where $\hat{\rho}$ represents here the distance of the point M' to the equator on the cylinder and λ the longitude.

The map scale factor along a parallel is given by:

$$\frac{a'd\lambda}{a \cos\varphi \, d\lambda},$$

while along a meridian it reads:

$$\frac{d\hat{\rho}}{a \, d\varphi}.$$

The value of $\hat{\rho}$ is then deduced from the definition of a conformal projection, which implies that the two map scale factors must be identical:

$$m = \frac{d\hat{\rho}}{a \, d\varphi} = \frac{a'}{a \cos\varphi}.$$

Integrating the differential equation $d\hat{\rho} = (a'/\cos\varphi)d\varphi$ allows us to determine $\hat{\rho}$:

$$\hat{\rho} = a'\int_0^\varphi \frac{d\varphi'}{\cos\varphi'}.$$

Re-writing this equation with $\theta' = \pi/2 - \varphi'$,

$$\hat{\rho} = -a'\int_{\frac{\pi}{2}}^\theta \frac{d\theta'}{\sin\theta'}$$

gives the relation:

$$\hat{\rho} = -a' \ln\left[\tan\left(\frac{\pi}{4} - \frac{\varphi}{2}\right)\right], \text{ also written } \varphi = \frac{\pi}{2} - 2 \arctan\left(e^{-\hat{\rho}/a'}\right).$$

These formulas enable us to compute the latitude and the map scale factor from the Cartesian coordinates on the projection plane.

2.4.3 The Lambert conical projection

The Lambert conical projection is very efficient for mapping a part of the Earth located at the mid-latitudes onto a plane. It consists in mapping the sphere onto a cone with an aperture angle α and an axis set on the polar axis. When developing this cone, the meridians are represented by half-lines drawn from the apex of the cone and the parallels are transformed into arcs of concentric circles (Figure 2.5). This transformation allows us to reduce the deformation of the spherical fields mapped onto the plane; however, its major drawback is that only limited domains in longitude can be represented because of the lack of continuity along a parallel.

When developing the cone, the circle on the cone located at the distance $\hat{\rho}$ from the apex, having a circumference of $2\pi\hat{\rho}\sin\alpha$, is transformed into an arc of angle Ψ_M from a circle of radius $\hat{\rho}$ (see Figure 2.5). The variation of the angle $\Delta\Psi$ on the projection corresponding to the variation $\Delta\lambda$ on the sphere is then given by:

$$\frac{\Delta\Psi}{\Delta\lambda} = \frac{\Psi_M}{2\pi} = \sin\alpha.$$

Conveniently choosing the axis system, with the origin located at the apex of the cone and the x-axis set on the origin meridian, the Cartesian coordinates of a point on the projection are obtained by:

$$\begin{aligned}
x &= \hat{\rho}\cos\left(\lambda\sin\alpha\right), \\
y &= \hat{\rho}\sin\left(\lambda\sin\alpha\right), \\
\hat{\rho}^2 &= x^2 + y^2.
\end{aligned}$$

Again, the relation between $\hat{\rho}$ and the co-latitude θ is then deduced from the definition of a conformal mapping, namely that the two map scale factors, along a parallel and along a meridian, need to be identical.

This leads to the relations:

$$m = \frac{d\hat{\rho}}{ad\theta} = \frac{\hat{\rho}\sin\alpha.d\lambda}{a\sin\theta.d\lambda},$$

and to the differential equation, $d\hat{\rho} = \hat{\rho}\sin\alpha\, d\theta/\sin\theta$, which, once integrated, gives:

$$\hat{\rho} = \hat{\rho}_0\left(\tan\frac{\theta}{2}\right)^{\sin\alpha} = \hat{\rho}_0\left[\tan\left(\frac{\pi}{4} - \frac{\varphi}{2}\right)\right]^{\sin\alpha}.$$

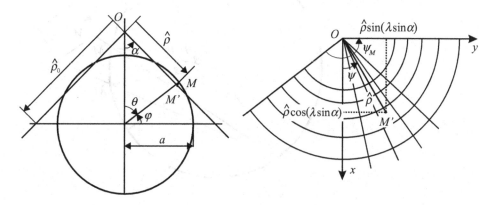

Figure 2.5   The Lambert conical projection.

The two constants, $\hat{\rho}_0$, the distance from the pole to the equator on the map, and $\sin \alpha$, can be determined by setting the co-latitudes θ_1 and θ_2 (intersections of the sphere with the cone, where the map scale factor is equal to 1 at a convenient location, taking into account the part of the sphere to be represented on the projection.

$$\sin \alpha = \ln \left(\frac{\sin \theta_1}{\sin \theta_2} \right) \div \ln \left(\frac{\tan \dfrac{\theta_1}{2}}{\tan \dfrac{\theta_2}{2}} \right),$$

$$\hat{\rho}_0 = \frac{a \sin \theta_1}{\sin \alpha} \left(\cot \frac{\theta_1}{2} \right)^{\sin \alpha} \quad \text{or} \quad \hat{\rho}_0 = \frac{a \sin \theta_2}{\sin \alpha} \left(\cot \frac{\theta_2}{2} \right)^{\sin \alpha}.$$

The expression of the map scale factor at any latitude φ is then given by:

$$m = \frac{\hat{\rho}_0 \sin \alpha}{a \, \cos \varphi} \left[\tan \left(\frac{\pi}{4} - \frac{\varphi}{2} \right) \right]^{\sin \alpha}.$$

2.4.4 The conformal transformation of the sphere onto itself

The principle of the conformal transformation of the sphere, described here, was introduced by Schmidt (1977). However, a clear interpretation of this transformation is credited to Courtier and Geleyn (1988). Let's consider the set of three successive transformations that are illustrated in Figure 2.6.

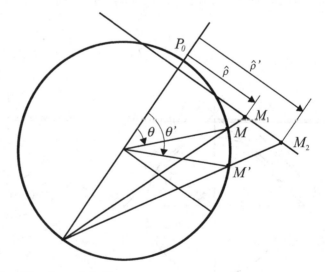

Figure 2.6   Conformal projection of the sphere onto itself.

First, there is a polar stereographic projection of the sphere on a plane, choosing any pole axis (not necessarily the North–South axis). Point M is then transformed into point M_1. Then in this plane, a dilatation of centre P_0 and ratio C is applied, which transforms point M_1 into point M_2. And finally an inverse polar stereographic projection is applied to transform the plane back to the sphere. And point M_2 is transformed into point M'.

Each of the three transformations is conformal, and so is their combination, $\mathcal{C}$ and it is even possible to demonstrate that this resulting transformation is the only non-trivial conformal mapping of the sphere onto itself. Thus we can transform any real sphere (on which geographical coordinates and meteorological fields are defined) into another sphere on which computations are carried out. It is especially valuable in that it allows us to zoom in on an area of interest on the real sphere while zooming out of the area located at the antipodes.

Figures 2.7a and 2.7b illustrate the effect of this transformation on the field representing the continental surfaces of the Earth. In these figures, the dilatation pole P_0 is located in the centre of France and the dilatation factor is of 3.5. It is important to understand that the transformed globe (Figure 2.7b) has been rotated to show the dilatation pole P_0 in the centre of the figure.

The inverse transformation $\mathcal{C}^{-1}$ carried out from a regular grid on the sphere (Figure 2.8a) provides a new grid with an increased resolution in the vicinity of the dilatation pole P_0 situated in the centre of France while resolution is reduced at the antipodes (Figure 2.8b).

Thus, working on a regular grid (Figure 2.8a) with transformed fields (Figure 2.7b) is equivalent to working on a stretched grid (Figure 2.8b) with real fields (Figure 2.7a).

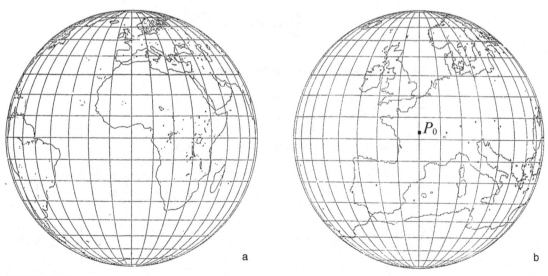

a   b

Figure 2.7 Zooming effect on the shape of the continents. (a) real field; (b) dilated field after rotation of the dilatation pole. (Météo-France image)

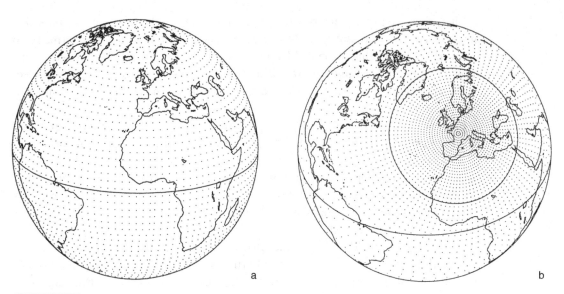

Figure 2.8   Stretching effect on the regular grid. (a) regular grid; (b) stretched grid. (Météo-France image)

In both cases, a structure of a given scale can be discretized with the same number of grid points. Obviously such zooming can be done for whichever part of the world is of interest.

Of course, when working with the stretched grid, it is necessary to use equations that take into account the metrics of the transformed sphere. These are easy enough to derive given that the map scale factor m is locally the ratio of the distances on the transformed sphere to the corresponding distances on the real sphere. By setting:

$$U = \frac{a\,u}{m}\cos\varphi \text{ and } V = \frac{a\,v}{m}\cos\varphi,$$

we get the following equations on the transformed sphere:

$$
\left.
\begin{aligned}
&\frac{\partial U}{\partial t} = (\zeta + f)V - \omega\frac{\partial U}{\partial p} - \frac{\partial}{\partial\lambda}(K + \Phi)\,, \\[2mm]
&\frac{\partial V}{\partial t} = -(\zeta + f)U - \omega\frac{\partial V}{\partial p} - (1 - \mu^2)\frac{\partial}{\partial\mu}(K \mid \Phi)\,, \\[2mm]
&\frac{\partial T}{\partial t} = -\frac{m^2}{a^2(1-\mu^2)}\left[U\frac{\partial T}{\partial\lambda} + (1-\mu^2)V\frac{\partial T}{\partial\mu}\right] - \omega\frac{\partial T}{\partial p} + \frac{R}{C_p}\frac{T}{p}\omega\,, \\[2mm]
&\nabla.\mathbf{V} + \frac{\partial\omega}{\partial p} = 0\,, \\[2mm]
&\frac{\partial\Phi}{\partial\ln p} = -RT
\end{aligned}
\right\}
\qquad (2.22)
$$

with: $\zeta = \dfrac{m^2}{a^2(1-\mu^2)}\left[\dfrac{\partial V}{\partial \lambda} - (1-\mu^2)\dfrac{\partial U}{\partial \mu}\right],$

$$\nabla.\mathbf{V} = \dfrac{m^2}{a^2(1-\mu^2)}\left[\dfrac{\partial U}{\partial \lambda} + (1-\mu^2)\dfrac{\partial V}{\partial \mu}\right],$$

and: $K = \dfrac{m^2\left(U^2+V^2\right)}{2a^2(1-\mu^2)}.$

Figure 2.6 shows that point M at the co-latitude θ, identified by $\cos\theta = \sin\varphi = \mu$ on the real sphere, is transformed into point M' at the co-latitude θ' identified by $\cos\theta' = \sin\varphi' = \mu'$ on the transformed sphere. The transformations allowing us to pass from one set of coordinates to the other are deduced from the basic relation defining the transformation, namely:

$$\tan\frac{\theta'}{2} = C\tan\frac{\theta}{2},$$

where C is the stretching factor.

The relation between μ' and μ (and therefore between φ' and φ) is obtained by expressing $\tan(\theta'/2)$ and $\tan(\theta/2)$ with respect to $\cos\theta'$ and $\cos\theta$:

$$\mu = \dfrac{(C^2-1)+\mu'\,(C^2+1)}{(C^2+1)+\mu'\,(C^2-1)}.$$

Differentiating this basic relation we get:

$$\left(1+\tan^2\frac{\theta'}{2}\right)\frac{d\theta'}{2} = C\left(1+\tan^2\frac{\theta}{2}\right)\frac{d\theta}{2},$$

giving us the map scale factor m:

$$m = \frac{d\theta'}{d\theta} = \frac{C\left(1+\tan^2\dfrac{\theta}{2}\right)}{\left(1+C^2\tan^2\dfrac{\theta}{2}\right)},$$

which can be re-written as a function of μ' defining the location of M' on the transformed grid:

$$m = \frac{(C^2+1)+\mu'(C^2-1)}{2C}. \tag{2.23}$$

The map scale factor is a first degree polynomial of the variable μ', a property which is of great computational interest when integrating these equations by the spectral method.

It is worth highlighting here that the transformed fields can be treated either by using a regular latitude-longitude-type grid or by expanding them into series of functions by using the spectral method (that will be explained later on). In both cases the meteorological fields are computed with a variable spatial resolution on the terrestrial sphere.

Finite differences

3.1 Introduction

The finite difference method is very widely used for numerically solving partial differential equations in science in general and in fluid mechanics in particular because it is simple to implement. In fact, it was the method Richardson (1922) used for the first attempt at weather forecasting by hand calculation. While methods using series expansion in terms of functions now compete with the finite difference method for the horizontal processing of meteorological fields, vertical discretization, to some extent, and time discretization still utilize finite differences. This presentation of finite differences is not designed to demonstrate the basic results that can be found in books on numerical analysis but rather to show in concrete terms how the finite difference method is applied to numerical weather prediction models.

3.2 The finite difference method

3.2.1 Computational principle, order of accuracy

The finite difference method consists in evaluating the partial derivatives of a function at one point from the differences between the values of that function at adjacent points. This procedure can be applied systematically insofar as a given continuous field is known in a discrete way. The values of the field to be processed are determined at the nodes of a regular grid (for example, rectangular grid for a two-dimensional field) characterized by its mesh size (Δx, Δy), applied over the spatial domain for which solutions are sought.

In the general case, we have to calculate the partial derivatives for a field depending on several variables $A(x, y, z, t)$, where x, y, z, and t denote the space and time variables. We look first at the derivative with respect to the variable x alone, and so, somewhat improperly, note as $A(x)$ the field $A(x, y, z, t)$ for which the values of variables y, z, and t are fixed.

To calculate the spatial derivative with respect to the variable x in a discrete way, we take a one-dimensional grid of mesh size Δx (Figure 3.1), at the nodes of which we know the values of the field $A(x)$.

The value of the derivative of A at point x can be calculated from the approximation:

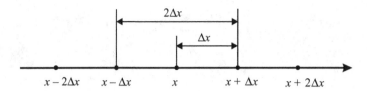

Figure 3.1   Use of a grid to calculate spatial derivatives.

$$\frac{\partial A}{\partial x} \approx \frac{A(x + \Delta x) - A(x)}{\Delta x}.$$

The Taylor expansion in the vicinity of x gives:

$$A(x + \Delta x) = A(x) + \frac{\Delta x}{1!} \frac{\partial A}{\partial x} + \frac{\Delta x^2}{2!} \frac{\partial^2 A}{\partial x^2} + \mathrm{O}(\Delta x^3).$$

We verify therefore that:

$$\frac{A(x + \Delta x) - A(x)}{\Delta x} = \frac{\partial A}{\partial x} + \frac{\Delta x}{2!} \frac{\partial^2 A}{\partial x^2} + \mathrm{O}(\Delta x^2). \tag{3.1}$$

This evaluation, called a *forward difference*, is a 1st-order accuracy approximation of the derivative of A at point x.

A similar evaluation may be made with a *backward difference* which gives:

$$\frac{A(x) - A(x - \Delta x)}{\Delta x} = \frac{\partial A}{\partial x} - \frac{\Delta x}{2!} \frac{\partial^2 A}{\partial x^2} + \mathrm{O}(\Delta x^2). \tag{3.2}$$

In this way we get 1st-order accuracy evaluations by approximating the behaviour of A in the vicinity of x to that of a 1st-degree polynomial function of x. Likewise, we get a 2nd-(resp. 4th) order accuracy approximation, by locally approximating the behaviour of A in the vicinity of x to that of a 2nd-(resp. 4th) degree polynomial function of x.

The *central difference* evaluation is obtained by adding (3.1) and (3.2) and gives a 2nd-order accuracy approximation:

$$\frac{A(x + \Delta x) - A(x - \Delta x)}{2\Delta x} = \frac{\partial A}{\partial x} + \frac{\Delta x^2}{3!} \frac{\partial^3 A}{\partial x^3} + \mathrm{O}(\Delta x^3). \tag{3.3}$$

By using a combination of central differences calculated over the intervals $2\Delta x$ and $4\Delta x$, we get a 4th-order accuracy approximation, which is written:

$$\frac{4}{3} \frac{A(x + \Delta x) - A(x - \Delta x)}{2\Delta x} - \frac{1}{3} \frac{A(x + 2\Delta x) - A(x - 2\Delta x)}{4\Delta x} = \frac{\partial A}{\partial x} - \frac{4\Delta x^4}{5!} \frac{\partial^5 A}{\partial x^5} + \mathrm{O}(\Delta x^5).$$

Generally, the truncation error of an approximation is defined as the difference between the value of an expression calculated numerically and its exact value. Notice that, for the calculation of the first derivative, the increased accuracy implies additional computation, since the 4th-order approximation comprises more terms than the 2nd-order approximation.

Of course, these approximations using differences can be employed for calculating higher order derivatives (second, third derivatives, etc.). We just apply the finite difference method to the discretized expressions of the lower order derivatives.

3.2.2 Common notations for finite differences

Writing out discretized equations in detail can be somewhat laborious. Accordingly some authors have come up with notations that shorten their formulation.

First we take the case of a one-dimensional field $A(x)$ that is known at discrete points x, $x + \Delta x$, $x + 2\Delta x$, ... of a grid with regular Δx spacing (points marked • in Figure 3.2).

We use two basic notations to identify an evaluation of a mean value of A and of its spatial derivative $\partial A/\partial x$ at the points located in the centre of the grid meshes (the points marked + in Figure 3.2). They are written:

$$\overline{A}^x = \frac{1}{2}\left[A\left(x + \frac{\Delta x}{2}\right) + A\left(x - \frac{\Delta x}{2}\right)\right], \tag{3.4}$$

$$A_x = \frac{1}{\Delta x}\left[A\left(x + \frac{\Delta x}{2}\right) - A\left(x - \frac{\Delta x}{2}\right)\right]. \tag{3.5}$$

These notations, proposed by Shuman (1962), are used to identify evaluations of the mean value of A and of its spatial derivative $\partial A/\partial x$ at the grid points marked + using the values A takes at the points •. These expressions may be considered as the result of the application of operators: a mean operator along x (noted $\overline{A}^x$), a difference operator along x (noted A_x, but often also $\delta_x A$ in the literature on finite differences).

These operators can be combined several times and after some elementary algebra they yield:

$$\overline{A}^{xx} = \overline{\left(\overline{A}^x\right)}^x = \frac{1}{4}\left[A(x + \Delta x) + 2A(x) + A(x - \Delta x)\right],$$

which is another way of calculating a mean of A on x.

Similarly, we can calculate:

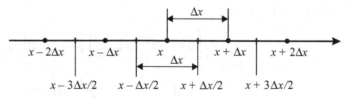

Figure 3.2   Location of the various sets of points on the one-dimensional grid.

$$A_{xx} = (A_x)_x = \frac{1}{\Delta x^2}\Big[A(x+\Delta x) + A(x-\Delta x) - 2A(x)\Big], \qquad (3.6)$$

which is an approximation of the second derivative of A in x, with 2nd-order accuracy.

By applying the mean operator after the difference operator (or vice versa), we obtain the central difference approximation of the first derivative of A in x:

$$\overline{A}_x^{\,x} = \frac{1}{2\Delta x}\Big[A(x+\Delta x) - A(x-\Delta x)\Big]. \qquad (3.7)$$

There are remarkable identities among certain finite difference expressions. For a given field $A(x)$ on a grid with mesh Δx (points marked • in Figure 3.2) and a field $B(x)$ defined on the grid offset by $\Delta x/2$ (points marked + in Figure 3.2), algebraic calculation enables us to verify the identity:

$$(\overline{A}^{\,x}B)_x \equiv AB_x + \overline{BA_x}^{\,x}. \qquad (3.8)$$

This identity is merely the straightforward transposition in finite differences of the formula for derivation of a product:

$$\frac{\partial}{\partial x}(AB) = A\frac{\partial B}{\partial x} + B\frac{\partial A}{\partial x};$$

it shall be utilized to make certain groupings in the construction of an evolution equation for total energy that shall be addressed in Chapters 6 and 7.

Notice that the successive application of both of the operators defined leads to quantities being evaluated at the same points as in the departure grid. By construction, it can be seen therefore that the number n_u of superscripts x gives the numerical coefficient that must be applied to the expression calculated; thus, the numerical coefficient $1/2^{n_u}$ corresponds to n_u. Likewise, the number n_d of subscripts x gives the mesh exponent for the denominator of the expression calculated; thus the denominator $(\Delta x)^{n_d}$ corresponds to n_d. When $n_u + n_d$ is odd, the quantity is evaluated on a grid that is offset with respect to the initial grid (points marked + in Figure 3.2). When $n_u + n_d$ is even, the quantity is evaluated at the same points as on the original grid (points marked • in Figure 3.2).

In the case of a field $A(x,y)$ that is a function of the two spatial variables x and y, the notations are easily extended to a two-dimensional grid and the mean and difference operators along x and y can be combined. Thus, the average of four grid points at the corners of the rectangle defining the grid, evaluated in the centre (point marked × in Figure 3.3), is written:

$$\overline{A}^{\,xy} = \frac{1}{4}\left[\begin{array}{l} A\left(x+\dfrac{\Delta x}{2}, y+\dfrac{\Delta y}{2}\right) + A\left(x+\dfrac{\Delta x}{2}, y-\dfrac{\Delta y}{2}\right) \\[2ex] + A\left(x-\dfrac{\Delta x}{2}, y+\dfrac{\Delta y}{2}\right) + A\left(x-\dfrac{\Delta x}{2}, y-\dfrac{\Delta y}{2}\right)\end{array}\right].$$

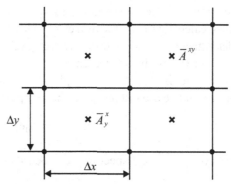

Figure 3.3 Location of quantities $\overline{A}^{xy}$ and $\overline{A}_y^{x}$ on a two-dimensional grid.

It can be verified that the mean and differences operators all commute with each other; thus we have:

$$\left(\overline{\overline{A}^x}\right)^y = \left(\overline{\overline{A}^y}\right)^x = \overline{A}^{xy}.$$

The total number of superscripts equals 2 and gives the multiplying factor 1/4; the number of indices relative to each variable is 1 and the quantity is therefore evaluated at the grid points marked × from the values taken by A at the points marked • (Figure 3.3). To get an average A evaluated at the original grid points (points marked •), we must apply the mean operator for a second time along x and y and calculate $\overline{A}^{xxyy}$.

We can also combine the mean operator in one direction and the difference operator in the other direction. Thus:

$$\overline{A}_y^x = \frac{1}{2\Delta y}\left[\begin{array}{l} A\left(x+\dfrac{\Delta x}{2},y+\dfrac{\Delta y}{2}\right) - A\left(x+\dfrac{\Delta x}{2},y-\dfrac{\Delta y}{2}\right) \\ + A\left(x-\dfrac{\Delta x}{2},y+\dfrac{\Delta y}{2}\right) - A\left(x-\dfrac{\Delta x}{2},y-\dfrac{\Delta y}{2}\right) \end{array}\right]$$

represents the derivative of A along y, calculated in the centre of the grid mesh.

The 2nd-order accuracy approximation of the Laplacian of a two-dimensional field for a regular grid ($\Delta x = \Delta y$) can be evaluated at the original grid points from the classical formula using the value inputs at five points:

$$A_{xx} + A_{yy} = \frac{1}{\Delta x^2}\left[A(x+\Delta x,y) + A(x,y+\Delta y) + A(x-\Delta x,y) + A(x,y-\Delta y) - 4A(x,y)\right]. \tag{3.9}$$

3.2.3 The accuracy of finite difference schemes

The truncation error of the approximation is defined as the difference between the value of the derivative calculated numerically and its true value. When the true value of the derivative of A can be calculated, it is easy to determine the truncation error. We can then calculate the truncation error made when using finite differences to evaluate the derivative of a sine function whose wavelength corresponds to the scale of the spatial structure we wish to represent.

Let $A(x)$ be the sine function of amplitude A_0 and wavenumber k, given by the complex exponential (we are interested in the real part of the function only):

$$A(x) = A_0 e^{ikx}.$$

The derivative of A with respect to x can be calculated analytically and is:

$$\frac{dA}{dx} = ikA(x).$$

The expression of the derivative calculated by 2nd-order accuracy central difference (3.8) can also be calculated analytically:

$$\overline{A}_x^x = A_0 \frac{e^{ikx}(e^{ik\Delta x} - e^{-ik\Delta x})}{2\Delta x} = A(x)\, i\, \frac{\sin(k\Delta x)}{\Delta x}.$$

The relative truncation error $|\varepsilon_T|_2$, which is the ratio of the truncation error to the analytical value of the function, is given by:

$$|\varepsilon_T|_2 = \left| \frac{\dfrac{dA}{dx} - \overline{A}_x^x}{\dfrac{dA}{dx}} \right| = \left| 1 - \frac{\sin(k\Delta x)}{k\Delta x} \right|.$$

The same calculation applied to the 4th-order accuracy central difference expression gives as the truncation error:

$$|\varepsilon_T|_4 = \left| 1 - \frac{\sin(k\Delta x)}{k\Delta x} \frac{[(4 - \cos(k\Delta x)]}{3} \right|.$$

It can be seen then that the truncation error varies with the wavenumber and mesh size, that is, with the characteristic scale of the field to be processed with respect to the size of the mesh used. Table 3.1 shows, by way of example, the truncation errors calculated for some wavelengths $\lambda = 2\pi/k$, that are multiples of mesh size Δx.

Examination of the table shows that the truncation errors of the finite difference approximations vary with the spatial scale to be handled. In all instances, for $\lambda = 2\Delta x$, the relative error is substantial (100%). In the case of a 2nd-order accuracy approximation, it is only from $\lambda \geq 8\Delta x$ onward that we get results that are acceptable in practice, albeit with a relative error of 10%. The 4th-order accuracy approximation, which requires more computation, performs better as the relative error is only 15% for $\lambda = 4\Delta x$ and reaches 1.2% for $\lambda = 8\Delta x$.

Table 3.1 Relative truncation errors for 2nd- and 4th-order schemes, for various wavelengths expressed as multiples of mesh size Δx.

λ	$k\Delta x$	$\|\varepsilon_T\|_2$ (%)	$\|\varepsilon_T\|_4$ (%)
$2\Delta x$	π	100	100
$4\Delta x$	$\pi/2$	36	15
$6\Delta x$	$\pi/3$	17	3,5
$8\Delta x$	$\pi/4$	10	1,2
$10\Delta x$	$\pi/5$	6,5	0,5
$12\Delta x$	$\pi/6$	4,4	0,2
$14\Delta x$	$\pi/7$	3,3	0,1
$16\Delta x$	$\pi/8$	2,6	0,08
$\downarrow$	$\downarrow$	$\downarrow$	$\downarrow$
∞	0	0	0

Accuracy is that much greater when mesh size is smaller with respect to wavelength, which is natural enough. This implies that seeking greater accuracy by reducing mesh size has as its counterpart an increase in the number of computations to be made. We need, then, given the available computing power, to find the right trade-off between accuracy and computation time.

We can now expect a few systematic effects arising from the finite differences, even for handling the simplest evolution problems. For example, if we use central differences to seek periodic solutions for the equation of advection with constant velocity U:

$$\frac{\partial A}{\partial t} = -U\frac{\partial A}{\partial x}, \tag{3.10}$$

the evaluation of the spatial derivative in central differences for a sine function of wavenumber k gives:

$$\frac{\partial A}{\partial t} = -U\frac{\sin(k\Delta x)}{k\Delta x}\frac{\partial A}{\partial x}. \tag{3.11}$$

This is an advection equation characterized by a propagation velocity:

$$U' = U\frac{\sin(k\Delta x)}{k\Delta x}.$$

As the multiplying factor is always less than 1, the effect of spatial discretization is reflected by a deceleration of the wave, which slows more for higher wavenumbers. Notice that velocity falls to zero for $k\Delta x = \pi$; this means that the wavelength $\lambda = 2\Delta x$ remains stationary.

The improved accuracy of finite difference approximations for spatial derivatives may be applied to time derivatives. The computation is formally identical, the wavenumber k being superseded by frequency ω and wavelength λ by period $T = 2\pi/\omega$ of the phenomenon under study.

Using the 2nd-order accuracy central difference approximation to evaluate the time derivative in the advection equation, assuming that the spatial derivative is evaluated exactly, leads to the equation:

$$\frac{\partial A}{\partial t} \frac{\sin(\omega \Delta t)}{\omega \Delta t} = U \frac{\partial A}{\partial x}. \tag{3.12}$$

This is an advection equation characterized by a propagation velocity:

$$U'' = U \left[\frac{\sin(\omega \Delta t)}{\omega \Delta t} \right]^{-1}.$$

Since the multiplying factor is always greater than 1, time discretization accelerates the wave. The higher the frequency or the shorter the period of the phenomenon under study, the greater the acceleration.

It is helpful to appreciate the implications of such constraints when designing the characteristics of a model. We assume, for instance, a small-scale synoptic structure akin to a sine wave (low pressure zone following a high pressure zone) typically with a wavelength of about 1000 km and moving at 20 m/s. The period of the phenomenon for a motionless observer is then 50 000 seconds, or about 14 hours.

To properly evaluate the 2nd-order accuracy spatial derivatives, we choose a mesh size of the order of 60 km ($\lambda/\Delta x \approx 16$). By the same reasoning, we take a time step of a little over an hour to evaluate the time derivatives with an accuracy equivalent to that for spatial derivatives ($T/\Delta t \approx 16$). However, with classical time integration schemes, the Courant-Friedrichs-Lewy condition (CFL for short), which shall be explained in detail in Chapter 5, forces us to choose Δt such that $U\Delta t/\Delta x < 1$, where U is the maximum speed of the waves handled by the model. By returning to the example of a mesh size of 60 km, the CFL constraint means we must take a time step of 4 minutes for a filtered model (processing slow waves only, of maximum speed of less than 50 m/s) or even of 30 seconds in the case of a primitive equation model (also handling gravity waves, the fastest of which may reach speeds of 300 m/s).

Thus the need to ensure the numerical scheme is stable compels us to adopt a time step such that the errors made in evaluating the time derivatives are far less than those made in evaluating the space derivatives. This has naturally led numerical analysts to look for time integration schemes with less restrictive constraints with respect to the choice of time step. This is how the *semi-implicit* and *semi-Lagrangian* techniques came to be proposed for constructing schemes for use with larger time steps.

3.3 The grids used and their properties

The equations used in numerical prediction involve several variables: wind components, temperature, and pressure. Accordingly numerical analysts have imagined a wide variety of grids differing in the disposition of the diverse variables so the various terms of the equations can be calculated with maximum accuracy.

One example of the use of these grids is given for the discretization of the shallow water model equations written for a conformal projection. This example illustrates how the problems related to horizontal discretization can be addressed concretely. The names proposed by Arakawa (1972) and taken up by Mesinger and Arakawa (1976) are used to identify the various grids.

3.3.1 The primitive equations in conformal projection

The system of equations (2.14) of the shallow water barotropic model, written for a conformal projection characterized by a local map scale factor m using the reduced wind variables $U = u/m$ and $V = v/m$ and the geopotential of the free surface Φ, takes the form:

$$
\left.
\begin{aligned}
\frac{\partial U}{\partial t} &= \ \ (\zeta + f)V - \frac{\partial}{\partial x}(K + \Phi), \\[4pt]
\frac{\partial V}{\partial t} &= -(\zeta + f)U - \frac{\partial}{\partial y}(K + \Phi), \\[4pt]
\frac{\partial \Phi}{\partial t} &= -m^2 \left[\frac{\partial}{\partial x}(\Phi U) + \frac{\partial}{\partial y}(\Phi V) \right], \\[4pt]
\text{with} \ \ K &= \frac{m^2}{2}(U^2 + V^2) \ \ \text{and} \ \ \zeta = m^2 \left(\frac{\partial V}{\partial x} - \frac{\partial U}{\partial y} \right).
\end{aligned}
\right\} \tag{3.13}
$$

This system comprises three prognostic equations; the variables K and ζ represent kinetic energy per unit mass and vorticity, respectively. The advection terms for velocities U and V appear in the right-hand sides of the equations of motion by combining the terms ζV and $-\partial K/\partial x$ on the one hand, and $-\zeta U$ and $-\partial K/\partial y$ on the other hand. The advection term for geopotential Φ is the first term of the expansion of divergence in the continuity equation. The Coriolis terms fV and $-fV$ appear in the right-hand sides of the equations of motion. The geopotential gradient in the equations of motion and the second term of the expansion of divergence in the continuity equation are the adaptation terms of the equations. They describe the exchanges between potential energy and kinetic energy, which are consequences of the fluctuations of the free surface of the barotropic fluid.

The technique for discretizing these equations consists in calculating the 2nd-order accuracy space derivatives by using certain types of grid on which the different variables occupy specific positions, the time derivatives being evaluated in central differences.

3.3.2 The A-type grid

The *A-type* grid is the simplest that can be imagined, with all the variables being expressed at the same place (Figure 3.4), together with the scale factor and the Coriolis parameter. The space derivatives are evaluated in central differences defined in (3.7) and the discretized system is written:

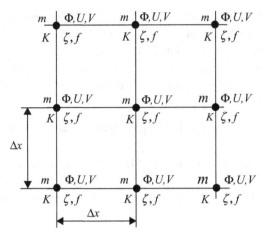

Figure 3.4 Location of variables on the A-type grid.

$$\overline{U}_t^t = \quad (\zeta + f)V - \overline{\left(\Phi + K\right)}_x^x,$$

$$\overline{V}_t^t = -(\zeta + f)U - \overline{\left(\Phi + K\right)}_y^y,$$

$$\overline{\Phi}_t^t = -m^2 \left[\overline{\left(\Phi U\right)}_x^x + \overline{\left(\Phi V\right)}_y^y \right], \tag{3.14}$$

$$\text{with} \quad K = \frac{m^2}{2}(U^2 + V^2) \quad \text{and} \quad \zeta = m^2 (\overline{V}_x^x - \overline{U}_y^y).$$

On the A grid, discretization (3.14) corresponds to evaluating the advection and adaptation terms in central differences over an interval of $2\Delta x$.

3.3.3 The B-type grid

A *B-type* grid may be used in which the velocities U and V are not calculated at the same locations as the geopotential Φ (Figure 3.5). The form of the gradient term suggests calculating the kinetic energy K and the map scale factor m at the same points of the grid as the geopotential Φ. The presence of terms involving the product of $(\zeta + f)$ by the velocities U and V leads to evaluating the vorticity ζ and the Coriolis parameter f at the same grid points as U and V. Under these circumstances, the discretized system is written:

$$\overline{U}_t^t = \quad (\zeta + f)V - \overline{\left(\Phi + K\right)}_x^y,$$

$$\overline{V}_t^t = -(\zeta + f)U - \overline{\left(\Phi + K\right)}_y^x,$$

$$\overline{\Phi}_t^t = -m^2 \left[(\overline{\Phi}^x \overline{U}^y)_x + (\overline{\Phi}^y \overline{V}^x)_y \right], \tag{3.15}$$

$$\text{with} \quad K = \frac{m^2}{2} \overline{(U^2 + V^2)}^{xy} \quad \text{and} \quad \zeta = \overline{m^2 (\overline{V}_x^y - \overline{U}_y^x)}^{xy}.$$

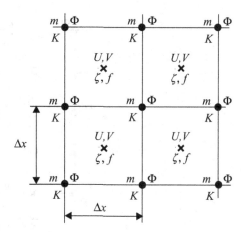

Figure 3.5   Location of variables on the B-type grid.

On the B grid, discretization (3.15) maintains evaluation in central differences over an interval $2\Delta x$ for the advection terms, whereas the discretization of the pressure gradient and wind divergence terms corresponds to an evaluation in central differences over an interval $\sqrt{2}\Delta x$ (as shall be shown in Chapter 5).

3.3.4 The C-type grid

This new *C-type* grid is characterized by the fact that the velocities U and V together with the geopotential Φ are evaluated at different points (Figure 3.6). It is natural to express the kinetic energy K and the map scale factor m at the same points as the geopotential Φ. The position of the velocities U and V suggests calculating vorticity and the Coriolis parameter f at the centre of the grid defined by the points where the geopotential Φ is evaluated.

With this arrangement of variables, it can be seen that the pressure gradient term and the divergence of vector $\Phi\mathbf{V}$, the right-hand side of the continuity equation, take on a particularly simple form by using central differences:

$$\left.\begin{aligned}
\overline{U}_t^t &= \overline{\overline{(\zeta+f)}^y \overline{V}^{xy}} - (\Phi+K)_x, \\
\overline{V}_t^t &= -\overline{\overline{(\zeta+f)}^x \overline{U}^{xy}} - (\Phi+K)_y, \\
\overline{\Phi}_t^t &= -m^2\left[(\overline{\Phi}^x U)_x + (\overline{\Phi}^y V)_y\right], \\
\text{with} \quad K &= \frac{m^2}{2}(\overline{U^2}^x + \overline{V^2}^y) \quad \text{and} \quad \zeta = \overline{m^2}^{xy}(V_x - U_y).
\end{aligned}\right\} \qquad (3.16)$$

On the C grid, discretization (3.16) can be used to evaluate adaptation terms in central differences over an interval Δx, so half as large as that used for the A grid. For the advection terms, the situation is no more favourable with the C grid than with the A or B grids.

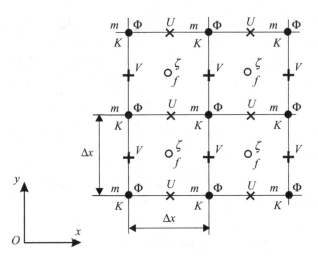

Figure 3.6   Location of the variables on the C-type grid.

3.3.5 The D′-type staggered grid (Eliassen grid)

The principle of this *D'-type* grid proposed by Eliassen (1956) is to switch around the position of the variables on passing from one time step to the next. For even time steps the position of the variables is as with the C grid, except that the positions of velocities U and V are reversed, so defining a grid D'. For odd time steps, all the variables switch position, the Us take the places of the Vs, the Vs of the Us and the geopotential Φ is now calculated in the centre of the grid, as in Figure 3.7.

In fact, the position of the variables at the odd time steps is obtained from the position of the same variables at even time steps by making them undergo a translation in amplitude $\Delta x / \sqrt{2}$ along the diagonal. The corresponding discretization is written:

$$
\left.
\begin{aligned}
\overline{U}_t^{\,t} &= \; (\overline{\zeta + f}^{\,x})V - (\Phi + K)_x, \\
\overline{V}_t^{\,t} &= -(\overline{\zeta + f}^{\,y})U - (\Phi + K)_y, \\
\overline{\Phi}_t^{\,t} &= -m^2 \left[(\overline{\Phi}^{\,y} U)_x + (\overline{\Phi}^{\,x} V)_y \right], \\
\text{with } \quad K &= \frac{m^2}{2}(\overline{U^2}^{\,x} + \overline{V^2}^{\,y}) \quad \text{and} \quad \zeta = m^2 (V_x - U_y).
\end{aligned}
\right\}
\tag{3.17}
$$

On this *staggered* grid, discretization (3.17) is used to obtain the same accuracy for the adaptation terms as with the C grid. Moreover, the advection terms are calculated with greater accuracy than with all other grids (corresponding to central differences over an interval $\sqrt{2}\Delta x$, as shall be shown in detail in Chapter 5), which is why it is so valuable. However, its use requires working with two different grids, one for even steps and one for odd steps, which does not make the scheme any easier to implement.

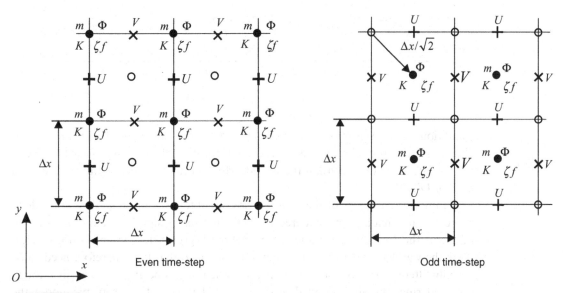

Figure 3.7 | Location of the variables on the D'-type staggered grid proposed by Eliassen.

3.3.6 The properties of the various grids

The different grids used for discretizing the equations of the *shallow water* barotropic model can be used to improve the accuracy of computation of the various terms. However, these discretizations are not neutral with respect to the linear stability of the systems obtained. In terms of efficiency, the improvement achieved with more accurate discretization is offset by extra computation because of the need to choose a shorter time step (meaning a larger number of iterations) for time discretization.

The properties of the various grids were discussed in detail by Winninghoff (1968), and then by Mesinger and Arakawa (1976). The effects of spatial discretization on wave propagation velocity and on the linear stability condition for the various time integration schemes are studied in detail in Chapter 5. Here we simply summarize the pros and cons of the various discretizations from various standpoints.

Greater accuracy in calculating advection terms means we can better represent transport, especially for short waves; from this standpoint, A, B, and C grids give equivalent results for processing advection and the D' staggered grid alone provides greater accuracy. Greater accuracy in calculating adaptation terms makes it possible to better evaluate gravity wave phase speed and to avoid false dispersion of these waves. For a packet of gravity waves, group velocity is also better evaluated, which improves the dynamic process of adaptation to geostrophic balance; from this standpoint, C and D' grids give better results than A or B grids.

For explicit models, for which the values of the parameters to be predicted are expressed directly from their known values at earlier points in time (which is the case for numerical schemes (3.14), (3.15), (3.16), and (3.17), the CFL stability criterion is related to the velocity of the gravity waves and therefore to the discretization of the

Table 3.2 Critical time steps $(\Delta t)_m$ for explicit models using various grids; here c is the speed of gravity waves.

Grid type	A	B	C, D'
$(\Delta t)_m$	$\dfrac{\Delta x}{c\sqrt{2}}$	$\dfrac{\Delta x}{2c}$	$\dfrac{\Delta x}{2c\sqrt{2}}$

adaptation terms. The maximum time steps $(\Delta t)_m$ that can be used with the explicit schemes studied over the various grids are set out in Table 3.2; they show that the A grid is more advantageous than the B grid, which in turn is more advantageous than the C or D'grids.

An important property of the various grids ought to be mentioned: their aptitude to provide results that are not rendered noisy by interfering signals. Practice shows that the use of the A grid and (to a lesser extent) the B grid yields extremely noisy fields, where the wavelength $\lambda = 2\Delta x$ dominates, so much so that they therefore need to be spatially filtered to be of any use. This phenomenon can be explained easily enough by considering a highly simplified version of the shallow water barotropic model: the one-dimensional linear model without advection terms. By assuming the scale factor is uniform and equal to 1, the system of equations is written:

$$\left.\begin{array}{l} \dfrac{\partial U}{\partial t} = -\dfrac{\partial \Phi}{\partial x}, \\[2ex] \dfrac{\partial \Phi}{\partial t} = -\Phi_0 \dfrac{\partial U}{\partial x}, \end{array}\right\} \tag{3.18}$$

Φ_0 being an average value of the geopotential that is assumed to be constant.

Spatial discretization of the system (3.18) with the A grid takes the form:

$$\dfrac{\partial U}{\partial t} = -\bar{\Phi}_x^x,$$

$$\dfrac{\partial \Phi}{\partial t} = -\Phi_0\, \bar{U}_x^x,$$

and leads to the equation:

$$\dfrac{\partial^2 \Phi}{\partial t^2} = \Phi_0 (\bar{\Phi}_{xx}^{xx}),$$

the expanded expression of which is written:

$$\left(\dfrac{\partial^2 \Phi}{\partial t^2}\right) = \Phi_0\, \bar{\Phi}_{xx}^{xx} = \dfrac{\Phi_0}{4\Delta x^2}\big(\Phi(x+2\Delta x) + \Phi(x-2\Delta x) - 2\Phi(x)\big). \tag{3.19}$$

The form of the right-hand side of this equation shows that the evolution of the even-index grid points depends only on the field values at the even-index grid points, just as the evolution of the odd-index grid points depends only on the field values at the odd-index grid points. It then follows that the use of the A grid leads to decoupling of the

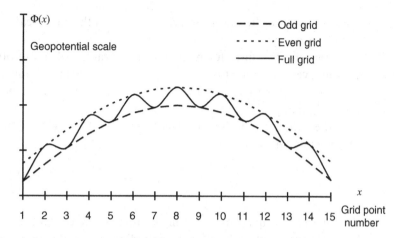

Figure 3.8   Decoupling of solutions and appearance of the stationary wave $\lambda = 2\Delta x$.

two systems, which evolve independently of one another (Figure 3.8), so favouring the creation of a wave $\lambda = 2\Delta x$, which has the property of being stationary, as seen before.

The remedy consists in filtering the fields obtained at the end of an integration period using a spatial filter with the right characteristics to eliminate this interfering wavelength while attenuating the other wavelengths as little as possible. The situation is, however, radically different with the C grid, since discretization of the system (3.18) on this grid yields:

$$\frac{\partial U}{\partial t} = -\Phi_x,$$

$$\frac{\partial \Phi}{\partial t} = -\Phi_0 \, U_x,$$

leading to the equation:

$$\frac{\partial^2 \Phi}{\partial t^2} = +\Phi_0 (\Phi_{xx}),$$

the expanded expression of which is written:

$$\left(\frac{\partial^2 \Phi}{\partial t^2}\right) = \Phi_0 \Phi_{xx} = \frac{\Phi_0}{\Delta x^2}\big(\Phi(x + \Delta x) + \Phi(x - \Delta x) - 2\Phi(x)\big). \tag{3.20}$$

In the case of the C grid, the evolution for the variable Φ involves the value of the field at each point and at its adjacent points and so avoids decoupling of the solutions. So the fields obtained with this grid do not need filtering.

3.3.7 Spatial filtering

In order to be able to use the fields predicted after integrating the primitive equations on the A grid, it is essential to carry out spatial filtering to remove the spurious short waves (and especially wavelength $\lambda = 2\Delta x$, which the numerical scheme processes as a stationary wave).

A very effective filter for removing short waves, which are often spurious waves, without overly affecting the other waves was suggested by Shuman (1957). It is applied in two stages in the x and y directions. It is actually a weighted mean calculated on the three adjacent points.

In the x-direction, we first calculate the quantity $\widetilde{A}^x(x)$ at all the grid points, except for the boundaries:

$$\widetilde{A}^x(x) = A(x)(1-v) + \left[A(x+\Delta x) + A(x-\Delta x)\right]\frac{v}{2};\qquad(3.21)$$

this intermediate quantity $\widetilde{A}^x(x)$ allows us in turn to compute the filtered value $\widetilde{\widetilde{A}}^x(x)$ at all the grid points, except for the boundaries:

$$\widetilde{\widetilde{A}}^x(x) = \widetilde{A}^x(x)(1-v') + \left[\widetilde{A}^x(x+\Delta x) + \widetilde{A}^x(x-\Delta x)\right]\frac{v'}{2},\qquad(3.22)$$

quantities v and v' being coefficients whose values are to be specified.

This two-stage procedure is then applied in the y-direction to complete the two-dimensional filtering.

The response of such a filter can be calculated for a sine function of wavenumber $k = 2\pi/\lambda$:

$$A(x) = A_0 e^{ikx}\mu.$$

Application of the two successive operations (3.21) and (3.22) in the x-direction gives:

$$\widetilde{A}^x(x) = (1-v)A(x) + vA(x)\cos k\Delta x,$$

$$\widetilde{\widetilde{A}}^x(x) = (1-v')\widetilde{A}^x(x) + v'\widetilde{A}^x(x)\cos k\Delta x.$$

By choosing $v = 0.5$ and $v' = -0.5$ we get:

$$\widetilde{A}^x(x) = A(x)\frac{(1+\cos k\Delta x)}{2}, \text{ which attenuates the field } A(x), \text{ then:}$$

$$\widetilde{\widetilde{A}}^x(x) = \widetilde{A}^x(x)\frac{(3-\cos k\Delta x)}{2}, \text{ which amplifies the field } \widetilde{A}^x(x).$$

The overall effect of filtering in the x-direction is given by:

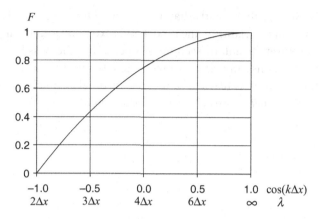

Figure 3.9 Effect of the Shuman filter for various wavelengths.

$$\overset{\approx x}{A}(x) = A(x)\frac{(1 + \cos k\Delta x)}{2}\frac{(3 - \cos k\Delta x)}{2}. \tag{3.23}$$

The Shuman filter response curve, that is, the value of the attenuation factor,

$$F = \frac{(1 + \cos k\Delta x)}{2}\frac{(3 - \cos k\Delta x)}{2},$$

for the various wavelengths expressed as multiples of mesh size Δx, is shown in Figure 3.9.

It can be observed that $F = 0$ for $\lambda = 2\Delta x$; the filter therefore totally removes this interfering wavelength that arises when integrating primitive equations on the A grid. The attenuation factor F takes the value 0.75 for $\lambda = 4\Delta x$ and 0.98 for $\lambda = 8\Delta x$. The filter's attenuation effect therefore declines rapidly for wavelengths λ of more than $2\Delta x$, becoming imperceptible whenever $\lambda > 8\Delta x$. This filter is therefore very effective for eliminating the spurious wavelength $\lambda > 2\Delta x$ without affecting the other wavelengths too much.

3.4 Conclusion

The finite difference method, which provides a straightforward means of computing space and time derivatives, can be used for numerically integrating the partial differential equations used in numerical weather prediction models. It has been shown that there are a large number of possible discretizations of these equations on various grids, the properties of which we have glanced at (they shall be examined in detail in Chapter 5). The choice of numerical scheme for making an operational prediction model is the outcome of a trade-off between accuracy and computation time. It shall be shown subsequently that the need to obtain discretized equations for conserving the physical invariants of the model adds extra constraints in the choice of discretization. Moreover, it shall be seen in Chapter 5 that the possibility of dealing with the adaptation terms implicitly and with the advection terms by a Lagrangian method alters the conclusions reached in the case of purely explicit schemes and leads to the use of *semi-implicit* and *semi-Lagrangian* schemes.

Spectral methods

4.1 Introduction

Here we present essentially the tools required for implementing the spectral method, which belongs more generally to the family of *Galerkin methods*. These are commonly used instead of the finite difference method to deal primarily with horizontal fields in weather forecasting models.

Galerkin methods, which are used for solving systems of partial differential equations numerically, do not directly use the values of fields at grid points but involve series expansions in terms of suitably chosen functions and lead to solving a system of ordinary differential equations. There are two types of methods with this procedure: the *finite element method* for which the functions are zero, except over a small domain where they are equal to low-order polynomials; and the *spectral method* for which the functions are eigenfunctions of a spatial operator defined on the entire working domain.

After a brief exposition of the principle behind the Galerkin method and how it is applied to solving the wave equation using the finite element method, we provide details of the use of the spectral method for processing fields defined on the sphere and for fields defined on a doubly periodic rectangular domain; these techniques are used for building *global* weather forecasting models (covering the entire terrestrial sphere) and for *limited area models*.

4.2 Using series expansions in terms of functions

4.2.1 General remarks on Galerkin methods

We examine how this method is applied for solving a partial differential equation describing the evolution of a field $A(x, y, t)$, defined on a spatial domain S:

$$\frac{\partial A}{\partial t} + F(A) = 0, \tag{4.1}$$

F being generally a nonlinear spatial operator.

After choosing N linearly independent functions $f_j(x, y)$ defined on S and forming a basis, projecting the field $A(x, y, t)$ onto that basis gives the approximation $\widetilde{A}(x, y, t)$:

$$\widetilde{A}(x, y, t) = \sum_{j=1}^{j=N} A_j(t) f_j(x, y). \tag{4.2}$$

Coefficients A_j depend here on time alone. The residual e_N obtained by replacing $A(x, y, t)$ by its expansion $\widetilde{A}(x, y, t)$ in equation (4.1) is written:

$$e_N = \frac{\partial}{\partial t} \sum_{j=1}^{j=N} A_j(t) f_j(x, y) + F\left[\sum_{j=1}^{j=N} A_j(t) f_j(x, y)\right]. \tag{4.3}$$

The Galerkin (or *weighted residual*) procedure consists in writing that the residual must be orthogonal to the basis functions for a given scalar product, that is:

$$\iint_S e_N f_i(x, y) \, dx \, dy = 0,$$

for each of the N functions $f_i(x, y)$, $i = 1, \ldots, N$. This constraint leads to system (4.4):

$$\sum_{j=1}^{j=N} \left\{\frac{d}{dt} A_j(t) \left[\iint_S f_i f_j \, dx \, dy\right]\right\} + \iint_S \left\{f_i F\left[\sum_{j=1}^{j=N} A_j(t) f_j\right]\right\} dx \, dy = 0, \, i = 1, \ldots N. \tag{4.4}$$

This system of N ordinary differential equations for coefficients $A_j(t)$, which are functions of variable t alone, is solved by evaluating the derivatives dA_j/dt using finite differences.

With the finite element method, the basis functions and their derivatives are nonzero over a small part of the working domain only. The inputs from nonzero terms to the sums appearing in system (4.4) relate to a limited number of coefficients only and lead to a fairly simple implicit system for determining the values of coefficients A_j at time $t + \Delta t$. With the spectral method, the functions used form an orthonormal basis and the sums are then reduced to a single term when the operator F is linear.

4.2.2 Using finite elements for the advection equation

Here we examine the use of the finite element method for processing the *advection equation at constant velocity* over a one-dimensional domain $D[-L, +L]$. The problem is to determine the solution $A(x, t)$ of the partial differential equation on D:

$$\frac{\partial A}{\partial t} + U\frac{\partial A}{\partial x} = 0, \tag{4.5}$$

U being a constant. We use trapezoidal functions defined on domain D as basis functions:

$$f_i(x) = 0 \qquad\qquad\qquad \text{if } x \in \left[-L,\, x_i - d/2 - \varepsilon\right],$$

$$f_i(x) = \frac{1}{\varepsilon}\left[x - (x_i - d/2 - \varepsilon)\right] \quad \text{if } x \in \left[x_i - d/2 - \varepsilon,\, x_i - d/2\right],$$

$$f_i(x) = 1 \qquad\qquad\qquad \text{if } x \in \left[x_i - d/2,\, x_i + d/2\right],$$

$$f_i(x) = -\frac{1}{\varepsilon}\left[x - (x_i + d/2 + \varepsilon)\right] \quad \text{if } x \in \left[x_i + d/2,\, x_i + d/2 + \varepsilon\right],$$

$$f_i(x) = 0 \qquad\qquad\qquad \text{if } x \in \left[x_i + d/2 + \varepsilon,\, +L\right],$$

d and ε being the positive real numbers defining the shape of the trapezium as in Figure 4.1.

The function $A(x, t)$ is written in expanded form:

$$A(x,t) = \sum_{j=1}^{j=N} A_j(t) f_j(x). \tag{4.6}$$

Given the shape of the functions f_j, the values of coefficients A_j coincide with the values of the function $A(x, t)$ at points x_i, since the expansion (4.6) implies the equality:

$$A(x_i,\, t) = A_i(t).$$

The Galerkin procedure leads to the system of equations:

$$\sum_{j=1}^{j=N} \frac{d}{dt} A_j(t) \int_D f_i f_j \, dx + \int_D \left\{ f_i U \left[\sum_{j=1}^{j=N} A_j(t) \frac{\partial}{\partial x} f_j(x) \right] \right\} dx = 0, \quad i = 1,...,\, N. \tag{4.7}$$

After calculating the integrals, in view of the shape of the functions $f_j(x)$ we finally have:

$$\frac{\varepsilon}{6}\left(\frac{dA_{i-1}}{dt} + \frac{dA_{i+1}}{dt}\right) + \left(\frac{2\varepsilon}{3} + d\right)\frac{dA_i}{dt} = -\frac{U}{2}(A_{i+1} - A_{i-1}). \tag{4.8}$$

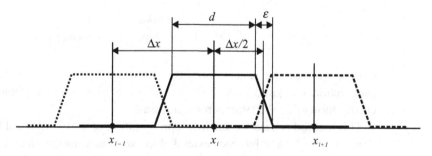

Figure 4.1   The trapezoidal functions.

Table 4.1 Comparison of truncation errors for the 4th-order accuracy finite difference method and the linear finite element method.

λ	$k\Delta x$	$\left|\varepsilon_T\right|_4$ (%)	$\left|\varepsilon_T\right|_{EF}$ (%)
$2\Delta x$	π	100	100
$4\Delta x$	$\pi/2$	36	15
$6\Delta x$	$\pi/3$	17	3.5
$8\Delta x$	$\pi/4$	10	1.2
$10\Delta x$	$\pi/5$	6.5	0.5
$\downarrow$	$\downarrow$	$\downarrow$	$\downarrow$
∞	0	0	0

In the case where $\varepsilon \to 0$ and $d \to \Delta x$, the trapezoidal function becomes a *crenel* function (that is equal to 1 on an interval Δx and 0 elsewhere) and the finite element method leads to the expression:

$$\frac{dA_i}{dt} = -\frac{U}{2\Delta x}(A_{i+1} - A_{i-1}). \tag{4.9}$$

Replacing the time derivatives by their evaluation in central differences yields, in view of the meaning of the coefficients A_j, an identical scheme to the 2nd-order accuracy finite difference method.

In the case where $d \to 0$ and $\varepsilon \to \Delta x$, the trapezoidal function becomes a *hat* function and leads for the coefficients to the expression:

$$\frac{\Delta x}{6}\left(\frac{dA_{i-1}}{dt} + 4\frac{dA_i}{dt} + \frac{dA_{i+1}}{dt}\right) = -\frac{U}{2}(A_{i+1} - A_{i-1}). \tag{4.10}$$

Replacing the time derivatives by central differences gives the values A_i at $t + \Delta t$ as solutions of a tridiagonal linear system.

The truncation error made with this scheme for a sine function of wavenumber $k = 2\pi/\lambda$, calculated by the same process as in Subsection 3.2.3, is written:

$$\left|\varepsilon_T\right|_{EF} = \left| 1 - \frac{\sin(k\Delta x)}{k\Delta x}\frac{3}{2 + \cos(k\Delta x)} \right|. \tag{4.11}$$

Comparison of the results from the 4th-order accuracy finite difference method (Subsection 3.2.3) and from the finite element method using hat functions (Table 4.1) clearly shows that the latter is more accurate.

This method is used for vertical representation of meteorological fields (Staniforth and Daley, 1977) and has been extended to two dimensions of space by choosing products of hat functions (Staniforth and Mitchell, 1978) as basis functions with a rectangular grid. In this case, the implicit system breaks down into tridiagonal subsystems.

This approach means a variable mesh support grid can be used, which is the principle employed for operational models in Canada.

We can also work with two spatial dimensions using a linear finite element (pyramidal element) on a triangular mesh grid (Cullen, 1973). The advantage of this approach is that we can adapt the triangulated grid to the geometry of the problem and to the scales we wish to represent locally. This explains why the method is commonly used for modelling hydrodynamic flows around obstacles or for simulating oceanic circulation.

4.3 Spectral method on the sphere

4.3.1 General remarks

The spectral method for processing atmospheric flow was introduced by Silberman (1954). However, the lengthy computation time required for the direct evaluation of the coefficients of nonlinear terms (computing interaction coefficients) limited its use to models with just a few degrees of freedom. The advent of the fast Fourier transformation (Cooley and Tukey, 1965) meant the *transform method* proposed at the same time by Eliassen et al. (1970) and Orszag (1970) could be implemented, so making the spectral method just as effective as the grid point method. The first successful uses of full spectral models in meteorology were achieved in the 1970s by Machenhauer and Daley (1972) and by Bourke (1974).

The principle of the spectral method is to expand a field defined on the sphere $A(\lambda, \mu)$ (λ denoting longitude and $\mu = \sin \varphi$ the sine of latitude) in terms of functions forming an orthonormal basis so as to minimize the truncation error and to provide a simple procedure for evaluating coefficients of the expansion. We confine ourselves here to presenting a few essential results for working with the basis of spherical functions chosen and the procedures for making the various calculations involved in weather forecasting models. For more information on the spectral method, it is worth reading Machenhauer (1979) and Rochas and Courtier (1992).

4.3.2 The basis of surface spherical harmonics

For representing fields defined on the sphere by means of series expansions in terms of functions, we use the surface spherical harmonics noted $Y_n^m(\lambda, \mu)$, the properties of which are to be explained. These surface spherical harmonics are complex functions of variables $\lambda \in [0, 2\pi]$ and $\mu \in [-1, +1]$; they are the eigenfunctions of the Laplace operator on the sphere that verify the relation:

$$\nabla^2 Y_n^m(\lambda, \mu) = -\frac{n(n+1)}{a^2} Y_n^m(\lambda, \mu), \qquad (4.12)$$

where a is the Earth's radius. When suitably normalized, these functions form an orthonormal basis of square-integrable functions on the sphere and verify the relation:

$$\frac{1}{4\pi} \int_{-1}^{+1} \int_{0}^{2\pi} Y_n^m(\lambda,\mu)\overline{Y_{n'}^{m'}}(\lambda,\mu) \, d\lambda d\mu = \begin{cases} 0 \text{ if } (n,m) \neq (n',m'), \\ 1 \text{ if } (n,m) = (n',m'), \end{cases} \quad (4.13)$$

where the operator $\overline{\quad \cdot \quad}$ denotes the complex conjugate.

It can be shown that surface spherical harmonics are expressed as the product of a function of the variable λ by a function of the variable μ in the form:

$$Y_n^m(\lambda,\mu) = P_n^m(\mu)e^{im\lambda}. \quad (4.14)$$

The complex nature of the function $Y_n^m(\lambda,\mu)$ appears clearly: on a circle of latitude (with fixed μ), $Y_n^m(\lambda,\mu)$ is periodic and the number m can therefore be interpreted as a zonal wavenumber (number of wavelengths on a circle of latitude).

$P_n^m(\mu)$ are the associated Legendre functions of first kind and of degree n and order m; they are real functions of μ, and solutions of the Legendre equation:

$$\frac{d}{d\mu}\left[(1-\mu^2)\frac{dP_n^m(\mu)}{d\mu}\right] + \left[n(n+1) - \frac{m^2}{(1-\mu^2)}\right]P_n^m(\mu) = 0 \quad \forall \mu \in [-1,+1], \quad (4.15)$$

which is introduced when looking for eigenvectors of the Laplace operator in spherical geometry. They are expressed analytically by the *Rodrigues formula*:

$$P_n^m(\mu) = \frac{(1-\mu^2)^{\frac{m}{2}}}{2^n n!}\frac{d^{n+m}}{d\mu^{n+m}}(\mu^2-1)^n. \quad (4.16)$$

They also verify the orthogonality relation:

$$\frac{1}{2}\int_{-1}^{+1} P_n^m P_{n'}^m d\mu = 0, \text{ for } n \neq n'. \quad (4.17)$$

From its expression, given by (4.16), it can be seen that the function $P_n^m(\mu)$ is only defined for $n \geq m$, leading us to take $P_n^m(\mu) \equiv 0$ for $n \leq m$. For positive values of m, by using (4.16) we get:

$$P_n^{-m}(\mu) = (-1)^m \frac{(n-m)!}{(n+m)!} P_n^m(\mu). \quad (4.18)$$

However, by convention, it is preferable to adopt the definition:

$$P_n^{-m}(\mu) \equiv P_n^m(\mu), \quad (4.19)$$

which allows us, for surface spherical harmonics, to verify the relation:

$$Y_n^{-m}(\mu) \equiv \overline{Y_n^m}(\mu). \quad (4.20)$$

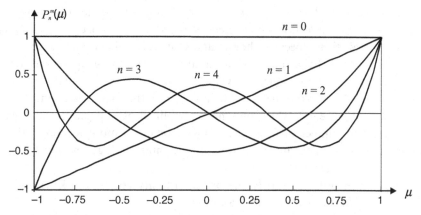

Figure 4.2   The first (non normalized) Legendre functions for $m = 0$.

This convention (4.19) allows us, when expanding a real field on the basis of spherical harmonics, to obtain complex conjugate coefficients corresponding to the conjugate basis functions (a property analogous to that obtained by expanding a real field in complex Fourier series).

Figure 4.2 represents graphically the associated Legendre functions for $m = 0$ and the first values of n.

4.3.3 The properties of spherical harmonics

Formula (4.16) shows that the function $P_n^m(\mu)$ can be written in the form:

$$P_n^m(\mu) = (1 - \mu^2)^{\frac{|m|}{2}} Q_{n-|m|}(\mu), \tag{4.21}$$

$Q_{n-|m|}(\mu)$ being a polynomial of degree $n - |m|$ and of parity $n - |m|$, which implies that:

$$P_n^m(\mu) = P_n^m(-\mu) \ \text{ if } n - |m| \text{ is even,}$$

$$P_n^m(\mu) = -P_n^m(-\mu) \text{ if } n - |m| \text{ is odd.}$$

As can be seen from Figure 4.2, the number $n-|m|$ corresponds to the number of zeros of the function $P_n^m(\mu)$ in the interval $]-1, +1[$ and so can be interpreted as a meridional wavenumber, with the number n still being called the total wavenumber.

Surface spherical harmonics $Y_n^m(\lambda, \mu)$ are subdivided into two families: even harmonics characterized by $n - |m|$ being even, which are symmetric with respect to the plane of the equator $(Y_n^m(-\mu) = -Y_n^m(\mu))$, and odd harmonics characterized by $n - |m|$ being odd, which are antisymmetric with respect to the plane of the equator $(Y_n^m(-\mu) = -Y_n^m(\mu))$.

So a field defined on the globe and that is symmetric with respect to the equatorial plane can be expanded by using even functions alone, just as an antisymmetric field can be expanded by using odd functions alone. This property has been used to construct *hemispheric models*, which are half as costly as *global models*, although less realistic too because they assume that meteorological fields are symmetric (or antisymmetric). Even functions are used to expand the geopotential, temperature, pressure, the zonal component of wind, and the velocity potential, whereas odd functions are used to expand the meridional wind and the streamfunction.

Figure 4.3, which presents a simplified plot (positive and negative values) of low wavenumber surface spherical harmonics (or more accurately the real part of them), gives a better understanding of their morphology and illustrates their properties of symmetry.

In order to satisfy the orthonormality condition (4.13), the Legendre functions $P_n^m(\mu)$ must verify the relation:

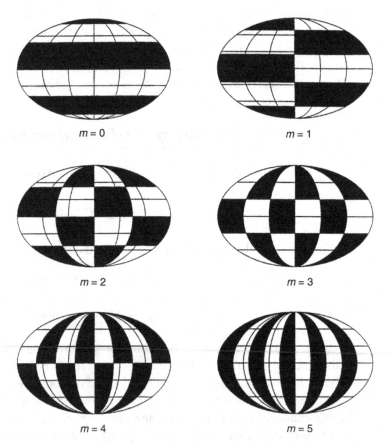

Figure 4.3   Schematic of the real part of spherical harmonics (positive values in black and negative values in white) for $n = 5$ on a Mollweide canvas. The harmonics corresponding to $m = 0$ (top left) are said to be zonal, those corresponding to $n = m$ (bottom right) are said to be sectorial, and the others are tesseral harmonics. (After Baer (1972). *J. Atmos. Sci.*, **29**, 651 © Amer. Met. Soc.)

$$\frac{1}{2} \int_{-1}^{+1} [P_n^m(\mu)]^2 d\mu = 1,$$

which leads to the expression of normalized Legendre functions:

$$P_n^m(\mu) = \sqrt{(2n+1)\frac{(n-m)!}{(n+m)!}} \frac{(1-\mu^2)^{\frac{m}{2}}}{2^n \, n!} \frac{d^{n+m}}{d\mu^{n+m}} (\mu^2-1)^n. \tag{4.22}$$

The Legendre functions and their derivatives satisfy the recurrence relations, a good number of which are featured in the working paper by Rochas and Courtier (1992). Here we give just the most useful ones.

The normalized Legendre functions verify the relation:

$$\mu P_n^m = \varepsilon_n^m P_{n-1}^m + \varepsilon_{n+1}^m P_{n+1}^m \text{ with: } \varepsilon_n^m = \sqrt{\frac{n^2-m^2}{4n^2-1}}. \tag{4.23}$$

Their derivatives with respect to μ are obtained by the relation:

$$(1-\mu^2)\frac{dP_n^m}{d\mu} = -n\mu P_n^m + (2n+1)\varepsilon_n^m P_{n-1}^m, \tag{4.24}$$

or again by using (4.23):

$$(1-\mu^2)\frac{dP_n^m}{d\mu} = (n+1)\varepsilon_n^m P_{n-1}^m - n\varepsilon_{n+1}^m P_{n+1}^m. \tag{4.25}$$

4.3.4 Expanding a spherical field

A field $A(\lambda,\mu)$ defined on the sphere can be written, by using the basis of normalized surface spherical harmonics, in the form:

$$A(\lambda,\mu) = \sum_{n=|m|}^{+\infty} \sum_{m=-\infty}^{+\infty} A_n^m Y_n^m(\lambda,\mu). \tag{4.26}$$

The expansion coefficients A_n^m, termed spectral coefficients, are complex numbers that, given the orthonormality relation (4.13), are calculated using the double integral:

$$A_n^m = \frac{1}{4\pi} \int_{-1}^{+1} \int_0^{2\pi} A(\lambda,\mu) \overline{Y_n^m} \, d\lambda d\mu. \tag{4.27}$$

These coefficients verify the Parseval-Plancherel relation:

$$\frac{1}{4\pi} \int_{-1}^{+1} \int_0^{2\pi} [A(\lambda,\mu)]^2 d\lambda d\mu = \sum_{n=|m|}^{\infty} \sum_{m=-\infty}^{+\infty} |A_n^m|^2, \tag{4.28}$$

which shows that the global quadratic quantities (e.g. kinetic energy) can be calculated directly in spectral space.

When the field $A(\lambda, \mu)$ is real, we verify the conjugation relation:

$$A_n^{-m} = \overline{A_n^m}, \tag{4.29}$$

by virtue of relation (4.20).

Notice too that the normalization adopted gives a precise meaning to the value of the first coefficient of the expansion A_0^0; since, $Y_0^0 = 1$, this coefficient represents the mean value of the field $A(\lambda, \mu)$ on the sphere, that is:

$$A_0^0 = \frac{1}{4\pi} \int_{-1}^{+1} \int_0^{2\pi} A(\lambda, \mu) \, d\lambda d\mu. \tag{4.30}$$

4.3.5 Truncating the expansion

In practice, given the limited capacity of computer memory, it is essential to truncate the expansion, which leads to a truncation error. The expansion truncation is defined as the set T of wavenumber n and m that must be used in the expansion:

$$A(\lambda, \mu) = \sum_{(n,m) \in T} A_n^m Y_n^m (\lambda, \mu). \tag{4.31}$$

Several types of truncation may be used to construct weather forecasting models. These are visualized by plotting the limits of the values taken by wavenumbers n and m on a Cartesian graph (Figure 4.4).

• Rhomboidal truncation is defined by:

$$T = \left\{ (n, m), 0 \le |m| \le M, \ |m| \le n \le |m| + N - M \right\};$$

its specific feature is that it gives the same number of degrees of freedom to each zonal wavenumber m. It provides an invariant representation for rotations around the axis of the poles (Machenhauer, 1979). The resolution can therefore be considered uniform in the zonal direction whereas it varies in the meridional direction.

• Triangular truncation is defined by:

$$T = \left\{ (n, m), 0 \le |m| \le M, \ 0 \le |m| \le n \le M \right\};$$

it provides an invariant representation for all arbitrary axis rotations passing through the centre of the sphere. This property arises because any spherical harmonic for a system of coordinates (λ, μ) is expressed in another system of coordinates (λ', μ') using a series of functions of the same total wavenumber n (Machenhauer, 1979). With this truncation, resolution can be considered uniform on the sphere.

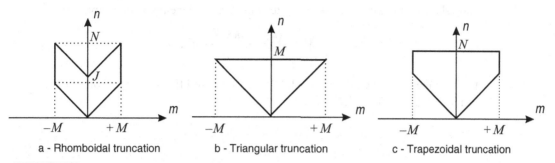

a - Rhomboidal truncation   b - Triangular truncation   c - Trapezoidal truncation

Figure 4.4   The various types of truncation used.

• Trapezoidal truncation is defined by:

$$T = \left\{ (n,m),\, 0 \le |m| \le M,\ |m| \le n \le N \right\},\text{ with } N > M;$$

it is analogous to triangular truncation and is used in very specific instances.

Triangular truncation is widely used for representing meteorological fields because of its isotropy property. This choice may also be vindicated by physical arguments, and Baer (1972) showed that if, on the Cartesian graph in coordinates m and n, we plot the lines of equal value of spectral coefficients of the global field of kinetic energy in the atmosphere, they are fairly close to a triangle; this shows that this type of truncation can be used to select scales providing the greatest contributions to kinetic energy.

The expansion of a spherical field $A(\lambda, \mu)$ using triangular truncation of maximum wavenumber M defines a new spherical field that we continue, improperly, to refer to as $A(\lambda, \mu)$ and that is written:

$$A(\lambda,\mu) = \sum_{m=-M}^{+M} \sum_{n=|m|}^{+M} A_n^m Y_n^m (\lambda,\mu) . \tag{4.32}$$

4.3.6 Calculating linear terms and application to wind calculation

The nature of the surface spherical harmonics and their property of being eigenfunctions of the Laplace operator on the sphere make it easy to calculate the coefficients of the result of applying linear operators on a spherical field $A(\lambda, \mu)$ known by its spectral coefficients A_n^m.

Thus definition (4.14) can be used to determine the coefficients of the derivative in λ:

$$\left(\frac{\partial A}{\partial \lambda} \right)_n^m = i\, m\, A_n^m , \tag{4.33}$$

and relation (4.12) can be used to determine the coefficients of the Laplacian:

$$\left(\nabla^2 A\right)_n^m = -\frac{n(n+1)}{a^2} A_n^m.$$ (4.34)

In the same way, given relations (4.23) and (4.24) and those arising from them for the spherical harmonics Y_n^m:

$$\mu Y_n^m = \varepsilon_n^m Y_{n-1}^m + \varepsilon_{n+1}^m Y_{n+1}^m,$$ (4.35)

$$(1-\mu^2)\frac{\partial Y_n^m}{\partial \mu} = (n+1)\varepsilon_n^m Y_{n-1}^m - n\varepsilon_{n+1}^m Y_{n+1}^m,$$ (4.36)

we can calculate the spectral coefficients of the results of multiplication by μ and derivation with respect to μ:

$$[\mu A]_n^m = \varepsilon_n^m A_{n-1}^m + \varepsilon_{n+1}^m A_{n+1}^m,$$ (4.37)

$$\left[(1-\mu^2)\frac{\partial A}{\partial \mu}\right]_n^m = -(n-1)\varepsilon_n^m A_{n-1}^m + (n+2)\varepsilon_{n+1}^m A_{n+1}^m.$$ (4.38)

The *Helmholtz relation*, $\mathbf{V} = \mathbf{k} \times \nabla\psi + \nabla\chi$, allows us to determine the wind vector from two scalar fields: the streamfunction $\psi(\lambda, \mu)$ and the velocity potential $\chi(\lambda, \mu)$. The two components U and V of the wind, already introduced in subsection 2.3.2, are then given by the formulas:

$$U = \frac{\partial \chi}{\partial \lambda} - (1-\mu^2)\frac{\partial \psi}{\partial \mu},$$
$$V = (1-\mu^2)\frac{\partial \chi}{\partial \mu} + \frac{\partial \psi}{\partial \lambda}.$$ (4.39)

Given relations (4.38) and (4.39), the coefficients χ_n^m and ψ_n^m of the velocity potential and the streamfunction enable us to calculate coefficients U_n^m and V_n^m as follows:

$$U_n^m = im\chi_n^m + [(n-1)\varepsilon_n^m \psi_{n-1}^m - (n+2)\varepsilon_{n+1}^m \psi_{n+1}^m],$$
$$V_n^m = im\psi_n^m - [(n-1)\varepsilon_n^m \chi_{n-1}^m - (n+2)\varepsilon_{n+1}^m \chi_{n+1}^m].$$ (4.40)

However, the synthesis must be made by summing from $n = |m|$ to $n = M + 1$ so as to take proper account of all the components χ_M^m and ψ_M^m in the expressions of U_{M+1}^m and V_{M+1}^m deduced from (4.40):

$$U(\lambda,\mu) = \sum_{m=-M}^{+M} \sum_{n=|m|}^{M+1} U_n^m Y_n^m(\lambda,\mu),$$
$$V(\lambda,\mu) = \sum_{m=-M}^{+M} \sum_{n=|m|}^{M+1} V_n^m Y_n^m(\lambda,\mu).$$ (4.41)

4.3.7 Calculating nonlinear terms

To calculate the coefficients of the product $A(\lambda, \mu)$ of the two fields defined on the sphere $B(\lambda, \mu)$ and $C(\lambda, \mu)$, we must return to the quadrature formula that gives the coefficients. It is assumed in what follows that all of the quantities are calculated using triangular truncation T_N. We get:

$$A_n^m = \frac{1}{4\pi} \int_{-1}^{+1} \int_0^{2\pi} A \overline{Y_n^m} d\lambda d\mu$$

$$= \sum_{n_1, m_1} \sum_{n_2, m_2} B_{n_1}^{m_1} C_{n_2}^{m_2} \frac{1}{4\pi} \int_{-1}^{+1} \int_0^{2\pi} Y_{n_1}^{m_1} Y_{n_2}^{m_2} \overline{Y_n^m} d\lambda d\mu ,$$

summation including all the wavenumbers included in the truncation T_N.

If B, C are defined with truncation T_N, the product $A(\lambda, \mu)$ is defined with truncation T_{2N}, for which only the terms included in truncation T_N need to be kept. It can be seen that the spectral coefficients of A are expressed as a weighted sum of the product of coefficients B and C, the weights being constituted by the integrals (termed *interaction coefficients*). This method is very demanding in computation time whenever the truncation becomes sizeable; it is therefore reserved for academic studies with very low truncations and the coefficients of nonlinear terms are calculated far more efficiently by the *transform method*, to be described.

The spectral coefficients of a product $A = BC$ can be calculated far more efficiently by running through the following stages:

• we calculate the value of the two terms $B(\lambda, \mu)$ and $C(\lambda, \mu)$ of the product at the points of longitude λ_j and of the sine of latitude μ_k on what is termed the transformation grid, G:

$$G : \left\{ \lambda_j, j = 1,...,J \; ; \; \mu_k, \, k = 1,..., K \right\}; \tag{4.42}$$

• the values are multiplied at the grid points:

$$A(\lambda, \mu) = B(\lambda, \mu) \, C(\lambda, \mu);$$

• from this result we calculate the corresponding spectral coefficients A_n^m.

In practice, the coefficients A_n^m are evaluated by calculating the double integral:

$$A_n^m = \frac{1}{2} \int_{-1}^{+1} \left[\frac{1}{2\pi} \int_0^{2\pi} A(\lambda, \mu) e^{-im\lambda} d\lambda \right] P_n^m(\mu) d\mu. \tag{4.43}$$

The first integral yields complex number $A_m(\mu)$ that is simply the Fourier coefficients of the function $A(\lambda, \mu)$ at fixed μ_k:

$$A_m(\mu_k) = \frac{1}{2\pi} \int_0^{2\pi} A(\lambda, \mu) e^{-im\lambda} d\lambda \tag{4.44}$$

This integral can be calculated using *trapezoidal* quadrature, written:

$$A_m(\mu_k) = \frac{1}{L}\sum_{j=1}^{L} A(\lambda_j,\mu_k)e^{-im\lambda_j}. \tag{4.45}$$

It can be shown (Krylov, 1962) that this finite sum on J regularly spaced points yields an exact result for the integral (4.44) if its integrand is a sum of sine functions of a degree less than or equal to $J-1$. In the case of the product $A = BC$, the integrand is written as a sum of sine functions of a degree less than or equal to $2M + M = 3M$. We therefore obtain an exact evaluation of the integral (4.44) by choosing $J \geq 3M + 1$.

This transformation is performed for a set of values $\mu_k = \sin\varphi_k$, $k = 1,\ldots, K$, characterizing the *Gaussian latitudes*; these values of μ_k are the roots of the equation:

$$P_K^0(\mu) = 0. \tag{4.46}$$

With each of these latitudes characterized by μ_k, distributed in a non-uniform manner over the interval $[-1,+1]$, a *Gaussian weight* w_k is associated, given by the formula:

$$w_k = \frac{2(1-\mu_k^2)}{\left[K P_{K-1}^0(\mu_k)\right]^2}, \tag{4.47}$$

from which the second integral can be calculated:

$$A_n^m = \frac{1}{2}\int_{-1}^{+1} A_m(\mu)P_n^m(\mu)d\mu \tag{4.48}$$

by using Gaussian numerical quadrature (Hildebrand, 1956), which is written:

$$A_n^m = \frac{1}{2}\sum_{k=1}^{K} w_k A_m(\mu_k)P_n^m(\mu_k). \tag{4.49}$$

It is shown (Krylov, 1962) that this weighted sum on K points yields an exact result for the integral (4.48) when the integrand is a polynomial of degree less than or equal to $2K-1$. In the case of the product $A = BC$, the integrand is written, by virtue of (4.21), as a sum of polynomials of degree less than or equal to $2M + M = 3M$. We therefore obtain an exact evaluation of the integral (4.48) by choosing $K \geq 3M + 1$.

By using triangular truncation T_M and choosing the number of points of the *Gaussian grid* (4.42) so as to comply with the constraints:

$$\left.\begin{array}{l} J \geq 3M +1, \\[2mm] K \geq \dfrac{3M+1}{2}, \end{array}\right\} \tag{4.50}$$

the spectral coefficients A_n^m of the product $A = BC$ belonging to truncation T_M can be calculated exactly using the quadrature indicated. This property is no longer true when the number of grid points is less than the specified values (in the case of a quadratic term) or when the product is of a higher order; the wavenumbers outside the truncation generated by nonlinear interaction when calculating the product are then falsely represented by wavenumbers belonging to the truncation, a phenomenon known as *aliasing* or *spectral folding* that is addressed in Chapter 6.

Notice that the calculations relating to physical parameterization are introduced simply on the grid point values: this is a further advantage of the transform method.

To sum up, beginning with the data evaluated at the points of the Gaussian grid, G, whose characteristics K and J have been suitably chosen (4.50), calculation of spectral coefficients by the transform method requires:

- a Fourier transform: (4.44) for passing from the value $A(\lambda, \mu)$ on the circles of latitude μ_k to the Fourier coefficients $A_m(\mu)$, which is done by means of the trapezoidal quadrature (4.45). This is done by using the *FFT* (*Fast Fourier Transform*) algorithm;
- a Legendre transform (4.48) for passing from Fourier coefficients $A_m(\mu)$ to spectral coefficients A_n^m, by using Gaussian quadrature (4.49).

Conversely, to obtain the field value $A(\lambda, \mu)$ from spectral coefficients A_n^m, it suffices to make the synthesis (4.32), which can be considered to be the succession of two transformations:

- an inverse Legendre transform for passing, by summation over the total wavenumber n, from the spectral coefficients value A_n^m to the Fourier coefficients $A_m(\mu)$ on the latitudes μ_k of the Gaussian grid:

$$X_m(\mu) = \sum_{n=|m|}^{M} X_n^m P_n^m(\mu);$$

(4.51)

- an inverse Fourier transform (Fourier synthesis) for passing, by summation over the zonal wavenumber m, from the value of Fourier coefficients $A_m(\mu)$ to the field $A(\lambda, \mu)$ over the latitude circles μ_k. This transform is done using the *IFFT* (*Inverse Fast Fourier Transform*) algorithm:

$$X(\lambda, \mu) = \sum_{m=-M}^{+M} X_m(\mu) e^{im\lambda}.$$

(4.52)

The Gaussian grid, as defined earlier, is regular over a circle of latitude. The convergence of meridians towards the poles therefore implies a gradual reduction in the distance between two consecutive grid points on a circle of latitude. Accordingly Machenhauer (1979) proposed a procedure for reducing the number of grid points as we approach the poles and tested it with encouraging results. It turns out that by using Gaussian quadrature, the contribution of polar latitudes for the tendency terms of spectral coefficients corresponding to the highest wavenumbers becomes extremely small (even smaller than the current accuracy of computers). Detailed study of this procedure by Hortal and Simmons (1991) has shown that the reduction in the number of grid points was a great advantage for saving computing time without significant loss of accuracy. The use of such a reduced grid has therefore become widespread.

It is important to keep in mind that the inequalities (4.50) apply when determining the characteristics of a Gaussian grid for an exact calculation of the quadratic terms. It shall be seen in Chapters 6 and 8 that advection can be handled in a Lagrangian way, eliminating the need to calculate the quadratic terms of the advection terms. In such a case, we can use a Gaussian grid allowing an accurate calculation for the linear terms only, the characteristics of which are provided by the inequalities:

$$\left. \begin{aligned} J &\geq 2M+1, \\ K &\geq \frac{2M+1}{2}. \end{aligned} \right\}$$

(4.53)

4.3.8 Practical implementation of the spectral method

The use of the spectral method therefore requires a certain working environment to be put in place before the model itself is constructed:

- choice of a truncation;
- computation of the characteristics of the relevant Gaussian grid (number of points, latitudes, and Gaussian weights) for computing the quadratic terms without aliasing;
- calculation of the normalized Legendre functions and their derivatives in μ, for all points on the Gaussian grid.

The Legendre functions are generally calculated by using recurrence relations like those proposed by Belousov (1962), cited by Rochas and Courtier (1992), and that are formulated for the normalized functions as below:

$$P_n^m(\mu) = c_n^m P_{n-2}^{m-2}(\mu) + d_n^m P_{n-1}^{m-2}(\mu) + e_n^m P_{n-1}^m(\mu),$$

with:

$$c_n^m = \left[\frac{(2n+1)(n+m-1)(n+m-3)}{(2n-3)(n+m)(n+m-2)} \right]^{1/2},$$

$$d_n^m = \left[\frac{(2n+1)(n+m-1)(n+m-1)}{(2n-1)(n+m)(n+m-2)} \right]^{1/2},$$

$$e_n^m = \left[\frac{(2n+1)(n-m)}{(2n-1)(n+m)} \right]^{1/2}.$$

Finally, it is essential to have effective codes for the fast Fourier transform and its inverse for implementing the transform method for computing the coefficients of the nonlinear terms.

The spectral method has a decisive advantage over the finite difference method. With finite differences, truncation error is greater for larger wavenumbers (small scales). With the spectral method, series expansion means spatial derivatives can be calculated with great accuracy (that given by the algorithms for calculating surface spherical harmonics, which in turn is dependent on the accuracy of the computer) for all the wavenumbers included in the truncation.

We can try to establish an equivalence between the truncation M of a spectral model and the mesh size Δx of a grid point model; the smallest wavelength represented with the spectral model is $2\pi a/M$ whereas the grid point method allows us to take into account $2\Delta x$ wavelength. We therefore write the relation:

$$2\Delta x \approx 2\pi a/M, \text{ or: } \Delta x_{(km)} \approx 20\,000/M.$$

This formula does not provide a strict equivalence, though, because unlike the spectral method, the finite difference method induces a truncation error that is larger when the characteristic wavelength of the structure considered is smaller. This practical equivalence was bolstered, though, by the comparative study by Girard and Jarraud (1982) between a spectral model with $T63$ truncation and a grid point model working with a 200 km mesh for a set of meteorological situations and that gave a certain advantage to the spectral method.

4.4 Spectral method on a doubly periodic domain

4.4.1 Constructing a doubly periodic domain

Given the advantages of the spectral method, it was tempting to apply it also to a limited area. Sine functions (or what amounts to the same thing, complex exponentials) form a basis of orthogonal functions on $[0, 2\pi]$ which can however be used for expanding only periodic functions over the plane domain under consideration.

Several solutions have been proposed for using a basis of sine functions effectively. A first technique is to use as variables on the limited area the deviations from the large-scale values provided by a model working on a larger domain (Hoyer, 1987); in this case the variables are zero at the lateral boundaries of the domain and the same hypothesis must be made for their space derivatives. In a second method, proposed by Tatsumi (1986), the basis of orthogonal sine functions is completed by nonorthogonal functions allowing the expansion of nonperiodic fields on the domain. Lastly, in a third simple and smart solution, proposed by Machenhauer and Haugen (1987), and taken up by Joly (1992), the working domain is enlarged (Figure 4.5) so as to extend the fields so that they, and their space derivatives, are

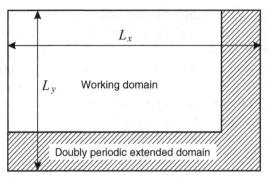

Figure 4.5 Extension from the initial domain to the biperiodic enlarged domain.

periodic in both directions over the new domain; the spectral method can then be applied to these *doubly periodic* fields.

4.4.2 Basis functions

Doubly periodic fields may be expanded by using the functions $H_n^m (x, y)$ which are products of sine functions, defined on the extended rectangular domain R $\{(x, y);$ $x \in [0, L_x], y \in [0, L_y]\}$ and that are written, in complex form:

$$H_n^m(x, y) = e^{im\left(\frac{2\pi}{L_x}\right)x} \, e^{in\left(\frac{2\pi}{L_y}\right)y}. \tag{4.54}$$

These functions, of modulus equal to 1, are eigenfunctions of the Laplace operator expressed in a system of Cartesian coordinates, since definition (4.54) leads to:

$$\nabla^2 H_n^m(x, y) = -4\pi^2 \left(\frac{m^2}{L_x^2} + \frac{n^2}{L_y^2}\right) H_n^m(x, y). \tag{4.55}$$

They form an orthonormal basis and confirm the relation:

$$\frac{1}{L_x L_y}\int_0^{L_x}\int_0^{L_y} H_n^m(x, y) \, \overline{H_{n'}^{m'}}(x, y) \, dydx = \begin{cases} 0 \text{ if } (n, m) \neq (n', m'), \\ 1 \text{ if } (n, m) = (n', m'), \end{cases} \tag{4.56}$$

where the operator $\overline{\cdot}$ denotes the complex conjugate.

They are suitable for expanding into series the doubly periodic functions on the domain R. Thus the expansion truncated at wavenumbers M and N of a doubly periodic function $A(x, y)$, that can now be termed $A(x, y)$ too so as to dispense with introducing additional specific notation, is written:

$$A(x, y) = \sum_{m=-M}^{+M} \sum_{n=-N}^{+N} A_n^m H_n^m(x, y). \tag{4.57}$$

The values of M and N, which must be specified, define the truncation of the expansion.

Given the relation (4.56), the coefficients A_n^m are easily calculated using the double integral:

$$A_n^m = \frac{1}{L_x L_y} \int_0^{L_x} \int_0^{L_y} A(x,y)\, \overline{H_n^m}(x,y)\, dy dx, \tag{4.58}$$

corresponding to the application of two Fourier transforms in succession.

The Parseval-Plancherel relation is also verified:

$$\frac{1}{L_x L_y} \int_0^{L_x} \int_0^{L_y} \left[A(x,y) \right]^2 dy dx = \sum_{m=-M}^{+M} \sum_{m=-N}^{+N} \left| A_n^m \right|^2. \tag{4.59}$$

The transition from coefficients A_m^n to the field $A(x, y)$ is made by performing the two Fourier syntheses in turn:

$$A_m(y) = \sum_{n=-N}^{+N} A_n^m\, e^{in\left(\frac{2\pi}{L_y}\right)y}, \tag{4.60}$$

$$A(x,y) = \sum_{m=-M}^{+M} A_m(y) e^{im\left(\frac{2\pi}{L_x}\right)x}. \tag{4.61}$$

If the field $A(x, y)$ is real, then $A_{-m} = \overline{A_m}$, and therefore also $\overline{A_{-m}} = A_m$.

By examining the expansion of these equalities and by identifying them term by term, it can be observed that the coefficients A_n^m are not independent and must verify the relations:

$$A_{-n}^{-m} = A_n^m, \text{ and: } A_{-n}^m = A_n^{-m}. \tag{4.62}$$

The existence of these relations shows that the real number of degrees of freedom defining a real doubly periodic field is halved with respect to the total number of complex coefficients.

4.4.3 Elliptical truncation

Elliptical truncation is defined by:

$$T_E: \left\{ (n,m) \,;\; -M(n) \leq m \leq +M(n), \; -N(m) \leq n \leq +N(m) \right\},$$

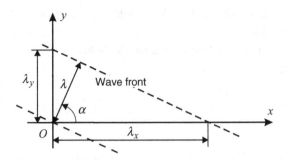

Figure 4.6 Plane wave.

with the quantities $M(n)$ and $N(m)$ verifying the relation:

$$\frac{N^2(m)}{N_{max}} + \frac{M^2(n)}{M_{max}} \leq 1, \tag{4.63}$$

N_{max} and M_{max} corresponding to the maximum wavenumbers handled in the direction of x and y. This truncation, the denomination of which becomes obvious (Figure 4.7), has the property of giving an invariant representation for a plane wave, whatever its direction of propagation with respect to the grid axes.

For a plane wave of wavelength λ propagating in a direction that makes an angle α with the x-axis, as shown in Figure 4.6, we verify the relations:

$$\lambda_x = \lambda / \cos \alpha \text{ and } \lambda_y = \lambda / \sin \alpha,$$

that lead to:

$$\frac{\lambda^2}{\lambda_x^2} + \frac{\lambda^2}{\lambda_y^2} = 1. \tag{4.64}$$

To obtain equivalent spatial resolution along the x- or y-axis, we must have for the smallest wavelength handled $\lambda_{\min}$:

$$\lambda_{\min} = L_x / M_{\max},$$
$$\lambda_{\min} = L_y / N_{\max}. \tag{4.65}$$

The wavelengths along x and y, for a plane wave characterized by the numbers $M(n)$ and $N(m)$, are given by:

$$\lambda_x = L_x / M,$$
$$\lambda_y = L_y / N. \tag{4.66}$$

Inserting relations (4.65) and (4.66) written for $\lambda = \lambda_{\min}$ into (4.64) gives the relation defining the elliptical truncation contour specified in (4.63) and illustrated in Figure 4.7.

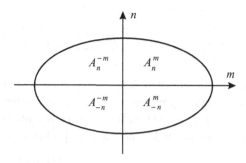

Figure 4.7   Elliptical truncation.

4.4.4 Calculating linear terms

Given the expression of the basis functions from (4.54), we get for the coefficients of
the space derivatives of a field A:

$$\left.\left[\frac{\partial}{\partial x}A(x,y)\right]^m_n = im\frac{2\pi}{L_x}A^m_n,\\[2mm] \left[\frac{\partial}{\partial y}A(x,y)\right]^m_n = in\frac{2\pi}{L_y}A^m_n,\right\} \tag{4.67}$$

and by using relation (4.55), we obtain the coefficients of the Laplacian:

$$\left(\nabla^2 A\right)^m_n = -\left[\left(m\frac{2\pi}{L_x}\right)^2 + \left(n\frac{2\pi}{L_y}\right)^2\right]A^m_n. \tag{4.68}$$

4.4.5 Calculating nonlinear terms

The coefficients of quadratic terms are calculated by the transform method, requiring
the calculation of nonlinear terms on a regular grid G_R applied over the domain R:

$$G_R\colon \left\{\, x_j,\, j = 1,...,\, J\,;\, y_k,\, k = 1,...,\, K \right\}. \tag{4.69}$$

The coefficients A^m_n of the quadratic expression $A(\lambda,\mu) = B(\lambda,\mu)\,C(\lambda,\mu)$ are obtained
by calculating the double integral (4.58), consisting in two successive Fourier trans-
forms using the trapezoidal quadrature formulas:

$$A_m(y_k) = \sum_{j=1}^{J} A(x_j,\, y_k)\,e^{-im\left(\frac{2\pi}{L_x}\right)x_j}, \tag{4.70}$$

$$A_n^m = \sum_{k=1}^{K} A_m(y_k)\, e^{-in\left(\frac{2\pi}{Ly}\right)y_k}. \tag{4.71}$$

As seen in Subsection 4.3.7, the expressions (4.70) and (4.71) yield accurate evaluations of the corresponding integrals if the number of grid points is suitably calculated given the degree of the sine functions of the integrand. The quadratic terms are evaluated without aliasing by choosing the grid characteristics G_R as below:

$$\left.\begin{aligned} J \geq 3M+1, \\ K \geq 3M+1. \end{aligned}\right\} \tag{4.72}$$

Limited area models often use equations written on a conformal projection. In this case, the nonlinear advection terms are not quadratic. The space derivatives are multiplied by a map scale factor which must be taken into account when calculating the characteristics of the grid, yielding a result without aliasing.

4.4.6 The advantage of the method

The procedure consisting in extending the domain and constructing doubly periodic fields on the domain R therefore provides a simple way of using the spectral method on a limited domain. This allows the space derivative to be calculated very accurately. A major advantage of the spectral method is that truncation means only the wavenumbers belonging to the truncation need be kept and filtering is automatic.

It must also be observed that the development of operational weather forecasting models requires a very big investment. We can consider using a single basis code to obtain both a spherical model and a limited area model. Apart from the process implemented to obtain doubly periodic fields, the two versions differ only in the processing of one of the two space dimensions (the Legendre transform in the spherical case and the Fourier transform in the limited area case).

5 The effects of discretization

5.1 Introduction

The effect of the various discretizations in space and time may be studied systematically in the context of a linearized model (Grotjhan and O'Brien, 1976). With such a model, it is possible to determine analytical solutions both for the system of equations under consideration and for the system obtained after discretization and therefore, by straightforward comparison, to appraise the effect of the chosen numerical schemes. The shallow water barotropic model is used as a study tool, as it has as solutions the two types of waves described by the primitive equation models: slow waves associated with advection terms and fast inertia-gravity waves associated with the Coriolis terms and the adaptation terms (pressure force in the equations of motion and divergence in the continuity equation). Moreover, it can be shown that a primitive equation model with N levels may be considered, once linearized, as the superposition of N shallow water barotropic models of decreasing equivalent heights.

5.2 The linearized barotropic model

5.2.1 The equations for the perturbations

The equations of the linearized shallow water model are obtained from the corresponding nonlinear equations by using the small perturbation method. This consists in writing that the model variables can be considered as the superposition of a time-independent basic state and of perturbations evolving over time. We must first therefore define a time-independent basic state that satisfies the nonlinear system of equations and then write the new system in which only the linear terms for perturbations are kept and which thus become the new variables.

The chosen working domain is a doubly periodic rectangular domain with a rotation characterized by a constant value of the Coriolis parameter f, that is assumed to be nonzero. To simplify, we place ourselves in a Cartesian geometry so that the scale factor m can be considered constant and equal to 1. We also assume the existence of topography of surface geopotential Φ_s whose characteristics shall be specified.

From the expanded equations of system (2.14), we obtain:

$$\frac{\partial U}{\partial t}+U\frac{\partial U}{\partial x}+V\frac{\partial U}{\partial y}-fV+\frac{\partial \Phi}{\partial x}=0,$$

$$\frac{\partial V}{\partial t}+U\frac{\partial V}{\partial x}+V\frac{\partial V}{\partial y}+fU+\frac{\partial \Phi}{\partial y}=0,$$

$$\frac{\partial \Phi}{\partial t}+U\frac{\partial (\Phi-\Phi_s)}{\partial x}+V\frac{\partial (\Phi-\Phi_s)}{\partial y}+(\Phi-\Phi_s)\left[\frac{\partial U}{\partial x}+\frac{\partial V}{\partial y}\right]=0.$$

(5.1)

We choose a geostrophic basic state characterized by a constant wind $\mathbf{V}_0$, with components U_0, V_0 related to the geopotential $\Phi_0(x, y)$ by the relations:

$$U_0 = -\frac{1}{f}\frac{\partial \Phi_0}{\partial y},$$

$$V_0 = \frac{1}{f}\frac{\partial \Phi_0}{\partial x}.$$

(5.2)

The choice of a constant wind $\mathbf{V}_0$ implies that the space derivatives of U_0 and V_0 are zero and that the surface $\Phi_0(x, y)$ is a plane. By choosing a topography Φ_s such that the horizontal gradient of $(\Phi_0 - \Phi_s)$ is zero (which means that the free surface $\Phi_0(x,y)$ that maintains the geostrophic flux constant is parallel to the topography, as shown in Figure 5.1), the basic state verifies the system of nonlinear equations (5.1).

The variables U, V and Φ can be written in the form of a sum of variables of the basic state U_0, V_0 and Φ_0 and of perturbations U', V' and Φ'. By conserving only the linear terms containing perturbations and given the hypothesis about the gradient of $(\Phi_0 - \Phi_s)$, system (5.1) leads, after introducing the notation $\Phi^* = (\Phi_0 - \Phi_s)$ (a constant characterizing the mean thickness of the fluid), to the system of linear equations (5.3 for the perturbations:

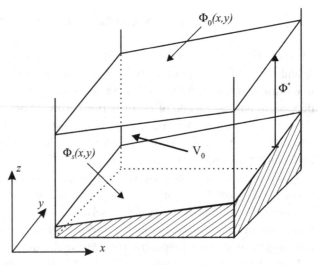

Figure 5.1   The geostrophic basic state.

$$\frac{\partial U'}{\partial t} + U_0 \frac{\partial U'}{\partial x} + V_0 \frac{\partial U'}{\partial y} - fV' + \frac{\partial \Phi'}{\partial x} = 0,$$

$$\frac{\partial V'}{\partial t} + U_0 \frac{\partial V'}{\partial x} + V_0 \frac{\partial V'}{\partial y} + fU' + \frac{\partial \Phi'}{\partial y} = 0, \qquad (5.3)$$

$$\frac{\partial \Phi'}{\partial t} + U_0 \frac{\partial \Phi'}{\partial x} + V_0 \frac{\partial \Phi'}{\partial y} + \Phi^* \left[\frac{\partial U'}{\partial x} + \frac{\partial V'}{\partial y} \right] = 0.$$

5.2.2 The analytical solutions of the linearized model

We look for analytical solutions of system (5.3) in the form of a plane wave characterized by its angular frequency (commonly simply termed frequency) ω and by its wavenumbers, k in the x-direction and l in the y-direction, and verifying the periodicity conditions at the boundaries of the chosen domain. The general solution can be obtained as a superposition of plane waves. We look for periodic solutions for the column vector $\mathbf{X} = (U', V', \Phi')^{\mathrm{T}}$ in the form:

$$\mathbf{X} = \mathbf{X}^* e^{\mathrm{i}(\omega t + kx + ly)} \qquad (5.4)$$

The system (5.3) is written in a short form by calling $\widetilde{R}$ the linear spatial operator:

$$\frac{\partial \mathbf{X}}{\partial t} + \widetilde{R}(\mathbf{X}) = 0. \qquad (5.5)$$

Substituting the expression (5.4) into system (5.3) leads to the homogeneous system:

$$\mathbf{MX} = 0, \qquad (5.6)$$

matrix $\mathbf{M}$ having the form:

$$\mathbf{M} = \begin{bmatrix} \mathrm{i}(\omega + kU_0 + lV_0) & -f & \mathrm{i}k \\ +f & \mathrm{i}(\omega + kU_0 + lV_0) & \mathrm{i}l \\ \mathrm{i}k\Phi^* & \mathrm{i}l\Phi^* & \mathrm{i}(\omega + kU_0 + lV_0) \end{bmatrix}. \qquad (5.7)$$

The nontrivial solutions ($\mathbf{X} \neq \mathbf{0}$) of the homogeneous system (5.6) are obtained when the determinant of the matrix $\mathbf{M}$ is zero, which leads to an equation of the third degree for the variable $v = \omega + kU_0 + lV_0$, which can be decomposed into an equation of the first degree and an equation of the second degree:

$$v \left\{ v^2 - \left[f^2 + \Phi^*(k^2 + l^2) \right] \right\} = 0. \qquad (5.8)$$

This equation admits as solutions:

$$v = 0,$$

$$v = \pm \left[f^2 + \Phi^*(k^2 + l^2) \right]^{1/2}. \qquad (5.9)$$

Three solutions are obtained for the frequencies ω:

$$\omega_1 = -(kU_0 + lV_0) = \omega_0, \tag{5.10}$$

characterizing a slow wave, and:

$$\omega_{2,3} = \omega_0 \pm \omega_g, \text{with } \omega_g = \left[f^2 + \Phi^*(k^2 + l^2) \right]^{1/2}, \tag{5.11}$$

characterizing the two inertia-gravity waves.

By calling α' and α the respective angles of the basic wind $\mathbf{V}_0$ and of the wave vector $\mathbf{K}$ with the x-axis, the quantities U_0, V_0, k and l verify the relations:

$$
\begin{aligned}
U_0 &= |\mathbf{V}_0| \cos \alpha', \quad V_0 = |\mathbf{V}_0| \sin \alpha', \\
k &= K \cos \alpha, \qquad l = K \sin \alpha, \qquad \text{with } K = \|\mathbf{K}\|.
\end{aligned}
\tag{5.12}
$$

The expressions (5.10) and (5.11) are then written in the simplified form:

$$
\begin{aligned}
\omega_1 &= -KV^*, \\
\omega_g &= \left[f^2 + \Phi^* K^2 \right]^{1/2} = K\Phi^{*1/2} \left(1 + \frac{1}{K^2 L_R^2} \right)^{1/2},
\end{aligned}
\tag{5.13}
$$

where V^* represents the projection of the basic wind on the wave vector and $L_R = \Phi^{*1/2}/f$, the Rossby radius of deformation. This quantity, which has the dimension of a length, separates the scales at which gravity predominates from those at which the Earth's rotation predominates.

The algebraic measurements of the wave propagation speeds (relative to a unit vector parallel to the wave vector $\mathbf{K}$ and oriented in the same direction), which are solutions of system (5.3), are therefore obtained by calculating $V = -\omega/K$, which gives:

$$
\begin{aligned}
V_1 &= V^* \\
V_{2,3} &= V^* \pm V_g
\end{aligned}
, \text{ with: } V_g = \Phi^{*1/2} \left(1 + \frac{1}{K^2 L_R^2} \right)^{1/2}.
\tag{5.14}
$$

5.3 Effect of horizontal discretization

5.3.1 General principle

When using the spectral method, it can be considered that the horizontal derivatives are evaluated exactly and that no error is made therefore (except for computer accuracy). Where finite differences are used on a grid, the horizontal derivatives are not calculated exactly and a discretization error is introduced, whose effects on wave propagation must be calculated.

Computation for the discretized system on a square mesh ($\Delta y = \Delta x$) follows the same lines as in Subsection 5.2.2. We simply take into account the effect of the finite difference operators defined in Subsection 3.2.2 for each of the components of the periodic solution **X**, noted generically as A:

$$A_x = iA\frac{\sin(k\Delta x/2)}{\Delta x/2}, \quad \overline{A}^x = A\cos(k\Delta x/2) \quad \text{and} \quad \overline{A}_x^x = iA\frac{\sin(k\Delta x)}{\Delta x},$$

$$A_y = iA\frac{\sin(l\Delta x/2)}{\Delta x/2}, \quad \overline{A}^y = A\cos(l\Delta x/2) \quad \text{and} \quad \overline{A}_y^y = iA\frac{\sin(l\Delta x)}{\Delta x}. \tag{5.15}$$

The elements of matrix **M** are modified by the presence of multiplicative factors from relations (5.15) and we obtain new equations of the third degree analogous to (5.8). We can therefore compare these new solutions for frequencies and phase speeds with 'exact' solutions given by the expressions (5.13) or (5.14).

5.3.2 Application to the various grids

We look first at the effect of the various forms of discretization on the propagation of slow waves. That effect depends exclusively on the way the advection terms are discretized. We therefore distinguish three cases (denoted a, b, and c, respectively) corresponding to the discretizations on the various grids studied in Chapter 3:

(a) central differences on the A, B, or C grids,
(b) discretization on the D' staggered grid,
(c) central differences of 4th-order accuracy on the A grid.

Computing the frequencies for discretization a, b, and c yields the expressions:

$$\left.\begin{array}{l} \omega_1^a = -\left[kU_0\dfrac{\sin(k\Delta x)}{k\Delta x} + lV_0\dfrac{\sin(l\Delta x)}{l\Delta x}\right], \\[3mm] \omega_1^b = -\left[kU_0\dfrac{\sin(k\Delta x/2)}{k\Delta x/2}\cos(l\Delta x/2) + lV_0\dfrac{\sin(l\Delta x/2)}{l\Delta x/2}\cos(k\Delta x/2)\right], \\[3mm] \omega_1^c = -\dfrac{1}{3}\left[kU_0\left(4\dfrac{\sin(k\Delta x)}{k\Delta x} - \dfrac{\sin(2k\Delta x)}{2k\Delta x}\right) + lV_0\left(4\dfrac{\sin(l\Delta x)}{l\Delta x} - \dfrac{\sin(2l\Delta x)}{2l\Delta x}\right)\right]. \end{array}\right\} \tag{5.16}$$

By using the relations (5.12), we finally get:

$$\omega_1^i = -KV_1^i = -KV^*F_1^i,$$

where F_1^i is the deceleration factor for the slow wave due to the horizontal discretization scheme i (i = a, b, or c). By choosing a basic wind of the same direction as the wave vector ($\alpha = \alpha'$), it is possible to evaluate F_1^i for various values of the angle α and wavenumber $K = 2\pi/\lambda$. The factors corresponding to discretizations a, b, and c are

given by the expressions (5.17) and are depicted in Figure 5.2 for some typical values
of wavelength λ:

$$\left.\begin{aligned}
F_1^{\,a} &= \frac{\cos\alpha\,\sin(K\Delta x\cos\alpha)+\sin\alpha\,\sin(K\Delta x\sin\alpha)}{K\Delta x}, \\[4pt]
F_1^{\,b} &= \frac{\cos\alpha\,\sin\left((K\Delta x/2)\cos\alpha\right)\cos\left((K\Delta x/2)\sin\alpha\right)+\sin\alpha\,\sin\left((K\Delta x/2)\sin\alpha\right)\cos\left((K\Delta x/2)\cos\alpha\right)}{K\Delta x/2}, \\[4pt]
F_1^{\,c} &= \frac{1}{3}\frac{\cos\alpha\,\sin(K\Delta x\cos\alpha)\left[4-\cos(K\Delta x\cos\alpha)\right]+\sin\alpha\,\sin(K\Delta x\sin\alpha)\left[4-\cos(K\Delta x\sin\alpha)\right]}{K\Delta x}.
\end{aligned}\right\}\quad (5.17)$$

The curves in Figure 5.2 show that the effects of discretization appear for the shorter
waves only and become negligible when $\lambda = 16\Delta x$. Notice the clear advantage of discretization of 4th-order accuracy, although it is true that this is offset by the greater
volume of computation required. Comparison of the curves shows that using the D'
grid enables us to achieve better accuracy than using the A, B, and C grids for small
values of α, but gives the same accuracy when $\alpha = \pi/4$.

Similar calculations can be made to study the effect of the various forms of discretization on the propagation of inertia-gravity waves. We confine ourselves to studying
the effect of the various discretizations on the frequency ω_g and speed V_g, since the
effect on ω_1 and V_1 has just been studied. It is also assumed that the Coriolis term is
always calculated exactly, regardless of grid type.

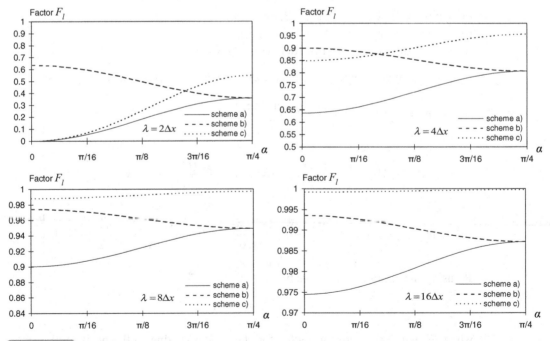

Figure 5.2   Deceleration factors for slow waves obtained with a, b, and c discretizations for wavelengths $\lambda = 2\Delta x$, $4\Delta x$, $8\Delta x$, and
$16\Delta x$. Because of the obvious symmetry for variables x and y, the curves are plotted for $\alpha \in '[0,\pi/4]$ only. Notice that
the y-axis scale differs for the various wavelengths.

Four types of discretization are distinguished for the adaptation terms:

(a) central differences on the A grid,
(b) discretization on the B grid,
(c) central differences on the C or D' grids,
(d) 4th-order accuracy central differences on the A grid.

Computing the frequencies for discretizations a, b, c, and d gives the expressions:

$$
\left.
\begin{aligned}
\omega_g^a &= \left\{ f^2 + \Phi^* \left[k^2 \frac{\sin^2(k\Delta x)}{(k\Delta x)^2} + l^2 \frac{\sin^2(l\Delta x)}{(l\Delta x)^2} \right] \right\}^{1/2}, \\
\omega_g^b &= \left\{ f^2 + \Phi^* \left[k^2 \frac{\sin^2(k\Delta x/2)\cos^2(l\Delta x/2)}{(k\Delta x/2)^2} + l^2 \frac{\sin^2(l\Delta x/2)\cos^2(k\Delta x/2)}{(l\Delta x/2)^2} \right] \right\}^{1/2}, \\
\omega_g^c &= \left\{ f^2 + \Phi^* \left[k^2 \frac{\sin^2(k\Delta x/2)}{(k\Delta x/2)^2} + l^2 \frac{\sin^2(l\Delta x/2)}{(l\Delta x/2)^2} \right] \right\}^{1/2}, \\
\omega_g^d &= \left\{ f^2 + \frac{\Phi^*}{9} \left[k^2 \frac{\sin^2(k\Delta x)[4-\cos(k\Delta x)]^2}{(k\Delta x)^2} + l^2 \frac{\sin^2(l\Delta x)[4-\cos(l\Delta x)]^2}{(l\Delta x)^2} \right] \right\}^{1/2}.
\end{aligned}
\right\}
\tag{5.18}
$$

By using the relations (5.12), we finally obtain:

$$
\omega_g^i = K V_g^i = K \Phi^{*1/2} \left[\left(F^i \right)^2 + \frac{1}{K^2 L_R^2} \right]^{1/2},
$$

where the factor F^i is the deceleration factor for pure gravity waves, due to the horizontal discretization scheme i (i = a, b, c, or d). The factors corresponding to discretizations a, b, c, and d are given by the expressions (5.19) and are depicted in Figure 5.3 for some typical values of wavelength λ:

$$
\left.
\begin{aligned}
F_g^a &= \left[\frac{\sin^2(K\Delta x \cos\alpha) + \sin^2(K\Delta x \sin\alpha)}{(K\Delta x)^2} \right]^{1/2}, \\
F_g^b &= \left[\frac{\sin^2((K\Delta x/2)\cos\alpha)\cos^2((K\Delta x/2)\sin\alpha) + \sin^2((K\Delta x/2)\sin\alpha)\cos^2((K\Delta x/2)\cos\alpha)}{(K\Delta x/2)^2} \right]^{1/2}, \\
F_g^c &= \left[\frac{\sin^2((K\Delta x/2)\cos\alpha) + \sin^2((K\Delta x/2)\sin\alpha)}{(K\Delta x/2)^2} \right]^{1/2}, \\
F_g^d &= \left\{ \frac{\sin^2(K\Delta x \cos\alpha)[4-\cos(K\Delta x \sin\alpha)]^2 + \sin^2(K\Delta x \sin\alpha)[4-\cos(K\Delta x \sin\alpha)]^2}{9(K\Delta x)^2} \right\}^{1/2}.
\end{aligned}
\right\}
\tag{5.19}
$$

The factors F_g^i correspond to pure gravity waves obtained for very large values of the Rossby radius of deformation ($L_R \to \infty$). In this case, examination of Figure 5.3

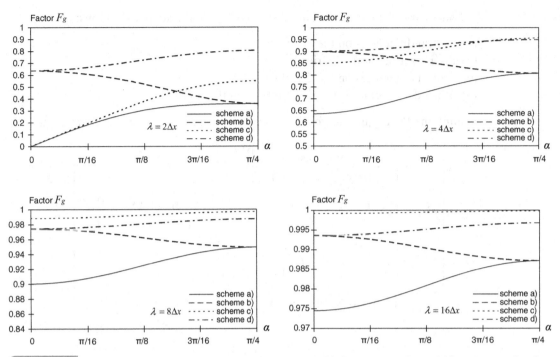

Figure 5.3 Deceleration factors for the inertia-gravity waves obtained with discretizations a, b, c, and d for various wavelengths $\lambda = 2\Delta x$, $4\Delta x$, $8\Delta x$, and $16\Delta x$. Because of the obvious symmetry for variables x and y, the curves are plotted for $\alpha' \in [0, \pi/4]$ only. Notice again that the y-axis scale differs for the various wavelengths.

shows that this factor is very close to 1 when $\lambda = 16\Delta x$ and beyond. For a mean geopotential height $\Phi^* = 10\ 000$ J/K (equivalent to a height of 1000 m) and a Coriolis parameter $f = 10^{-4}$ s^{-1}, the Rossby radius of deformation is $L_R = 1000$ km. The term $1/K^2 L_R^2$ becomes large for long wavelengths only, which corresponds for F_g^i to values close to 1. So the deceleration factor for inertia-gravity waves, which is given by the formula:

$$V_g^i / V^* = \left[\frac{\left(F_g^i\right)^2 + \dfrac{1}{K^2 L_R^2}}{1 + \dfrac{1}{K^2 L_R^2}} \right]^{1/2}, \tag{5.20}$$

can be replaced in practice by the approximation:

$$V_g^i / V^* \cong F_g^i.$$

The attenuation factors obtained for inertia-gravity waves, depicted in Figure 5.3, show that the effect of spatial discretization is fairly similar to what is observed for slow waves. It is observed that the C or D' grids provide greater accuracy than the B grid, which in turn provides greater accuracy than the A grid. Here again it is seen that at the cost of additional computation, the 4th-order accuracy discretization outclasses the 2nd-order accuracy discretizations.

5.4 Various time integration schemes

Having examined the effects of spatial discretization taken in isolation, we study the effect of the various time integration schemes, confining ourselves, though, to the centred schemes, which have the property of preserving wave amplitude. We examine in turn the forward uncentred scheme, then the centred explicit scheme, the centred semi-implicit scheme, and lastly, the semi-Lagrangian semi-implicit scheme. To study the properties of time discretization only, we assume therefore in the remainder of this chapter that the horizontal derivatives are computed exactly by the spectral method for example.

The system (5.5) is written in the form:

$$\frac{\partial \mathbf{X}}{\partial t} + \widetilde{A}(\mathbf{X}) + \widetilde{F}(\mathbf{X}) + \widetilde{G}(\mathbf{X}) = 0, \tag{5.21}$$

$\widetilde{A}$, $\widetilde{F}$ and $\widetilde{G}$ representing the advection, Coriolis, and adaptation operators, respectively.

For periodic solutions of the type (5.4) sought, the effect of these various operators is reflected by a multiplication by matrices with complex coefficients:

$$\frac{\partial \mathbf{X}}{\partial t} \equiv i\omega\, \mathbf{IX}, \quad \widetilde{A}(\mathbf{X}) \equiv \mathbf{AX} \equiv i(kU_0 + lV_0)\mathbf{IX}, \quad \widetilde{F}(\mathbf{X}) \equiv \mathbf{FX} \text{ and } \widetilde{G}(\mathbf{X}) \equiv \mathbf{GX}, \tag{5.22}$$

I being the identity matrix, **F** and **G** matrices and being of the form:

$$\mathbf{F} = \begin{bmatrix} 0 & -f & 0 \\ +f & 0 & 0 \\ 0 & 0 & 0 \end{bmatrix}, \qquad \mathbf{G} = \begin{bmatrix} 0 & 0 & ik \\ 0 & 0 & il \\ ik\Phi^* & il\Phi^* & 0 \end{bmatrix}. \tag{5.23}$$

5.4.1 The Euler explicit scheme

The *forward uncentred* or *Euler scheme* consists in evaluating the time derivative by using a forward difference so that system (5.21) is replaced by:

$$\frac{\mathbf{X}(t+\Delta t)-\mathbf{X}(t)}{\Delta t}+\widetilde{A}[\mathbf{X}(t)]+\widetilde{F}[\mathbf{X}(t)]+\widetilde{G}[\mathbf{X}(t)]=0. \tag{5.24}$$

When applied to solution (5.4), the discretization of the time derivative is reflected by a multiplication of vector $\mathbf{X}$ by a complex quantity:

$$\frac{\mathbf{X}(t+\Delta t)-\mathbf{X}(t)}{\Delta t}=\frac{e^{i\omega\Delta t}-1}{\Delta t}\mathbf{X}(t). \tag{5.25}$$

The solutions for this new system are calculated by following the same process as in Subsection 5.2.2 and we obtain for the slow wave, instead of (5.10):

$$\frac{e^{i\omega\Delta t}-1}{\Delta t}=-i\left(kU_0+lV_0\right), \tag{5.26}$$

which, in view of the relations (5.12), gives for the frequency:

$$e^{i\omega\Delta t}=1-iKV^*\Delta t. \tag{5.27}$$

As the right-hand side of (5.27) is a complex number with a modulus greater than 1, the frequency must be a complex number that is written in the form:

$$\omega=\omega'+i\omega'',$$

ω' and ω'' being real numbers.

The quantity ω' is the true frequency corresponding to the slow wave speed and is obtained from relation (5.24):

$$\omega'\Delta t=-KV_1\Delta t=\arctan(-KV^*\Delta t). \tag{5.28}$$

The ratio Q_1 of the speed calculated numerically to the speed calculated analytically for the slow wave is then written:

$$Q_1=\frac{\arctan(KV^*\Delta t)}{KV^*\Delta t}. \tag{5.29}$$

The quantity ω'' is then given by:

$$R_1=e^{-\omega''\Delta t}=\left[1+(KV^*\Delta t)^2\right]^{1/2}, \tag{5.30}$$

which is the modulus of the amplification factor of the solution between times t and $t+\Delta t$, since we verify:

$$\|\mathbf{X}(t+\Delta t)\|=\left|e^{i\omega\Delta t}\right|.\|\mathbf{X}(t)\|. \tag{5.31}$$

Relation (5.29) shows that the Euler scheme slows down the slow wave and relation (5.30) that it increases its amplitude from one time step to the next, since the modulus of the amplification factor R_1 is greater than 1.

For the inertia-gravity waves, the analogous relation to (5.11) is given by:

$$\frac{e^{i\omega\Delta t} - 1}{\Delta t} = -i\left\{ \left(kU_0 + lV_0 \right) \pm \left[f^2 + \Phi^*(k^2 + l^2) \right]^{1/2} \right\}. \tag{5.32}$$

By the same reasoning as for slow waves, we calculate the characteristics of inertia-gravity wave propagation, resulting from the use of the numerical scheme. The ratio $Q_{2,3}$ of the speed calculated numerically to the speed calculated analytically for inertia-gravity waves is then written:

$$Q_{2,3} = \frac{\arctan\left[K(V^* \pm V_g)\Delta t \right]}{K(V^* \pm V_g)\Delta t}, \tag{5.33}$$

and the modulus of the amplification factor of the solution:

$$R_{2,3} = e^{-\omega''\Delta t} = \left[1 + \left[K(V^* \pm V_g)\Delta t \right]^2 \right]^{1/2}. \tag{5.34}$$

Figure 5.4 shows the values taken by factors Q and R for the slow wave depending on $KV^*\Delta t$ and those same factors for inertia-gravity waves depending on $K(V^* + V_g)\Delta t$, when these quantities vary from 0 to 1. Assuming $V^*/V_g = 1/5$, it can be seen that the variation of quantity $KV^*\Delta t$ is limited to 1/6 of the variation of $K(V^* + V_g)\Delta_t$. It can be observed that the Euler scheme, while decelerating waves, increases their amplitude from one time step to the next, the most prominent effect being for fast inertia-gravity waves. The Euler scheme is *unconditionally unstable*; its use in numerical prediction is reserved strictly to the first time step.

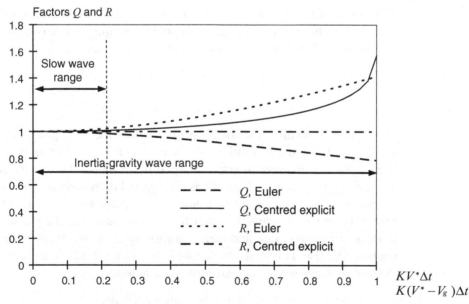

Figure 5.4   Effect on slow waves and on inertia-gravity waves of the (Euler) forward uncentred explicit scheme and the (leapfrog) centred explicit scheme.

5.4.2 The centred explicit scheme

The centred explicit scheme consists in evaluating the time derivative using central differences in time and therefore replacing the system (5.21) by:

$$\frac{\mathbf{X}(t+\Delta t)-\mathbf{X}(t-\Delta t)}{2\Delta t} + \widetilde{A}\big[\mathbf{X}(t)\big] + \widetilde{F}\big[\mathbf{X}(t)\big] + \widetilde{G}\big[\mathbf{X}(t)\big] = 0. \tag{5.35}$$

This *leapfrog* scheme, as it is called, allows us to obtain the variables explicitly at time $t+\Delta t$ from their known values at the previous times t and $t-\Delta t$; accordingly it can be applied only from the second time step onwards, with the values at the first time step usually being computed using the Euler scheme.

When applied to solution (5.4), this discretization of the time derivative is reflected by a multiplication of the vector $\mathbf{X}$ by a complex quantity:

$$\frac{\mathbf{X}(t+\Delta t)-\mathbf{X}(t-\Delta t)}{2\Delta t} = i\frac{\sin(\omega\Delta t)}{\Delta t}\mathbf{X}(t) \tag{5.36}$$

The equivalent of relation (5.10) for the slow wave is written:

$$\frac{\sin(\omega_1\Delta t)}{\Delta t} = -(kU_0 + lV_0), \tag{5.37}$$

and the equivalent of relation (5.11) for inertia-gravity waves is:

$$\frac{\sin(\omega_{2,3}\Delta t)}{\Delta t} = -(kU_0 + lV_0) \pm \Big[f^2 + \Phi^*(k^2 + l^2) \Big]^{1/2}. \tag{5.38}$$

If the value of the right-hand side lies between -1 and $+1$, the solutions for ω are real and the modulus of the amplification factor $e^{i\omega\Delta t}$ is equal to 1. To obtain a stable solution, we must therefore necessarily verify the inequalities:

$$-1 \leq KV^* \Delta t \leq +1, \tag{5.39}$$

$$-1 \leq K(V^* \pm V_g)\Delta_t \leq +1. \tag{5.40}$$

The second inequality is more constraining than the first, given the usual orders of magnitude for the quantities V^*, mean velocity of advection (barely exceeding 50 m/s), and V_g, the speed of gravity waves (up to 330 m/s). These sufficient conditions for ensuring the stability of equations with partial derivatives processed by finite differences were brought to light by Courant, Friedrichs, and Lewy (1928) and are now known as the *CFL conditions*. If they hold, we can calculate the propagation characteristics of the slow wave and inertia-gravity waves. However, close examination of relations (5.37) and (5.38) shows that we in fact obtain for both the slow wave and for the inertia gravity waves, two categories of solutions:

$$\omega_1\Delta t = -\arcsin(KV^*\Delta t), \tag{5.41}$$

$$\omega_{2,3}\Delta t = -\arcsin\left[K(V^* \pm V_g)\Delta t\right], \tag{5.42}$$

which correspond to the *physical* solutions, and:

$$\omega_1\Delta t = \pi + \arcsin(KV^*\Delta t), \tag{5.43}$$

$$\omega_{2,3}\Delta t = \pi + \arcsin\left[K(V^* \pm V_g)\Delta t\right] \tag{5.44}$$

which are additional solutions introduced by the chosen discretization and are called *computational solutions.*

For the physical solutions, the ratios Q_1 and $Q_{2,3}$ of the speeds calculated numerically to the speeds calculated analytically are written, for the slow wave:

$$Q_1 = \frac{\arcsin(KV^*\Delta t)}{KV^*\Delta t}, \tag{5.45}$$

and for inertia-gravity waves:

$$Q_{2,3} = \frac{\arcsin\left[K(V^* \pm V_g)\Delta t\right]}{K(V^* \pm V_g)\Delta t}, \tag{5.46}$$

whereas the moduli R_1 and $R_{2,3}$ of the amplification factors of the solutions remain equal to 1.

The values taken by the factors Q and R as a function of $KV^*\Delta t$ for the slow wave, and of $K(V^* + V_g)\Delta t$ for inertia-gravity waves, are shown in Figure 5.4. The curves obtained in this case show that, compared with the analytical solutions, the numerical solutions are accelerated but maintain their amplitude. As already noted for the Euler scheme, the faster the inertia-gravity waves, the more prominent the effect.

The form of frequencies obtained for the computational solutions (5.43) and (5.44) show that these solutions, while maintaining their amplitude, oscillate from one time step to the next because of the factor $e^{in\pi} = (-1)^n$ in front of the solution corresponding to the n-th time step.

In the general case, the solution of the discretized system is expressed as a linear combination of the two independent solutions we have just obtained (physical solutions and computational solutions). The amplitude of the computational solutions is determined entirely by the values of the variables at the initial time ($t = 0$) and at the end of the first time step (i.e. at $t = \Delta t$). We then obtain a time-oscillating solution as indicated in Figure 5.5. In the academic case of the *linear model*, it is possible to totally eliminate the computational solution from the general solution by choosing the analytical solutions of the discretized system as values of variables at Δt. In the case of *nonlinear real models*, the amplitude of the time-oscillating computational solution can be reduced by calculating the variables more precisely at the end of the first time step (Hoskins and Simmons, 1975); even so, a time filter, such as the one proposed by Robert (1966) and described by Asselin (1972), has to be applied to reduce the amplitude of the computational solutions while preserving the physical solutions as much as possible.

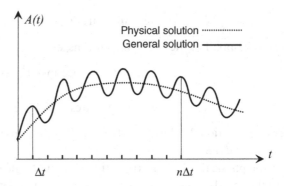

Figure 5.5 Effect of the computational solution on the evolution over time of a variable $A(t)$ with a central difference scheme.

In short, the centred explicit scheme is *conditionally stable*, the constraint on the time step depending essentially on the speed of the inertia-gravity waves. When it is stable, it accelerates the waves slightly, but preserves their amplitude. However, because it utilizes three time levels, it introduces in addition to physical solutions, spurious solutions known as *computational solutions* that need to be brought under control in a model when it is integrated.

5.4.3 The centred semi-implicit scheme

It has been recognized with simple problems (constant velocity advection equations in one spatial dimension) that slightly more complex schemes, termed *implicit schemes* (solution at time $t + \Delta t$ generally requiring the solution of a linear system of equations), are unconditionally stable (Kurihara, 1965). As the tightest constraint on the time step for atmospheric models comes from the processing of inertia-gravity waves, the origin of which is due to adaptation terms, it has been imagined (Robert, 1969) that these terms can be processed implicitly, while continuing to process advection terms explicitly.

The *centred semi-implicit scheme* is implemented by evaluating the time derivative in central differences, the advection, and Coriolis terms at time t (as in the centred explicit scheme) and by calculating the adaptation terms (the geopotential gradient in the equations of motion and the divergence term in the continuity equation) as the arithmetic mean of their values at time levels $t + \Delta t$ and $t - \Delta t$, thereby implying the need to solve a linear system to obtain the values of the variables at time $t + \Delta t$.

In the context of the study undertaken here with the linear model, the solution can however be determined directly.

The semi-implicit scheme is discretized by replacing system (5.21) by:

$$\frac{\mathbf{X}(t+\Delta t) - \mathbf{X}(t-\Delta t)}{2\Delta t} + \widetilde{A}\big[\mathbf{X}(t)\big] + \widetilde{F}\big[\mathbf{X}(t)\big] + \frac{1}{2}\widetilde{G}\big[\mathbf{X}(t+\Delta t) + \mathbf{X}(t-\Delta t)\big] = 0. \quad (5.47)$$

For a solution of the form (5.4), we obtain the homogeneous system analogous to (5.6):

$$i\frac{\sin(\omega\Delta t)}{\Delta t}\mathbf{X}+\mathbf{AX}+\mathbf{FX}+\cos(\omega\Delta t)\mathbf{GX}=0. \tag{5.48}$$

The equivalent of relation (5.10) for the slow wave is written:

$$\frac{\sin(\omega_1\Delta t)}{\Delta t}=-(kU_0+lV_0), \tag{5.49}$$

and the equivalent of relation (5.11) for inertia-gravity waves is:

$$\frac{\sin(\omega_{2,3}\Delta t)}{\Delta t}=-(kU_0+lV_0)\pm\left[f^2+\cos^2(\omega\Delta t)\,\Phi^*(k^2+l^2)\right]^{1/2}. \tag{5.50}$$

We find for the slow wave the same results as in the case of the explicit model with the same stability condition:

$$-1\le KV^*\Delta t\le+1. \tag{5.51}$$

For inertia-gravity waves, relation (5.50) leads to the second-degree equation in $\sin(\omega\Delta t)$:

$$(1+K^2\Phi^*\Delta t^2)\sin^2(\omega\Delta t)+2V^*K\Delta t\,\sin(\omega\Delta t)+K^2\Delta t^2(V^{*2}-V_g^2)=0. \tag{5.52}$$

We obtain stable solutions insofar as the roots of this equation are real and lie between -1 and $+1$. The discriminant of the equation must therefore be positive, which leads to the relation:

$$K^2\Delta t^2V^{*2}\le K^2\Delta t^2V_g^2\left(1+\frac{1}{K^2\Phi^*\Delta t^2}\right). \tag{5.53}$$

This relation is generally verified for current values of speeds V^* and V_g. For large values of the Rossby radius of deformation ($L_R\to\infty$), this inequality comes down to:

$$K^2\Delta t^2V^{*2}\le K^2\Phi^*\Delta t^2+1, \tag{5.54}$$

which is the stability condition given by Robert (1969).

We must also make sure the roots lie between -1 and $+1$. This is verified if the values of the trinomial of equation (5.52) calculated for $\sin(\omega\Delta t)=\pm1$ are positive. We then come to the inequalities:

$$-1\le f\Delta t\pm KV^*\Delta t\le+1. \tag{5.55}$$

Considering the mean value of f (10^{-4} s^{-1}) and of the values that can be used for the time step (1000 s), this stability condition provides a constraint analogous to condition (5.51) seen previously.

As for the centred explicit scheme, we obtain two categories of solutions.

For physical solutions, ratios Q_1 and $Q_{2,3}$ of numerically computed to analytically computed speeds for the slow wave are then written:

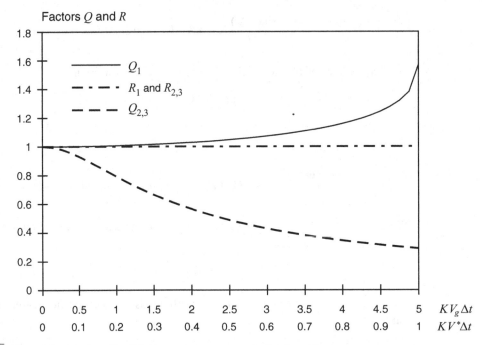

Factors Q and R

Figure 5.6 Effect of the centred semi-implicit scheme on slow waves and on inertia-gravity waves.

$$Q_1 = \frac{\arcsin(KV^*\Delta t)}{KV^*\Delta t}, \tag{5.56}$$

and for inertia-gravity waves:

$$Q_{2,3} = \frac{\arcsin\left[K\Delta t\left(V^* \pm \sqrt{V_g^2 + K^2\Phi^*\Delta t^2(V_g^2 - V^{*2})}\right)\right]}{(1 + K^2\Phi^*\Delta t^2)\left[K(V^* \pm V_g)\Delta t\right]}, \tag{5.57}$$

while the amplification factor moduli R_1 and $R_{2,3}$ of the solutions remain equal to 1.

Figure 5.6 shows the values taken by factors Q and R for the slow wave as a function of $KV^*\Delta t$, this quantity varying from 0 to 1 and factors Q and R for inertia-gravity waves (calculated for a basic wind $V^* = 0$ and for $f\Delta t = 0.1$, corresponding to a time step of the order of 1000 s) as a function of $KV_g\Delta t$, which varies from 0 to 5 ($V^*/V_g = 1/5$). It can be seen that the semi-implicit scheme accelerates the slow wave and slows down the inertia-gravity waves (the effect being more prominent the higher the wavenumber and the higher the wave speed, with amplitudes being maintained.

For the computational solutions, we find the same behaviour as with the centred explicit scheme.

In short, the semi-implicit scheme is *conditionally stable*, the constraint on the time step arising essentially from the velocity of the basic wind. When it is stable, it slightly accelerates the speed of slow waves, but considerably reduces the speed of inertia-gravity waves while conserving their amplitude. It also introduces, in addition to physical solutions, spurious computational solutions. Considering the speed ratio $V*/V_g = 1/5$ found with numerical prediction models, we can then generally use a time step Δt five times greater than with an explicit scheme.

5.4.4 The centred semi-Lagrangian semi-implicit scheme

The success of the implicit method prompted the search for an unconditionally stable scheme for processing advection terms. Krishnamurti (1962) and Sawyer (1963) proposed Lagrangian advection schemes to improve the accuracy in calculating advection terms. But it is Robert (1981) who must be credited with showing the advantage, not in accuracy but in computing time, that could be obtained with this method by associating it with implicit processing of adaptation terms.

5.4.4.1 Implementation with perfect interpolation

The principle of the Lagrangian method is to evaluate the total (or Lagrangian) derivative numerically for the different variables. This involves determining the trajectories of particles reaching the grid points and their starting points, and interpolating variables at those points. From that stage on, we can contemplate extending the semi-implicit scheme.

The total derivative for vector $\mathbf{X}$ is calculated as follows:

$$\frac{d\mathbf{X}}{dt} \cong \frac{\partial \mathbf{X}}{\partial t} + \widetilde{A}(\mathbf{X}) \cong \frac{\mathbf{X}_G(t+\Delta t) - \mathbf{X}_O(t-\Delta t)}{2\Delta t}, \tag{5.58}$$

where $\mathbf{X}_G(t + \Delta t)$ denotes vector $\mathbf{X}$ at grid point G at time $t + \Delta t$ and $\mathbf{X}_O(t - \Delta t)$ vector $\mathbf{X}$ at the starting point O of the trajectory at time $t - \Delta t$ (Figure 5.7).

The adaptation terms are handled implicitly by computing the arithmetic mean of their value at point G at time $t + \Delta t$ and their value at point O at time $t - \Delta t$, to maintain

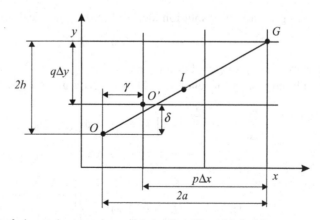

Figure 5.7   Location of variables for Lagrangian processing and interpolation at the point of origin.

the symmetry with what is done for advection (Robert, 1981). Only the Coriolis term is evaluated at time t and at midpoint I of the trajectory (we shall see, though, that this can also be processed implicitly).

The semi-Lagrangian semi-implicit scheme is discretized by replacing system (5.21) by:

$$\frac{\mathbf{X}_G(t+\Delta t) - \mathbf{X}_O(t-\Delta t)}{2\Delta t} + \widetilde{F}\left[\mathbf{X}_I(t)\right] + \frac{1}{2}\widetilde{G}\left[\mathbf{X}_G(t+\Delta t) + \mathbf{X}_O(t-\Delta t)\right] = 0. \qquad (5.59)$$

In the *linear model*, it is trivial to calculate the trajectory and determine its starting point, since the basic wind is constant. The coordinates of the starting point O of the trajectory at time $t - \Delta t$ for a particle that ends up at time $t + \Delta t$ at grid point G of coordinates (x, y) are then:

$$\begin{aligned} x_O &= x - 2a, & a &= U_0 \Delta t, \\ y_O &= y - 2b, & b &= V_0 \Delta t. \end{aligned} \right\} \qquad (5.60)$$

The value of vector $\mathbf{X}$ at point O at $t - \Delta t$ is therefore calculated exactly and is written:

$$\mathbf{X}_O(t - \Delta t) = \mathbf{X}^* e^{i[\omega(t-\Delta t)+k(x-2a)+l(y-2b)]}, \qquad (5.61)$$

and at median point I of the trajectory at t, it is written:

$$\mathbf{X}_I(t) = \mathbf{X}^* e^{i[\omega t+k(x-a)+l(y-b)]}. \qquad (5.62)$$

System (5.59) then leads to the homogeneous system analogous to (5.6) in the form:

$$i\frac{\sin\left[(\omega+kU_0+lV_0)\Delta t\right]}{\Delta t}\mathbf{X}_I + \mathbf{F}\mathbf{X}_I + \cos\left[(\omega+kU_0+lV_0)\Delta t\right]\mathbf{G}\mathbf{X}_I = 0. \qquad (5.63)$$

The equivalent of relation (5.10) is then written, for the slow wave:

$$\sin[(\omega_1+kU_0+lV_0)\Delta t] = 0, \qquad (5.64)$$

which gives a *perfect* solution identical to the analytical solution:

$$\omega_1 = -KV^*, \qquad (5.65)$$

corresponding to $Q_1 = 1$ and $R_1 = 1$.

The equivalent of relation (5.11) for inertia-gravity waves is written:

$$\frac{\sin[(\omega_{2,3}+kU_0+lV_0)\Delta t]}{\Delta t} = \pm\left[f^2 + \cos^2[(\omega_{2,3}+kU_0+lV_0)\Delta t]\,\Phi^*(k^2+l^2)\right]^{1/2}, \qquad (5.66)$$

or:

$$\tan[(\omega_{2,3}+kU_0+lV_0)\Delta t)] = \pm\left\{\frac{\left[f^2+\Phi^*(k^2+l^2)\right]\Delta t^2}{1-f^2\Delta t^2}\right\}^{1/2}. \qquad (5.67)$$

The expression in the denominator of the square root must be positive, which leads to the inequalities:

$$-1 \leq f\Delta t \leq +1, \tag{5.68}$$

which, considering the order of magnitude of the Coriolis parameter f (10^{-4} s^{-1}), do not induce a tight constraint for the time step.

The physical solutions for inertia-gravity waves are therefore written:

$$\omega_{2,3} = -KV^* \pm \frac{1}{\Delta t}\arctan\left[\frac{KV_g\Delta t}{(1-\Delta t^2 f^2)^{1/2}}\right], \tag{5.69}$$

which gives as the deceleration factor:

$$Q_{2,3} = \frac{V^* \pm \dfrac{1}{K\Delta t}\arctan\left[\dfrac{KV_g\Delta t}{(1-\Delta t^2 f^2)^{1/2}}\right]}{V^* \pm V_g}, \tag{5.70}$$

whereas the amplification factor of solution $R_{2,3}$ is equal to 1.

Figure 5.8 shows the values taken by factors Q and R for the slow wave (constant value 1) as a function of $KV^*\Delta t$, varying from 0 to 5, and the same factors for the inertia-gravity waves (calculated for a basic wind $V^* = 0$ and for $f\Delta t = 0.1$, corresponding to a time step of the order of 1000 s) as a function of $KV_g\Delta t$, varying from 0 to 25. It can be seen that the semi-Lagrangian semi-implicit scheme deals perfectly well with the slow wave, whereas it considerably slows down the inertia-gravity waves, amplitudes being maintained.

In short, the semi-Lagrangian semi-implicit scheme with *perfect interpolation* is *conditionally stable*, albeit with a relatively weak constraint on the time step. It gives an exact evaluation of the slow wave and noticeably slows inertia-gravity waves while conserving their amplitudes. It also introduces, in addition to physical solutions, spurious

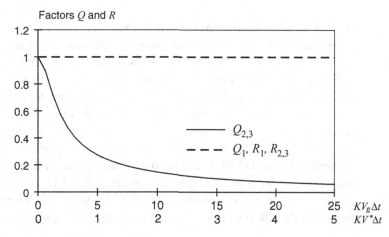

Figure 5.8   Effect of the various Lagrangian schemes on slow waves and inertia-gravity waves.

computational solutions. In practice, it enables us to increase the time step by a factor of three to four compared with the semi-implicit scheme.

5.4.4.2 Implicit treatment of the Coriolis parameter

The Coriolis parameter can also be handled implicitly by replacing system (5.59) by:

$$\frac{\mathbf{X}_G(t+\Delta t)-\mathbf{X}_O(t-\Delta t)}{2\Delta t}+\frac{1}{2}\widetilde{F}\left[\mathbf{X}_G(t+\Delta t)+\mathbf{X}_O(t-\Delta t)\right]+\frac{1}{2}\widetilde{G}\left[\mathbf{X}_G(t+\Delta t)+\mathbf{X}_O(t-\Delta t)\right]=0 . \qquad (5.71)$$

System (5.71) applied to solution (5.4) leads to a homogeneous system analogous to (5.6), which is written:

$$i\frac{\sin(\omega\Delta t+kU_0+lV_0)}{\Delta t}\mathbf{X}_I+\cos(\omega\Delta t+kU_0+lV_0)\left[\mathbf{F}+\mathbf{G}\right]\mathbf{X}_I=0. \qquad (5.72)$$

We find the same perfect solution (5.65) for the slow wave as before and for inertia-gravity waves, instead of (5.767) we get the relation:

$$\tan[(\omega_{2,3}+kU_0+lV_0)\Delta t]=\pm\left[f^2+\Phi^*(k^2+l^2)\right]^{1/2}\Delta t \qquad (5.73)$$

which does not entail any constraint on the time step. We have thus obtained an unconditionally stable scheme.

The physical solutions for inertia-gravity waves are therefore written:

$$\omega_{2,3}=-KV^*\pm\frac{1}{\Delta t}\arctan(KV_g\Delta t). \qquad (5.74)$$

For the deceleration factor we get:

$$Q_{2,3}=\frac{V^*\pm\arctan(KV_g\Delta t)/K\Delta t}{V^*\pm V_g}, \qquad (5.75)$$

whereas the amplification factor of solution $R_{2,3}$ remains equal to 1.

The results for the deceleration factor Q and attenuation factor R, processing the Coriolis term implicitly, are practically identical (taking $f\Delta t = 0.1$) to the results with explicit processing of the Coriolis term. The curves in Figure 5.8 are identical for quantities R_1, $R_{2,3}$, and Q_1 and can only be distinguished for the quantities $Q_{2,3}$.

We notice too that this discretization no longer involves variables at time level t, which thus gives a *two time level scheme*. An interesting consequence of this perfectly symmetrical formulation is the disappearance of computational solutions. Although relations (5.64) and (5.74) lead to two determinations for the arcs, these actually correspond to the same physical solution: the modulus of the amplification factor of the solution must be calculated over the interval $2\Delta t$ and the computational solutions then coincide with the physical solutions, as they only differ from them by the factor $e^{in2\pi}=1$.

In short, the semi-Lagrangian semi-implicit scheme (with implicit treatment of the Coriolis parameter and perfect interpolation at the origin point) is an *unconditionally stable scheme*. In the absence of any constraint on the time step, we can choose a value that is compatible with the accuracy requirements. The scheme provides an accurate evaluation for the speed of the slow wave and markedly slows down inertia-gravity waves while preserving their amplitudes. In practice, semi-Lagrangian schemes required the variables to be interpolated at the point of origin of the trajectories, the effects of which can be calculated. Moreover, accurately computing the trajectories of particles in any velocity field is a crucial problem for using the semi-Lagrangian method. This is addressed in Chapter 6.

5.4.4.3 The effects of interpolation in the semi-Lagrangian scheme

To study the effects of interpolation of variables at the point of origin O, the exact values $\mathbf{X}_O$ must be replaced by the interpolated values $\hat{\mathbf{X}}_O$ and system (5.71) becomes:

$$
\frac{\hat{\mathbf{X}}_G(t+\Delta t)-\hat{\mathbf{X}}_O(t-\Delta t)}{2\Delta t}+\frac{1}{2}\widetilde{F}\Big[\mathbf{X}_G(t+\Delta t)+\hat{\mathbf{X}}_O(t-\Delta t)\Big]
$$
$$
+\frac{1}{2}\widetilde{G}\Big[\mathbf{X}_G(t+\Delta t)+\hat{\mathbf{X}}_O(t-\Delta t)\Big]=0.
$$
(5.76)

By noting $E(x)$ the integer part of the real number x, we determine the coordinates of the grid point O' closest to O (Figure 5.7):

$$
\left.\begin{aligned}
x_{O'} &= x-p\Delta x, && p = E(2a/\Delta x), \\
y_{O'} &= y-q\Delta x, && q = E(2b/\Delta x).
\end{aligned}\right\}
$$
(5.77)

The interpolated value $\hat{\mathbf{X}}_O$ at point O can then be written:

$$
\hat{\mathbf{X}}_O = C\mathbf{X}_{O'},
$$
(5.78)

where C is a complex number of modulus ρ and θ argument characterizing the interpolator used.

By bilinear interpolation from the four grid points surrounding the origin point O, the value of C is given by:

$$
C=\Big[(1-\gamma)+\gamma e^{-ik\Delta x}\Big]\Big[(1-\delta)+\delta e^{-il\Delta x}\Big],
$$
(5.79)

with $\gamma = 2a/\Delta x-p$ and $\delta = 2b/\Delta x-q$, as in Figure 5.7.

The system leads, for determining the slow wave, to the relation:

$$
\Big[e^{i\omega\Delta t}-Ce^{-i(\omega\Delta t+kp\Delta x+lq\Delta x)}\Big]\Big/2\Delta t = 0,
$$
(5.80)

or, considering the complex nature of C:

$$
e^{i2\omega\Delta t} = \rho e^{-i(kp\Delta x+lq\Delta x-\theta)}.
$$
(5.81)

As the left-hand side of equality (5.81) is a complex number with a modulus different from 1, the frequency must be a complex number, that can be written:

$$\omega = \omega' + i\omega'',$$

ω' being ω'' real numbers.

The quantity ω' is the true frequency giving the speed of the slow wave and is obtained from relation (5.81):

$$2\omega'\Delta\Delta t = -KV_1\Delta t = -(kp\Delta x + lq\Delta x) + \theta, \text{ with:} \tag{5.82}$$

$$\theta = -\arctan\frac{\gamma\sin(k\Delta x)}{1 - \gamma + \gamma\cos(k\Delta x)} - \arctan\frac{\delta\sin(l\Delta x)}{1 - \delta + \delta\cos(l\Delta x)}. \tag{5.83}$$

The ratio Q_1 of the speed calculated numerically to the speed calculated analytically for the slow wave is therefore written:

$$Q_1 = \frac{1}{KV^*\Delta t}\frac{1}{2}\left[(kp\Delta x + lq\Delta x) + \arctan\frac{\gamma\sin(k\Delta x)}{1 - \gamma + \gamma\cos(k\Delta x)} + \arctan\frac{\delta\sin(l\Delta x)}{1 - \delta + \delta\cos(l\Delta x)}\right]. \tag{5.84}$$

The quantity ω'' is given by:

$$R_1 = e^{-2\omega''\Delta t} = \rho, \tag{5.85}$$

which is the modulus of the amplification factor of the solution between times $t-\Delta t$ and $t+\Delta t$, since we verify:

$$\|\mathbf{X}(t + \Delta t)\| = \left| e^{i2\omega\Delta t} \right| \cdot \|\mathbf{X}(t - \Delta t)\|. \tag{5.86}$$

In the case of bilinear interpolation defined in (5.79), we get:

$$\rho = \left\{ \left[1 - 2\gamma(1 - \gamma)(1 - \cos(k\Delta x))\right]\left[1 - 2\delta(1 - \delta)(1 - \cos(k\Delta x))\right] \right\}^{1/2}. \tag{5.87}$$

So that the modulus ρ is less than 1 and the scheme is therefore stable, conditions:

$$\left.\begin{array}{l} 0 \le \gamma \le 1, \\ 0 \le \delta \le 1. \end{array}\right\} \tag{5.88}$$

must be verified.

These relations mean that the value at O must be interpolated from the four grid points immediately around it.

To appraise the effect of interpolation on the slow wave quantitatively, we assume $\alpha = \alpha' = \pi/4$, which also leads to $\gamma = \delta$. In this case, considering the definition of p, q, γ, and δ, relation (5.84) takes the form:

$$Q_l = 1 + \frac{1}{KV^*\Delta t}\arctan\left[\frac{\gamma\sin(K\Delta x/\sqrt{2})}{1 - \gamma + \gamma\cos(K\Delta x/\sqrt{2})} - \gamma(K\Delta x/\sqrt{2})\right], \tag{5.89}$$

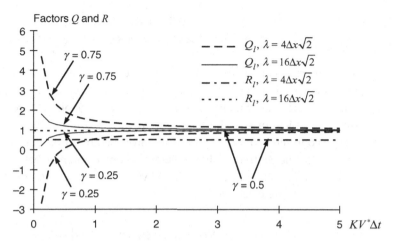

Factors Q and R

Figure 5.9 Effects of interpolation at the origin point on the speed and amplitude of the slow wave for values of two wavelengths and for particular values of the normalized distance γ.

whereas relation (5.87) is written:

$$R_l = 1 - 2\gamma(1-\gamma)[1 - \cos(K\Delta x/\sqrt{2})]. \tag{5.90}$$

The values taken by the factor Q_1 vary with the quantity $KV^*\Delta t$, the wave number K, and the normalized distance γ. A detailed examination of the quantity in square brackets in expression (5.89) shows that it reaches its maximum values for short waves (large values of K); for a given wavenumber K, it varies as a function of γ, cancels out for $\gamma = 0$, $\gamma = 0.5$, and $\gamma = 1$ is minimum for values of γ close to 0.25, and is maximum for values close to 0.75. We therefore choose in Figure 5.9 to represent the values of factor Q_1 for two wavelengths ($\lambda = 4\Delta x/\sqrt{2}$ and $\lambda = 16\Delta x/\sqrt{2}$), with parameter γ taking the values 0.25 and 0.75. It is observed that the changes in speed of the slow wave due to interpolation become negligible when $\lambda = 16\Delta x/\sqrt{2}$.

Examination of expression (5.90) giving the attenuation factor R_1 shows that this is only significantly less than 1 for very short waves. For a fixed wavenumber K, it takes the value 1 for $\gamma = 0$, $\gamma = 1$ and is minimum for $\gamma = 0.5$. Value R_1 (independent of $KV^*\Delta t$) is plotted in Figure 5.9 for two wavelengths ($\lambda = 4\Delta x/\sqrt{2}$ and $\lambda = 16\Delta x/\sqrt{2}$) by choosing $\gamma = 0.5$. As with factor Q_1, the effects of attenuation become negligible when $\lambda = 16\Delta x/\sqrt{2}$.

These interpolation effects that arise essentially for short wavelengths may be minimized by a higher order (quadratic or cubic) interpolation. Its effect can readily be calculated by the approach used above after determining the modulus ρ and argument θ of the complex number C corresponding to the chosen interpolator.

In conclusion, insofar as interpolation is actually done, the semi-Lagrangian semi-implicit scheme remains stable. However, the wave amplitude is reduced slightly and the wave speed is modified in comparison with a scheme where interpolation is perfect. These effects can be minimized by choosing high-order interpolations which, while improving accuracy, are more costly in computing time.

5.5 Time filtering

Time filtering is used to attenuate the decoupling between even time steps and odd time steps brought about by the existence of the computational solution in central difference schemes, which uses the values of the variables at three successive time levels (like the explicit or semi-implicit centred schemes).

The continuous filtering procedure, proposed by Robert (1966) and applied here to the case of the centred explicit scheme, is implemented as below.

Supposing that we know the filtered values at time $t - \Delta t$, noted $\mathbf{X}\,(t - \Delta t)$, the values at time $t + \Delta t$ are calculated first:

$$\mathbf{X}(t+\Delta t) = \check{\mathbf{X}}(t-\Delta t) + 2\Delta t \left(\frac{\partial \mathbf{X}}{\partial t} \right)_t ; \tag{5.91}$$

the filtering is then applied to the values at time t:

$$\check{\mathbf{X}}(t) = \mathbf{X}(t) + \frac{\nu}{2}[\mathbf{X}(t+\Delta t) + \check{\mathbf{X}}(t-\Delta t) - 2\,\mathbf{X}(t)]. \tag{5.92}$$

These values will be used for the next time step, giving the values of the variables at $t + 2\Delta t$:

$$\mathbf{X}(t+2\Delta t) = \check{\mathbf{X}}(t) + 2\Delta t \left(\frac{\partial \mathbf{X}}{\partial t} \right)_{t+\Delta t} . \tag{5.93}$$

The properties of this continuous filtering process were studied by Asselin (1972) within the framework of the constant velocity U advection equation in one spatial dimension:

$$\frac{\partial \mathbf{X}}{\partial t} = -U \frac{\partial \mathbf{X}}{\partial x}.$$

For periodic solution of wavenumber k, the spatial derivatives are calculated exactly: $\partial \mathbf{X}/\partial x = ik\mathbf{X}$, and eliminating the filtered values $\check{\mathbf{X}}_t$ and $\check{\mathbf{X}}_{t-\Delta t}$ between equations (5.91), (5.92), and (5.93) leads to the relation:

$$\mathbf{X}(t+2\Delta t) = \mathbf{X}(t) + \frac{\nu}{2}[\mathbf{X}(t+\Delta t) + \mathbf{X}(t-\Delta t) + 2ikU\Delta t\,\mathbf{X}_t - 2\,\mathbf{X}_t] - 2ikU\,\mathbf{X}(t+\Delta t) . \tag{5.94}$$

The amplification factor of the solution, which is defined by $\|\mathbf{X}(t+\Delta t)\| = |R_f|\|\mathbf{X}(t)\|$, is then given by the relation:

$$R_f^2 - (\nu - 2ikU\Delta t)R_f + \nu - \nu ikU\Delta t = 0,$$

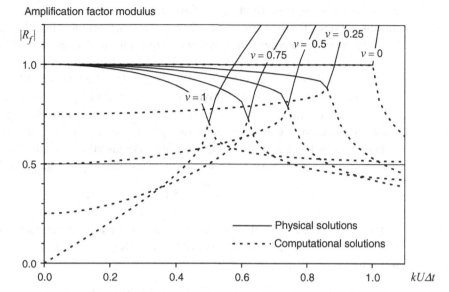

Amplification factor modulus

Figure 5.10   Effect of time filtering for different values of the filter parameter v.

an equation of the second degree whose roots, corresponding to the physical solution for one and to the computational solution for the other, are given by:

$$R_f = v/2 - \mathrm{i}Uk\Delta t \pm \sqrt{\left(v/2-1\right)^2 - (kU\Delta t)^2}. \tag{5.95}$$

When $-(1 - v/2) \le kU\Delta t \le (1 - v/2)$, the discriminant is positive and it can then be verified that the modulus of the amplification factor is less than 1. By accepting negative values for the discriminant, we find as the stability condition:

$$(kU\Delta t)^2 + kU\Delta t \sqrt{(kU\Delta t)^2 - (1 - v/2)^2} - (1 - v/2) \le 0. \tag{5.96}$$

The plot of $\left|R_f\right|$ as a function of $kU\Delta t$ is shown in Figure 5.10. It clearly shows that introducing filtering gives a more restrictive CFL stability condition (Déqué and Cariolle, 1986). The higher the wavenumber k (and therefore the shorter the wavelength), the more attenuated is the physical solution and the computational solution is itself attenuated much more than the physical solution, which was the object of the exercise.

5.6 Effect of spatial discretization on stability

5.6.1 The case of finite difference models

The effect of discretization has been addressed by clearly separating the effects of horizontal discretization (Section 5.3) from the effects of time discretization (Section 5.4). Of course, both aspects must be taken into consideration to appraise the overall effects of discretization. In the case of the use of a grid point model, the stability conditions obtained for the various time-integration schemes must be calculated by taking into account too the attenuation factors corresponding to each of the finite difference schemes used.

Thus, for *explicit schemes*, the stability conditions (5.39) and (5.40) must be modified for each of the grids by multiplying velocities V^* and V_g by the attenuation factors resulting from spatial discretization for the various grids. Introducing these attenuation factors therefore modifies the stability conditions laid down in Section 5.4.

In the case of the A grid, the attenuation factors obtained in (5.17) and (5.19) reach their maximum values for $\alpha = \pi/4$ and the most restrictive stability condition is of the form:

$$-1 \le \sqrt{2}(V^* \pm V_g)\Delta t / \Delta x \le +1. \tag{5.97}$$

In the case of the B grid, the maximum values of the attenuation factor for the adaptation terms are obtained for $\alpha = 0$ and the most restrictive stability condition is then written:

$$-1 \le (V^* \pm 2V_g)\Delta t / \Delta x \le +1. \tag{5.98}$$

For the C and D' grids, the maximum values of the coefficient for the adaptation terms are obtained for $\alpha = \pi/4$ and the most restrictive condition then becomes:

$$-1 \le \sqrt{2}(V^* \pm 2V_g)\Delta t / \Delta x \le +1. \tag{5.99}$$

Lastly, in the case of discretization on the A grid with 4th-order accuracy, we get:

$$-\frac{3}{4} \le \sqrt{2}(V^* \pm V_g)\Delta t / \Delta x \le +\frac{3}{4}. \tag{5.100}$$

Upon examination of inequalities (5.97) to (5.100), it can be seen that the schemes that provide the best accuracy for processing inertia-gravity waves are also those that require the shortest time step.

We can reason similarly for *semi-implicit schemes* knowing that the constraint on the time step is related to discretization of the advection terms.

For the A, B, and C grids, the maximum value of the attenuation factor is obtained for $\alpha = \pi/4$ and we get:

$$-1 \le \sqrt{2}\, V^* \Delta t / \Delta x \le +1. \tag{5.101}$$

For the D' staggered grid, the maximum value of the attenuation factor for the advection term is obtained for $\alpha = 0$, which gives as the condition:

$$-1 \le 2V^* \Delta t / \Delta x \le +1. \tag{5.102}$$

The use of 4th-order accuracy central differences on the A grid leads to the condition:

$$-\frac{3}{4} \le \sqrt{2}\, V^* \Delta t / \Delta x \le +\frac{3}{4}. \tag{5.103}$$

Here again it can be seen that the gain in accuracy entails additional computation because of the constraint on the time step. It is important to point out too that the use of finite differences of 4th-order accuracy in a semi-implicit scheme increases the complexity of the implicit system, giving the values of variables at the end of the time step.

5.6.2 The case of spectral models

The computations made in this chapter on a doubly periodic domain may be reworked on the sphere by determining the solutions of the linearized equation of the spectral model truncated after having established a basic state (Dickinson and Williamson, 1972; Machenhauer, 1977). For spectral models processed explicitly, the stability condition obtained for the fastest gravity waves, assuming a basic state at rest, is written:

$$-1 \leq \frac{\sqrt{N(N+1)}}{a} \sqrt{\Phi^*} \Delta t \leq +1, \qquad (5.104)$$

where a denotes the radius of the Earth and N the maximum total wavenumber of the truncation used.

For semi-implicit processing, the linear analysis by Kwizak and Robert (1971) shows that for realistic atmospheric conditions, the time-step condition is related to the speed of propagation of Rossby waves. This speed can be approximated with a quasi-geostrophic linearized model having as its basic state a solid rotation of angular velocity Ω. The stability condition obtained for the fastest Rossby waves is written (Machenhauer, 1979):

$$-1 \leq M\Omega\Delta t \leq +1,$$

where M denotes the largest zonal wave of the truncation used.

It can be seen, then, that the stability conditions obtained in the case of the explicit scheme and the semi-implicit scheme, and therefore the performance gain obtained, depend on the type of truncation.

6   Barotropic models

6.1   Barotropic models using the vorticity equation

We examine the barotropic model based on the vorticity equation (or filtered barotropic model) for historical reasons: the first successful numerical prediction (Charney, Fjörtoft, and von Neumann, 1950) was based on this type of model and has been the subject of many theoretical studies. After looking at the practical implementation of this model and the improvements that can be made to it, we explain how nonlinear instability arises and how it can be corrected.

6.1.1   The zero divergence model

We begin by considering the zero divergence model (6.1), which comes down to a single evolution equation for absolute vorticity, entrained by a nondivergent (purely rotational) wind; by using a conformal projection, this equation is written:

$$\frac{\partial \zeta}{\partial t} = -m^2 \left[U \frac{\partial}{\partial x}(\zeta + f) + V \frac{\partial}{\partial y}(\zeta + f) \right]. \tag{6.1}$$

By introducing the streamfunction ψ as a variable, the reduced wind components $U = u/m$; $V = v/m$, and the vorticity ζ are written:

$$U = -\frac{\partial \psi}{\partial y}, \quad V = \frac{\partial \psi}{\partial x} \quad \text{and} \quad \zeta = \nabla^2 \psi.$$

After permuting the Laplacian and the time derivative operators, the evolution equation is written:

$$\nabla^2 \frac{\partial \psi}{\partial t} = -J(\psi, \nabla^2 \psi + f), \; J \text{ being the Jacobian.} \tag{6.2}$$

Such a model is numerically integrated from streamfunction ψ data at time t by the following operations:

- calculate the right-hand side of the Poisson equation (6.2): we first calculate the Laplacian of ψ then the Jacobian $J(\psi, \nabla^2\psi + f)$ by evaluating the spatial derivatives using finite differences;
- solve the Poisson equation using a tried and tested method (e.g. over-relaxation) to obtain the tendency $\partial \psi / \partial t$ at time t;

- extrapolate ψ in time using a *forward difference* for the first time step and then a central difference for the next time steps:

$$\psi(t + \Delta t) = \psi(t - \Delta t) + 2\Delta t \frac{\partial \psi(t)}{\partial t}. \tag{6.3}$$

Over a rectangular working domain, the vorticity can be calculated by evaluating spatial derivatives in central differences for the points within the domain only and not on its boundaries; similarly the Jacobian can only be calculated within the domain of definition of vorticity. The boundary conditions must therefore be prescribed on the two outer rows of grid points. A condition that is simple to implement consists in specifying constant values of ψ, and so zero values of $\partial \psi / \partial t$ for the two outer rows of the grid, and so solving the Poisson equation.

By replacing the rotational wind and the vorticity by their geostrophic approximations:

$$U = U_g = -\frac{1}{\overline{f}} \frac{\partial \Phi}{\partial y}, \ V = V_g = \frac{1}{\overline{f}} \frac{\partial \Phi}{\partial x} \ \text{and} \ \zeta_g = \frac{1}{\overline{f}} \nabla^2 \Phi,$$

where $\overline{f}$ is a mean value of f over the domain (so a nondivergent wind can be kept), we obtain an evolution equation for the geopotential Φ:

$$\nabla^2 \frac{\partial \Phi}{\partial t} = -J\left(\Phi, \frac{\nabla^2 \Phi}{\overline{f}} + f\right). \tag{6.4}$$

This equation for the variable Φ is solved in the same way as equation (6.3) for the variable ψ.

Following the historical experiment by Charney, Fjörtoft, and von Neumann, this model was used with some success for short-range forecasting of the 500 hPa geopotential height over a limited area. The option to ignore isobaric divergence of the horizontal wind at 500 hPa was then based on the assumption of a maximum vertical velocity ($\partial \omega / \partial p = 0$) in the vicinity of this level.

6.1.2 Introducing a divergence term

When using zero divergence barotropic models operationally over quasi-hemispheric domains, it was observed (Wolff, 1958) that very long waves move westwards (retrogression) while observation of the real atmosphere shows they are stationary or slow moving. This behaviour is explained after examining solutions of the evolution equation (6.4) linearized with respect to geostrophic basic state with a constant wind U^* parallel to the x-axis, and for a perturbation $\Phi(x)$ independent of y; this equation is written (assuming a scale factor $m = 1$):

$$\nabla^2\left(\frac{\partial \Phi}{\partial t}\right) + \overline{f} U^* \frac{\partial}{\partial x}\left(\frac{\nabla^2 \Phi}{\overline{f}}\right) + \beta\left(\frac{\partial \Phi}{\partial x}\right) = 0, \tag{6.5}$$

the quantity $\beta = \partial f/\partial y$ characterizing the meridional variation of the Coriolis parameter f.

We look for periodic solutions in the form:

$$\Phi(x) = \Phi_0 \, e^{i(kx - ct)}, \tag{6.6}$$

where k denotes the wavenumber $k = 2\pi/\lambda$ and c the phase speed of the wave.

Replacing $\Phi(x)$ by its expression given in (6.6), equation (6.5) yields the wave speed expression:

$$c_R = U^* - \frac{\beta\lambda^2}{4\pi^2}. \tag{6.7}$$

The quantity c_R is the speed of Rossby waves, which are the only solutions of equation (6.5); the inertia-gravity wave solutions of the linearized shallow water model have vanished. The zero divergence condition eliminates any propagation of inertia-gravity waves. Hence we speak of *filtered* models (meaning having filtered out inertia-gravity waves). This property allows us to use a comparatively long time step with explicit time schemes (6.3), since the CFL stability condition involves only velocity U^* which is approximated locally by the advecting wind.

Relation (6.7) explains the retrogression of long waves that is found for:

$$\lambda > 2\pi\sqrt{U^*/\beta}.$$

For $U^* = 25$ m/s and $\beta = 1{,}5.10^{-11}$ (value for latitude 45° North), we obtain a wavelength $\lambda > 8\,000$ km.

This behaviour, owing to the zero divergence assumption, can be attenuated by taking into account a mean divergence term in the vorticity equation:

$$\frac{\partial \zeta_g}{\partial t} = -\mathbf{V}_g \cdot \nabla(\zeta_g + f) - \bar{f}\,\nabla \cdot \mathbf{V}, \tag{6.8}$$

as shown by Bolin (1955) and Cressman (1958).

This mean divergence term can be calculated by using a geostrophic evaluation of the advection terms in the continuity equation. As the geostrophic wind is perpendicular to the geopotential gradient, by introducing a mean value of the geopotential Φ^*, we get:

$$\frac{\partial \Phi}{\partial t} = -\Phi\nabla \cdot \mathbf{V} \cong -\Phi^*\nabla \cdot \mathbf{V}. \tag{6.9}$$

The evolution equation for geopotential thus becomes:

$$\nabla^2 \frac{\partial \Phi}{\partial t} - \frac{\bar{f}^2}{\Phi^*}\frac{\partial \Phi}{\partial t} = -J(\Phi, \frac{\nabla^2\Phi}{\bar{f}} + f). \tag{6.10}$$

Equation (6.10) for determining the tendency $\partial\Phi/\partial t$ is a Helmholtz equation, which is solved using the same methods as for a Poisson equation.

The term $M^* = \overline{f}/\Phi^*$ is the *long-wave stabilization* term; it increases as Φ^* decreases, which corresponds to marked divergence. The linearized equation taking account of the divergence term is written:

$$(\nabla^2 - M^*)\frac{\partial\Phi}{\partial t} + \overline{f}\,U^* \frac{\partial}{\partial x}\left(\frac{\nabla^2\Phi}{\overline{f}}\right) + \beta\left(\frac{\partial\Phi}{\partial x}\right) = 0. \qquad (6.11)$$

For a perturbation of the form (6.6), we obtain the Rossby wave speed solutions of this new equation:

$$c_R = \frac{U^* - \beta\lambda^2/4\pi^2}{1 + M^*\lambda^2/4\pi^2}. \qquad (6.12)$$

Introducing the stabilization term does not prevent retrogression of long waves, but reduces it as M^* increases. In practice, this term has been evaluated empirically using equation (6.10) applied to the geopotential at 500 hPa; it corresponds approximately to adopting a mean geopotential height $\Phi^* \cong 15\,000$ J/kg.

The evolution equation (6.10) therefore describes the behaviour of the 500 hPa geopotential height relatively realistically, provided the mean divergence is taken as that of a homogenous atmospheric layer that is Φ^*/g, i.e. 1500 m thick.

6.1.3 Nonlinear instability and how to prevent it

Experience shows that in the models under study, which are reputedly stable (in the sense that the linear stability criterion is satisfied), small-scale vortices develop, in the event of long integrations, that may lead locally to extremely large wind values, so entailing linear instability. This type of instability was first identified by Phillips (1959) and is due to the nonlinearity of the equations and to it being impossible to represent correctly all of the small scales generated by nonlinear interaction on a grid containing a finite number of points.

We consider a periodic variable $\psi(x)$(of wavenumber k) defined over a domain L, known from its values at the $2M + 1$ points of a grid with mesh $\Delta x = L/(2M)$. The values of the variable at the different grid points are given by:

$$\psi(j\Delta x) = \psi_0 \sin(kj\Delta x). \qquad (6.13)$$

The shortest wavelength that can be represented on the grid is $\lambda_{min} = L/M = 2\Delta x$, since the values at the $2M+1$ points make it possible to determine a Fourier expansion up to wavenumber $k_{max} = 2\pi M/L$, that is, $k_{max} = \pi/\Delta x$. When integrating a nonlinear model, the occurrence of product terms generates wavenumbers $k > k_{max}$.

For $k_{max} < k < 2k_{max}$, we can write $k^* = 2k_{max} - k$, with $0 < k^* < k_{max}$; expression (6.13) becomes:

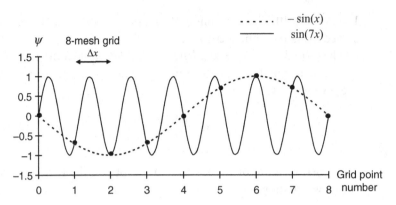

Figure 6.1 Illustration of aliasing.

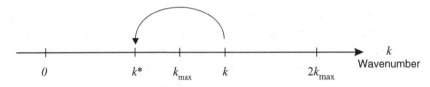

Figure 6.2 Spectrum folding.

$$\psi(j\Delta x) = \psi_0 \sin(kj\Delta x) = \psi_0 \sin(2k_{max}j\Delta x - k^* j\Delta x). \qquad (6.14)$$

Because $k_{max} = \pi/\Delta x$, $2k_{max}j\Delta x = 2\pi j$, and the sine development gives:

$$\psi(j\Delta x) = \psi_0 \left[\sin(2\pi j)\cos(k^* j\Delta x) - \cos(2\pi j)\sin(k^* j\Delta x) \right] = -\psi_0 \sin(k^* j\Delta x). \quad (6.15)$$

The plot in Figure 6.1 shows that at the grid points, in accordance with the relation (6.15), it is impossible to spot any difference between $\psi_0 \sin(kj\Delta x)$ and $-\psi_0 \sin(k^* j\Delta x)$. The wavenumber k cannot be distinguished from its *alias*: the wavenumber k^*. This poor representation of wavenumbers greater than the cutoff (the largest wavenumber allowed) goes by the name of *aliasing*, or *spectrum folding*, because of the simple interpretation made of it and illustrated in Figure 6.2.

When integrating a numerical model, because the equations are nonlinear, the product of two waves of wavenumbers k_1 and k_2 generates waves of wavenumbers $k_1 - k_2$ and $k_1 + k_2$, which can easily exceed the cutoff value k_{max} (whether a grid point model or a spectral model). Aliasing therefore arises by the progressive buildup of small-scale structures (corresponding to large wavenumbers) that can locally generate large wind speed values and so lead to linear instability.

This phenomenon is specific to the numerical processing of nonlinear equations; there is, however, a way to attenuate its effect using a specific discretization of the Jacobian, in the case of the zero divergence model.

The zero divergence model is written in short form by writing $\xi = \zeta + f$:

$$\frac{\partial \xi}{\partial t} = -J(\psi, \xi). \tag{6.16}$$

By calling $\mathbf{V}_\psi = \mathbf{k} \times \nabla \psi$ the rotational wind, the Jacobian takes the form:

$$\begin{aligned} J(\psi, \xi) &= \mathbf{V}_\psi \cdot \nabla \xi, \\ &= \nabla \cdot (\xi \mathbf{V}_\psi) - \xi \nabla \cdot \mathbf{V}_\psi, \\ &= \nabla \cdot (\xi \mathbf{V}_\psi), \end{aligned} \tag{6.17}$$

since the divergence of the rotational wind is zero.

Integrating equation (6.16) over a working domain S with zero flux conditions on the edges (or conditions such that the outgoing flux is equal to the incoming flux) gives by virtue of the Green-Ostrogradsky theorem:

$$\frac{\partial}{\partial t} \iint_S \xi \, ds = -\iint_S J(\psi, \xi) \, ds = 0, \tag{6.18}$$

indicating that the absolute vorticity integral is conserved.

Likewise, we can demonstrate the relation:

$$\frac{\partial}{\partial t} \iint_S \frac{\xi^2}{2} \, ds = -\iint_S \xi J(\psi, \xi) \, ds = -\iint_S J(\psi, \frac{\xi^2}{2}) \, ds = 0, \tag{6.19}$$

meaning that the integral of the half-square of the vorticity (a quantity termed *absolute enstrophy*) is conserved.

Multiplying both sides of equation (6.16) by ψ leads to the relations:

$$\left. \begin{aligned} \psi \frac{\partial \xi}{\partial t} &= \psi \frac{\partial}{\partial t} \nabla^2 \psi = \nabla \cdot \left(\psi \nabla \frac{\partial \psi}{\partial t} \right) - \nabla \psi \cdot \nabla \frac{\partial \psi}{\partial t}, \\ -\psi J(\psi, \xi) &= -\psi \nabla \cdot (\xi \mathbf{V}_\psi) = -\nabla \cdot (\psi \xi \mathbf{V}_\psi) + \nabla \psi \cdot \xi \mathbf{V}_\psi. \end{aligned} \right\} \tag{6.20}$$

The final term of the right-hand side of the second equality is zero since the rotational wind is perpendicular to the gradient of the streamfunction. After observing that:

$$\nabla \psi \cdot \nabla \frac{\partial \psi}{\partial t} = \frac{\partial}{\partial t} \frac{(\nabla \psi)^2}{2} = \frac{\partial K}{\partial t}$$

represents the time derivative of kinetic energy K, it can be seen that integrating the vorticity equation multiplied by the streamfunction ψ over the working domain (under the same conditions as specified previously) leads to the relation:

$$\frac{\partial}{\partial t} \iint_S K \, ds = +\iint_S \psi J(\psi, \xi) \, ds = \iint_S \nabla \cdot (\psi \xi \mathbf{V}_\psi) \, ds = 0, \tag{6.21}$$

meaning that the integral of kinetic energy (or more specifically of the half-square of velocity) is preserved.

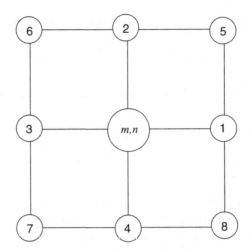

Figure 6.3   Arrangement of the nine points for calculating the Jacobian at the central point (m, n).

The absolute vorticity, kinetic energy, and absolute enstrophy summed over the whole of the working domain are therefore *integral invariants* of the zero divergence model. A discretization scheme will therefore be more realistic insofar as these quantities are also conserved when the integrals are replaced by finite sums.

It can be shown that, among all the possible discretizations of the Jacobian with 2nd-order accuracy obtained by using nine points (Figure 6.3), a small number only allow us to satisfy the integral invariant conservation constraints. Such discretizations were determined systematically by Arakawa (1966).

A Jacobian $J(\psi, \xi)$ evaluated over nine grid points is a bilinear function of the values of ψ and of ξ at each of the nine points and is therefore written in a general form:

$$J_{mn} = \sum_{l=-1}^{+1} \sum_{k=-1}^{+1} \sum_{j=-1}^{+1} \sum_{i=-1}^{+1} a_{ijkl} \; \psi_{m+i,n+j} \; \xi_{m+k,n+l}.$$

Coefficients a_{ijkl} can be determined by taking account of the Jacobian invariance properties for a rotation of $\pi/2$ and its sign change for symmetry around the x-axis. It can be shown that there are six independent 2nd-order approximations of the Jacobian; by making a weighted average of these Jacobians we get further expressions of the Jacobian with 2nd-order accuracy.

Of the six evaluations of the 2nd-order accuracy Jacobian, three (J_1, J_2 and J_3) have interesting properties with respect to the conservation of integral invariants.

J_1 is a plain discretization in central differences of the formula:

$$J = \frac{\partial \psi}{\partial x} \frac{\partial \xi}{\partial y} - \frac{\partial \psi}{\partial y} \frac{\partial \xi}{\partial x},$$

and is written, given the distribution of the points:

$$J_1 = \frac{(\psi_1 - \psi_3)(\xi_2 - \xi_4) - (\psi_2 - \psi_4)(\xi_1 - \xi_3)}{4\Delta x^2}. \tag{6.22}$$

J_2 is a central difference discretization of the alternative formula:

$$J = \frac{\partial}{\partial x}\left(\psi\frac{\partial\xi}{\partial y}\right) - \frac{\partial}{\partial y}\left(\psi\frac{\partial\xi}{\partial x}\right),$$

and is written:

$$J_2 = \frac{\left[\psi_1(\xi_5 - \xi_8) - \psi_3(\xi_6 - \xi_7)\right] - \left[\psi_2(\xi_5 - \xi_6) - \psi_4(\xi_8 - \xi_7)\right]}{4\Delta x^2}. \tag{6.23}$$

J_3, lastly, is a discretization of the symmetric formula of the previous one (by permuting both ψ and ξ, and x and y):

$$J = -\frac{\partial}{\partial x}\left(\xi\frac{\partial\psi}{\partial y}\right) + \frac{\partial}{\partial y}\left(\xi\frac{\partial\psi}{\partial x}\right),$$

and is written:

$$J_3 = -\frac{\left[\xi_1(\psi_5 - \psi_8) - \xi_3(\psi_6 - \psi_7)\right] - \left[\xi_2(\psi_5 - \psi_6) - \xi_4(\psi_8 - \psi_7)\right]}{4\Delta x^2}. \tag{6.24}$$

The discrete homologues of integrals (6.18), (6.19), and (6.21) are written:

$$\left.\begin{array}{l}\sum_m\sum_n J_{mn} = 0, \\[2mm] \sum_m\sum_n \xi_{mn}J_{mn} = 0, \\[2mm] \sum_m\sum_n \psi_{mn}J_{mn} = 0,\end{array}\right\} \tag{6.25}$$

summation over m and n extending so as to cover the working domain S.

From close scrutiny of the proposed discretizations it can be shown that:

• the Jacobian J_1 ensures the total absolute vorticity is conserved;
• Jacobians J_3 and $(J_1 + J_2)/2$ ensure the total kinetic energy is conserved;
• Jacobians J_2 and $(J_1 + J_3)/2$ ensure the total absolute enstrophy is conserved.

Lastly, the Jacobian:

$$J_A = \frac{J_1 + J_2 + J_3}{3} \tag{6.26}$$

ensures the total absolute vorticity, kinetic energy, and absolute enstrophy are conserved.

This last expression is termed the Arakawa Jacobian. It allows the integral invariants of the zero divergence model to be conserved when making finite summations under suitably chosen boundary conditions. This formulation is relatively complex, but can thus prevent the occurrence of small-scale structures as a result of aliasing. Substantial aliasing manifests itself by the development of small vortices with positive and negative values of vorticity, the mean value of the vorticity having to be conserved. Such vortices therefore contribute to increasing absolute enstrophy (half-square of vorticity) and so it can be understood why conserving this quantity protects against over intense aliasing.

6.2 The shallow water barotropic model

We now study the properties of the shallow water model and the implementation of the various time integration schemes applied to a discretized model using the finite difference method.

6.2.1 The properties of the shallow water model

For a rectangular domain in conformal projection of scale factor m, the equations (2.14) of the shallow water model with zero orography ($\Phi_s = 0$) are written:

$$\left.\begin{aligned}
\frac{\partial U}{\partial t} &= (\zeta + f)V - \frac{\partial}{\partial x}(K + \Phi), \\
\frac{\partial V}{\partial t} &= -(\zeta + f)U - \frac{\partial}{\partial y}(K + \Phi), \\
\frac{\partial \Phi}{\partial t} &= -m^2\left[\frac{\partial}{\partial x}(\Phi U) + \frac{\partial}{\partial y}(\Phi V)\right]
\end{aligned}\right\} \quad (6.27)$$

with $K = m^2\dfrac{U^2 + V^2}{2}$, the kinetic energy per unit mass and $\zeta = m^2\left(\dfrac{\partial V}{\partial x} - \dfrac{\partial U}{\partial y}\right)$ the vorticity.

These equations allow us to check that in the absence of any flux on the domain boundaries (or with incoming flux equal to outgoing flux), the shallow water model ensures that a certain number of integral invariants are conserved: total mass $\mathfrak{M}$, total energy $\mathfrak{E}$, total potential absolute vorticity $\mathfrak{Z}$, and *total potential absolute enstrophy* $\mathfrak{H}$ (integral of the half-square of potential vorticity), the expressions for which are set out below.

$$\left.\begin{aligned}
\mathfrak{M} &= \frac{\rho}{g}\iint_S \Phi\,\frac{dx\,dy}{m^2}, \\
\mathfrak{E} &= \frac{\rho}{g}\iint_S \Phi\left(\frac{\Phi}{2} + K\right)\frac{dx\,dy}{m^2}, \\
\mathfrak{Z} &= \frac{\rho}{g}\iint_S \Phi\left(\frac{\zeta + f}{\Phi}\right)\frac{dx\,dy}{m^2}, \\
\mathfrak{H} &= \frac{\rho}{g}\iint_S \frac{\Phi}{2}\left(\frac{\zeta + f}{\Phi}\right)^2\frac{dx\,dy}{m^2},
\end{aligned}\right\} \quad (6.28)$$

ρ being the density of the atmosphere which is assumed to be homogeneous (constant ρ), S the spatial domain, and g acceleration due to gravity.

Note that total energy comprises the potential energy E_p and the kinetic energy E_c of the column of fluid of height z and surface area $dx\,dy/m^2$:

$$dE_P = \rho\int_0^z gz'dz'\frac{dx\,dy}{m^2} = \rho\frac{\Phi^2}{2g}\frac{dx\,dy}{m^2}, \tag{6.29}$$

$$dE_c = \rho\int_0^z Kdz'\frac{dx\,dy}{m^2} = \rho\frac{\Phi K}{g}\frac{dx\,dy}{m^2}. \tag{6.30}$$

Since the time derivation operator commutes with the integration operator over the spatial domain, the evolution equations for these quantities can be determined. The computation consists in deriving with respect to time the two sides of equations (6.28) and replacing the tendencies of U, V and Φ by their expressions derived from (6.27). We thus obtain, after suitable grouping of the various terms:

$$\left.\begin{array}{l}\dfrac{d\mathfrak{M}}{dt} = -\dfrac{\rho}{g}\iint_S \left[\dfrac{\partial(\Phi U)}{\partial x}+\dfrac{\partial(\Phi V)}{\partial y}\right]dx\,dy,\\[3mm] \dfrac{d\mathfrak{E}}{dt} = -\dfrac{\rho}{g}\iint_S \left\{\dfrac{\partial}{\partial x}[\Phi(\Phi+K)U]+\dfrac{\partial}{\partial y}[\Phi(\Phi+K)V]\right\}dx\,dy,\\[3mm] \dfrac{d\mathfrak{Z}}{dt} = -\dfrac{\rho}{g}\iint_S \left\{\dfrac{\partial}{\partial x}[(\zeta+f)U]+\dfrac{\partial}{\partial y}[(\zeta+f)V]\right\}dx\,dy,\\[3mm] \dfrac{d\mathfrak{H}}{dt} = -\dfrac{\rho}{g}\iint_S \dfrac{1}{2}\left\{\dfrac{\partial}{\partial x}\left[\Phi\left(\dfrac{\zeta+f}{\Phi}\right)^2 U\right]+\dfrac{\partial}{\partial y}\left[\Phi\left(\dfrac{\zeta+f}{\Phi}\right)^2 V\right]\right\}dx\,dy.\end{array}\right\} \tag{6.31}$$

The right-hand sides of these equations bring out integrals whose integrands are horizontal divergence terms. By virtue of the Green-Ostrogradsky theorem, with suitably chosen boundary conditions, the right-hand sides cancel out, proving that the quantities $\mathfrak{M}$, $\mathfrak{E}$, $\mathfrak{Z}$, and $\mathfrak{H}$ are conserved by equations (6.27). It is logical, then, to look for discretization schemes conserving the same integral invariants, so as to ensure that these physical quantities remain bounded when the model is numerically integrated.

6.2.2 Discretization of the equations on a C grid

Discretization on the C grid has already been addressed in Subsection 3.3.4; here, though, we opt for a different solution of (3.16) for the terms $(\zeta+f)V$ and $-(\zeta+f)U$, so as to introduce the potential absolute vorticity variable ξ^* and thereby ensure the conservation of certain integral invariants:

$$\left.\begin{array}{l} \overline{U}_t^t = \overline{\xi^*}^y \overline{\overline{\Phi}^y V}^{xy} - (\Phi + K)_x, \\[2mm] \overline{V}_t^t = -\overline{\xi^*}^x \overline{\overline{\Phi}^x U}^{xy} - (\Phi + K)_y, \\[2mm] \overline{\Phi}_t^t = -m^2 \left[\left(\overline{\Phi}^x U \right)_x + \left(\overline{\Phi}^y V \right)_y \right], \end{array}\right\} \qquad (6.32)$$

with $\xi^* = \dfrac{\overline{m^2}^{xy}(V_x - U_y) + f}{\overline{\Phi}^{xy}}$ and $K = \dfrac{m^2}{2} \left(\overline{U^2}^x + \overline{V^2}^y \right).$

Figure 6.4 shows the positions of the prognostic variables U, V and Φ and the diagnostic variables ξ^* (potential absolute vorticity) and K (kinetic energy) on the C grid, of mesh size Δx. It also shows the locations of the map scale factor m and the Coriolis parameter f.

The evolution equation for total mass $\mathfrak{M}$ is written, in view of this discretization:

$$\frac{\partial \mathfrak{M}}{\partial t} = -\frac{\rho}{g} \sum \sum \left[(\overline{\Phi}^x U)_x + (\overline{\Phi}^y V)_y \right] \Delta x^2 = 0, \qquad (6.33)$$

summation being extended to the whole of the spatial domain.

This result follows from the discretized form of the term $\nabla \cdot (\Phi V)$ on the C grid; when summed over neighbouring grid points, the divergence terms calculated numerically cancel out stepwise and only the contribution from the boundary terms remains, which is zero for suitably chosen boundary conditions. Given the embedded locations

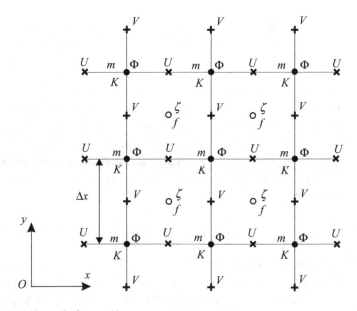

Figure 6.4   Location of the variables on the C-type grid.

of Φ (with integer subscripts) and U variables (with fractional subscripts) along the x-axis, we verify:

$$\sum (\overline{\Phi}^x U)_x = \frac{1}{\Delta x}\left[\frac{\Phi_2+\Phi_3}{2}U_{5/2} - \frac{\Phi_1+\Phi_2}{2}U_{3/2} + \frac{\Phi_3+\Phi_4}{2}U_{7/2} - \frac{\Phi_2+\Phi_3}{2}U_{5/2} + \dots \right.$$

$$\left. \dots + \frac{\Phi_{j-2}+\Phi_{j-1}}{2}U_{j-3/2} - \frac{\Phi_{j-3}+\Phi_{j-2}}{2}U_{J-5/2} + \frac{\Phi_{j-1}+\Phi_j}{2}U_{j-1/2} - \frac{\Phi_{j-2}+\Phi_{j-1}}{2}U_{j-3/2} \right]$$

$$= \frac{1}{\Delta x}\left[\frac{\Phi_{j-1}+\Phi_j}{2}U_{j-1/2} - \frac{\Phi_1+\Phi_2}{2}U_{3/2} \right].$$

This calculation is in a way a practical demonstration of the Green-Ostrogradsky theorem applied to a divergence expressed in finite differences. This property of cancellation when summing occurs whenever we can bring out a *numerical divergence* on the C grid of the form:

$$\delta(X\mathbf{V}) = (XU)_x + (XV)_y, \tag{6.34}$$

X being a scalar quantity in this instance.

The evolution equation for total energy $\mathfrak{E}$ is written:

$$\frac{\partial \mathfrak{E}}{\partial t} = \frac{\rho}{g}\sum\sum\left[(\Phi+K)\frac{\partial \Phi}{\partial t} + m^2\left(\overline{\overline{\Phi}^x U \frac{\partial U}{\partial t}}^x + \overline{\overline{\Phi}^y V \frac{\partial V}{\partial t}}^y \right) \right]\frac{\Delta x^2}{m^2}. \tag{6.35}$$

By replacing the time derivatives of U, V, and Φ by their values given in (6.32) and by using the identity in finite differences (3.8), seen in Subsection 3.2.2, to group terms, we get:

$$\frac{\partial \mathfrak{E}}{\partial t} = -\frac{\rho}{g}\sum\sum\left\{ \left[\overline{(\overline{\Phi+K}^x)\overline{\Phi}^x U}\right]_x + \left[\overline{(\overline{\Phi+K}^y)\overline{\Phi}^y V}\right]_y - \overline{\overline{\Phi}^x U\ \overline{\xi^*}^y \overline{\overline{\Phi}^y V}^{xy}}^x + \overline{\overline{\Phi}^y V\ \xi^*\ \overline{\overline{\Phi}^x U}^{xy}}^y \right\}\Delta x^2. \tag{6.36}$$

The first two terms cancel when summing over the entire domain, but not so the last two terms. Contrary to what we saw with the analytical formulation, the chosen discretization cannot conserve total energy.

The evolution equation for total potential absolute vorticity $\mathfrak{Z}$ is written:

$$\frac{\partial \mathfrak{Z}}{\partial t} = \frac{\rho}{g}\sum\sum\left[\left(\frac{\partial V}{\partial t}\right)_x - \left(\frac{\partial U}{\partial t}\right)_y \right]\Delta x^2 \tag{6.37}$$

and yields, after an analogous calculation to that done above:

$$\frac{\partial \mathfrak{Z}}{\partial t} = -\frac{\rho}{g}\sum\sum\left[\left(\overline{\xi^*}^x \overline{\overline{\Phi}^x U}^{xy}\right)_x + \left(\overline{\xi^*}^y \overline{\overline{\Phi}^y V}^{xy}\right)_y \right]\Delta x^2, \tag{6.38}$$

which expression shows that $\mathfrak{Z}$ is conserved.

The evolution equation for total potential absolute enstrophy $\mathfrak{H}$ is written:

$$\frac{\partial \mathfrak{H}}{\partial t} = \frac{\rho}{g} \sum \sum \frac{\partial}{\partial t} \left[\frac{(\xi^*)^2}{2} \overline{\Phi}^{xy} \right] \frac{\Delta x^2}{\overline{m^2}^{xy}}$$

$$= \frac{\rho}{g} \sum \sum \left[\xi^* \frac{\partial \xi^*}{\partial t} \overline{\Phi}^{xy} + \frac{(\xi^*)^2}{2} \frac{\partial \overline{\Phi}^{xy}}{\partial t} \right] \frac{\Delta x^2}{\overline{m^2}^{xy}}. \tag{6.39}$$

By replacing $\xi^*, \dfrac{\partial \xi^*}{\partial t}, \Phi$, and $\dfrac{\partial \Phi}{\partial t}$ by their finite difference expressions, we get:

$$\frac{\partial \mathfrak{H}}{\partial t} = -\frac{\rho}{g} \sum \sum \frac{1}{2} \left[\widetilde{\xi^* 2}^x \left(\overline{\overline{\Phi}^x U}^{xy} \right)_x + \widetilde{\xi^* 2}^y \left(\overline{\overline{\Phi}^y V}^{xy} \right)_y \right] \Delta x^2, \tag{6.40}$$

the notations ~x and ~y indicating a geometric mean by analogy with the notations used for the arithmetic mean.

The discretization used allows us therefore to conserve the vorticity and potential absolute enstrophy, but does not conserve kinetic energy.

Alongside this, another discretization of the terms $\xi^* V$ and $-\xi^* U$, written:

$$\overline{(\xi^* \overline{\overline{\Phi}^x V})^x}^y \quad \text{and} \quad -\overline{(\xi^* \overline{\overline{\Phi}^y V})^y}^x,$$

allows us to conserve the absolute vorticity and the kinetic energy, but not the potential absolute enstrophy; however, from a practical point of view, to avoid the appearance of nonlinear instability, it is preferable to conserve the potential absolute enstrophy rather than the kinetic energy.

There is also a scheme for conserving both the kinetic energy and the total potential absolute enstrophy for a nondivergent flow (Sadourny, 1975). Such a scheme is, however, quite cumbersome to implement and considerably increases the number of calculations to be made at each time step. Computation of the discretized equation of the zero divergence model obtained from this specific discretization of the equations of the shallow water model brings out, as might be expected, the Arakawa Jacobian. In practice, it can therefore be remembered that the use of a scheme conserving potential absolute enstrophy is sufficient for making long-term integrations.

6.2.3 The centred explicit scheme

We henceforth term A, B and C the right-hand sides of the evolution equations for U, V and Φ, the discretized expressions of which are given in (6.32). Having examined how to calculate the tendency terms, we now look at the different time integration schemes.

For the first time step, we use the Euler scheme, which allows us to determine the value of variables at time Δt after having calculated the tendencies at time 0.

The extrapolation formulas are written, at the first time step:

$$\left.\begin{aligned} U(\Delta t) &= U(0) + \Delta t\, A(0), \\ V(\Delta t) &= V(0) + \Delta t\, B(0), \\ \Phi(\Delta t) &= \Phi(0) + \Delta t\, C(0). \end{aligned}\right\} \tag{6.41}$$

When we have the value of variables at two successive times $t - \Delta t$ and t, we can calculate the tendencies at t and apply the central difference scheme for the following time steps, which gives the extrapolation formulas:

$$\left.\begin{aligned} U(t + \Delta t) &= U(t - \Delta t) + 2\Delta t\, A(t), \\ V(t + \Delta t) &= V(t - \Delta t) + 2\Delta t\, B(t), \\ \Phi(t + \Delta t) &= \Phi(t - \Delta t) + 2\Delta t\, C(t). \end{aligned}\right\} \tag{6.42}$$

As seen in Subsection 5.6.1, the calculation will be stable by choosing a time step Δt so as to satisfy the CFL condition (5.99), which is written:

$$-1 \le \sqrt{2}(V^* \pm 2V_g)\Delta t / \Delta x \le +1, \tag{6.43}$$

where V^* denotes the basic wind speed and $V_g \cong \sqrt{\Phi^*}$, the gravity wave speed.

The Euler scheme is numerically unstable and is used for the first time step only. Accuracy for calculating variables at time Δt can be improved by making progressive time steps (Hoskins and Simmons, 1975): the Euler scheme is applied first with a time step $\Delta t/2$, then the leapfrog scheme provides the values of the variables at time Δt.

So as to be able to make the calculations for each time step, the values of the variables at $t + \Delta t$ must be prescribed on the edge of the domain: a simple boundary condition consists in conserving the initial values on the edges of the domain.

6.2.4 The centred semi-implicit scheme

Semi-implicit processing consists in implicit processing of the linearized adaptation terms, which are responsible for gravity waves, while conserving explicit processing for the advection and Coriolis terms. We must therefore prescribe a mean value of the geopotential Φ^* so as to bring out a linear part in the continuity equation.

The arithmetic mean of variable X at times $t + \Delta t$ and $t - \Delta t$ being noted:

$$\overline{X}^{2t} = [X(t + \Delta t) + X(t - \Delta t)]/2,$$

and the discretized divergence $\delta = U_x + V_y$, the discretized equations (6.32) are modified as follows:

$$
\left.
\begin{aligned}
\bar{U}_t^t &= \overline{\xi^*}^y\,\overline{\overline{\Phi}^y V}^{xy} - K_x - \bar{\Phi}_x^{2t}, \\
\bar{V}_t^t &= -\overline{\xi^*}^x\,\overline{\overline{\Phi}^x U}^{xy} - K_y - \bar{\Phi}_y^{2t}, \\
\bar{\Phi}_t^t &= -m^2\left\{\left[\overline{(\Phi-\Phi^*)}^x U\right]_x + \left[\overline{(\Phi-\Phi^*)}^y V\right]_y\right\} - m^2\Phi^*\,\bar{\delta}^{2t}.
\end{aligned}
\right\}
\qquad (6.44)
$$

The expansion of these discretized equations can still be written by isolating the variables to be calculated at time $t + \Delta t$:

$$
\left.
\begin{aligned}
U(t+\Delta t) &= U_T - \Delta t\,[\Phi(t+\Delta t)]_x\,, \\
V(t+\Delta t) &= V_T - \Delta t\,[\Phi(t+\Delta t)]_y\,, \\
\Phi(t+\Delta t) &= \Phi_T - m^2\Delta t\,\Phi^*\,\delta(t+\Delta t),
\end{aligned}
\right\}
\qquad (6.45)
$$

where U_T, V_T, Φ_T represent transient variables (quantities obtained explicitly and involving only data evaluated at t and $t - \Delta t$), which are written:

$$
\left.
\begin{aligned}
U_T &= U(t-\Delta t) + 2\Delta t A(t) + \Delta t\big[2\Phi(t) - \Phi(t-\Delta t)\big]_x, \\
V_T &= V(t-\Delta t) + 2\Delta t B(t) + \Delta t\big[2\Phi(t) - \Phi(t-\Delta t)\big]_y, \\
\Phi_T &= \Phi(t-\Delta t) + 2\Delta t C(t) + m^2\Delta t\,\Phi^*\big[2\delta(t) - \delta(t-\Delta t)\big].
\end{aligned}
\right\}
\qquad (6.46)
$$

We can therefore calculate the divergence at time $t + \Delta t$:

$$
\delta(t+\Delta t) = [(U_T)_x + (V_T)_y] - \Delta t\left\{\,[\Phi(t+\Delta t)]_{xx} + [\Phi(t+\Delta t)]_{yy}\right\};
\qquad (6.47)
$$

by substituting this value into the third equation of system (6.45), we get:

$$
\Phi(t+\Delta t) = \Phi_T - m^2\Phi^*\Delta t\big[(U_T)_x + (V_T)_y\big] + m^2\Phi^*\Delta t^2\left\{\,[\Phi(t+\Delta t)]_{xx} + [\Phi(t+\Delta t)]_{yy}\right\}.
\qquad (6.48)
$$

This Helmholtz equation gives the value of $\Phi(t + \Delta t)$ implicitly. The corresponding linear system can be solved by using an iterative over-relaxation method (Dady, 1969). At each time step, the choice of the last known value $\Phi(t)$ as *background* for the value sought $\Phi(t + \Delta t)$ accelerates the iterative process.

Knowing $\Phi(t + \Delta t)$, we can obtain $U(t + \Delta t)$ and $V(t + \Delta t)$ by using the first two equations of system (6.45):

$$
\left.
\begin{aligned}
U(t+\Delta t) &= U_T - \Delta t\,[\Phi(t+\Delta t)]_x\,, \\
V(t+\Delta t) &= V_T - \Delta t\,[\Phi(t+\Delta t)]_y\,.
\end{aligned}
\right\}
\qquad (6.49)
$$

Given the chosen discretization, it can be seen that the right-hand sides of equations (6.45) cannot be calculated for the outermost rows of each of the variables. These must therefore be prescribed on the outer rows; the values of the variable Φ on the boundary also provide the boundary conditions required for solving the Helmholtz equation (6.48).

This method is fairly easy to implement but increases the number of calculations to be made for a time step; however, the time step may be augmented significantly since the constraint given by the CFL criterion (5.101) seen in Subsection 5.6.1 is henceforth:

$$-1 \leq \sqrt{2}\, V^* \Delta t / \Delta x \leq +1. \tag{6.50}$$

Comparison of this constraint (6.50) with constraint (6.43) highlights the advantage of the semi-implicit method for processing the shallow water model. The gain achieved depends on the ratio of velocities $\sqrt{\Phi^*}/V^*$, which may attain a factor of 6, and justifies the increased amount of computation for each time step.

These conclusions are based on the results from Subsection 5.4.3. Processing the nonlinear shallow water model, as proposed in (6.44), implies explicit processing of the nonlinear divergence term. This may become large if the values of the geopotential Φ depart too far from the value of the mean geopotential Φ^*, and may cause instability (Simmons et al., 1978; Bénard, 2003); this may, however, be avoided by choosing a suitable value for Φ^*.

6.2.5 Semi-Lagrangian schemes

6.2.5.1 The centred scheme and determination of the particle origin point

Lagrangian processing of advection requires us to bring out the Lagrangian derivatives. Given the choice of variables $U = u/m$ and $V = v/m$, the Lagrangian derivative takes the form:

$$\frac{d}{dt} \equiv \frac{\partial}{\partial t} + m^2 \left(U \frac{\partial}{\partial x} + V \frac{\partial}{\partial x} \right). \tag{6.51}$$

The equations of system (6.27) are then written by isolating the linear divergence term in the continuity equation:

$$\left. \begin{array}{l} \dfrac{dU}{dt} = -\dfrac{U^2 + V^2}{2} \dfrac{\partial(m^2)}{\partial x} + fV - \dfrac{\partial \Phi}{\partial x}, \\[3mm] \dfrac{dV}{dt} = -\dfrac{U^2 + V^2}{2} \dfrac{\partial(m^2)}{\partial y} - fU - \dfrac{\partial \Phi}{\partial y}, \\[3mm] \dfrac{d\Phi}{dt} = -m^2(\Phi - \Phi^*)\left(\dfrac{\partial U}{\partial x} + \dfrac{\partial V}{\partial y} \right) - m^2 \Phi^* \left(\dfrac{\partial U}{\partial x} + \dfrac{\partial V}{\partial y} \right). \end{array} \right\} \tag{6.52}$$

The total derivatives are evaluated as differences in values at grid point G at $t + \Delta t$ and in the values at the particle origin point O at $t - \Delta t$ (Figure 6.5); the linear terms are expressed as arithmetic means of the values at G at $t + \Delta t$ and of the values at O at $t - \Delta t$; finally, the nonlinear residual terms are calculated at time t at the median point of the trajectory, noted I (Figure 6.5).

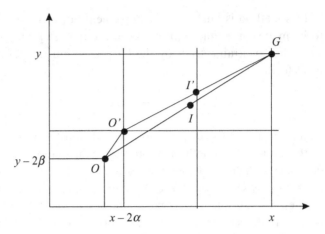

Figure 6.5   Location of the various points for Lagrangian processing.

By using the notations:

$$\hat{X}_t^t \equiv \frac{[X(t+\Delta t)]_G - [X(t-\Delta t)]_O}{2\Delta t} \quad \text{and} \quad \hat{X}^{2t} \equiv \frac{[X(t+\Delta t)]_G + [X(t-\Delta t)]_O}{2},$$

and by conserving as far as possible the principles already adopted for the C grid in (6.32), the discretized equations are written:

$$
\begin{aligned}
\hat{U}_t^t &= \left[-\overline{\left(\overline{U^2}^x + \overline{V^2}^y \right)}^x \left(\frac{m^2}{2} \right)_x + \overline{\frac{\overline{f}^y}{\overline{\Phi}^{xy}} \overline{\Phi}^y V}^{xy} \right]_I - \hat{\Phi}_x^{2t}, \\[2ex]
\hat{V}_t^t &= \left[-\overline{\left(\overline{U^2}^x + \overline{V^2}^y \right)}^y \left(\frac{m^2}{2} \right)_y - \overline{\frac{\overline{f}^x}{\overline{\Phi}^{xy}} \overline{\Phi}^x U}^{xy} \right]_I - \hat{\Phi}_y^{2t}, \\[2ex]
\hat{\Phi}_t^t &= \left[-m^2 (\Phi - \Phi^*) \delta \right]_I - m^2 \Phi^* \hat{\delta}^{2t}.
\end{aligned}
\tag{6.53}
$$

The coordinates of the median point I are obtained from its coordinates (α, β), calculated in the local reference frame the origin of which is point G, using the iterative algorithm proposed by Robert (1983):

$$
\left.
\begin{aligned}
\alpha_0 &= 0, \\
\beta_0 &= 0,
\end{aligned}
\right\}
\quad
\left.
\begin{aligned}
\alpha_{n+1} &= m^2 \Delta t U(x_G - \alpha_n, y_G - \beta_n), \\
\beta_{n+1} &= m^2 \Delta t V(x_G - \alpha_n, y_G - \beta_n).
\end{aligned}
\right\}
\tag{6.54}
$$

It is noticed that this calculation requires spatial interpolation of the value of velocities U and V at time t for each iteration.

When the process converges, the solution (α, β) then satisfies equations (6.55):

$$
\left.
\begin{aligned}
\alpha &= m^2 \Delta t U(x - \alpha, y - \beta), \\
\beta &= m^2 \Delta t V(x - \alpha, y - \beta).
\end{aligned}
\right\}
\tag{6.55}
$$

Convergence of the iterative process (6.54), which was studied by Pudykiewicz et al. (1985), is set out in detail below.

By applying the mean-value theorem over the interval ($[x_G - \alpha, x]$, $[y_G - \beta, y]$), we obtain for $U(x, y)$:

$$
\begin{aligned}
U(x, y) = {} & U(x_G - \alpha, y_G - \beta) \\
& + [x - (x_G - \alpha)]\frac{\partial U}{\partial x}\big(x_G - \alpha + \theta_1[x - (x_G - \alpha)], y_G - \beta + \theta_2[y - (y_G - \beta)]\big) \\
& + [y - (y_G - \beta)]\frac{\partial U}{\partial y}\big(x_G - \alpha + \theta_1[x - (x_G - \alpha)], y_G - \beta + \theta_2[y - (y_G - \beta)]\big) ,
\end{aligned}
\tag{6.56}
$$

with $0 < \theta_1, \theta_2 < 1$.

This formula, applied to the coordinates of point I at iteration n, ($x = x_G - \alpha$, $y = y_G - \beta$), gives:

$$
\begin{aligned}
U(x_G - \alpha_n, y_G - \beta_n) = {} & U(x_G - \alpha, y_G - \beta) \\
& + (\alpha - \alpha_n)\frac{\partial U}{\partial x}\big(x_G - \alpha + \theta_1(\alpha - \alpha_n), y_G - \beta + \theta_2[(\beta - \beta_n)]\big) \\
& + (\beta - \beta_n)\frac{\partial U}{\partial y}\big(x_G - \alpha + \theta_1(\alpha - \alpha_n), y_G - \beta + \theta_2[(\beta - \beta_n)]\big) .
\end{aligned}
\tag{6.57}
$$

By replacing $U(x_G - \alpha_n, y_G - \beta_n)$ by its value drawn from the first relation of (6.54) and by using the first relation of (6.55) to eliminate $U(x_G - \alpha, y_G - \beta)$, we finally obtain a first recurrence relation for the deviations from the solution (α, β). These deviations being noted:

$$
F_n = \alpha_n - \alpha \quad \text{and} \quad G_n = \beta_n - \beta,
$$

this first relation is written:

$$
F_{n+1} = -m^2 \Delta t \left[\left(\frac{\partial U}{\partial x}(x_A, y_A) \right) F_n + \left(\frac{\partial U}{\partial y}(x_A, y_A) \right) G_n \right],
\tag{6.58}
$$

with:

$$
x_A = x_G - \alpha + \theta_1 F_n, \quad y_A = y_G - \beta + \theta_2 G_n.
$$

By performing a similar calculation for the variable $V(x, y)$, we get the second recurrence relation:

$$
G_{n+1} = -m^2 \Delta t \left[\left(\frac{\partial V}{\partial x}(x_B, y_B) \right) F_n + \left(\frac{\partial V}{\partial y}(x_B, y_B) \right) G_n \right],
\tag{6.59}
$$

with:

$$
x_B = x_G - \alpha + \theta_3 F_n, \quad y_B = y_G - \beta + \theta_4 G_n.
$$

The quantity $\mathbf{E}_n = (F_n, G_n)^T$ represents the error of the solution obtained at iteration n. Convergence of the process requires that $\|\mathbf{E}_{n+1}\| < \|\mathbf{E}_n\|$. By using the norm of the maximum, defined by $\|\mathbf{E}\| = \max(|F|,|G|)$, we finally find:

$$\|\mathbf{E}_{n+1}\| = m^2 \Delta t \max \left\{ \left| \left(\frac{\partial U}{\partial x}(x_A, y_A) \right) F_n + \left(\frac{\partial U}{\partial y}(x_A, y_A) \right) G_n \right|, \left| \left(\frac{\partial V}{\partial x}(x_B, y_B) \right) F_n + \left(\frac{\partial V}{\partial y}(x_B, y_B) \right) G_n \right| \right\}$$

$$\leq 2m^2 \Delta t \max \left(\left| \frac{\partial U}{\partial x} \right|, \left| \frac{\partial U}{\partial y} \right|, \left| \frac{\partial V}{\partial x} \right|, \left| \frac{\partial V}{\partial y} \right| \right) \cdot \|\mathbf{E}_n\|,$$

which yields as a sufficient stability condition:

$$2m^2 \Delta t \max \left(\left| \frac{\partial U}{\partial x} \right|, \left| \frac{\partial U}{\partial y} \right|, \left| \frac{\partial V}{\partial x} \right|, \left| \frac{\partial V}{\partial y} \right| \right) < 1. \tag{6.60}$$

For a given time step, the convergence of the iterative process (6.55) is not ensured unless condition (6.60) is satisfied; this remains, though, far less of a constraint than the CFL condition for the semi-implicit scheme.

Once the coordinates of the median point I: (α, β) are known, we obtain those of the point of origin O: $(2\alpha, 2\beta)$ by assuming point I is located at the midpoint of segment OG.

The process of searching for the origin point requires interpolating velocities U and V, so as to determine their values at point $(x - \alpha_k, y - \beta_k)$ at iteration k. It has been shown (Robert, 1981) that a straightforward bilinear type interpolation is sufficient. Conversely, to avoid the attenuation effects mentioned in Paragraph 5.4.4.3, we have to interpolate the variables at the origin point O and the nonlinear terms at the median point I very accurately, by using a higher order interpolator (e.g. quadratic or cubic).

The discretized equations are in a form analogous to (6.45):

$$\left. \begin{aligned} U(t + \Delta t) &= U_T - \Delta t \left[\Phi(t + \Delta t) \right]_x, \\ V(t + \Delta t) &= V_T - \Delta t \left[\Phi(t + \Delta t) \right]_y, \\ \Phi(t + \Delta t) &= \Phi_T - m^2 \Delta t \Phi^* \delta(t + \Delta t), \end{aligned} \right\} \tag{6.61}$$

with the quantity $\delta = U_x + V_y$ representing the divergence calculated numerically and the expressions of the transient quantities U_T, V_T, Φ_T now being:

$$\left. \begin{aligned} U_T &= \left[U(t - \Delta t) \right]_O + 2\Delta t \left[-\left(\overline{U^{2^x}} + \overline{V^{2^y}} \right)^x \left(\frac{m^2}{2} \right)_x + \frac{\overline{f}^y}{\overline{\Phi}^{xy}} \overline{\Phi}^y \overline{V}^{xy} \right]_I - \left[\Phi_x(t - \Delta t) \right]_O, \\ V_T &= \left[U(t - \Delta t) \right]_O + 2\Delta t \left[-\left(\overline{U^{2^x}} + \overline{V^{2^y}} \right)^y \left(\frac{m^2}{2} \right)_y - \frac{\overline{f}^x}{\overline{\Phi}^{xy}} \overline{\Phi}^x \overline{U}^{xy} \right]_I - \left[\Phi_y(t - \Delta t) \right]_O, \\ \Phi_T &= \left[\Phi(t - \Delta t) \right]_O + 2\Delta t \left[-m^2 (\Phi - \Phi^*)\delta \right]_I - m^2 \Phi^* \left[\delta(t - \Delta t) \right]_O. \end{aligned} \right\} \tag{6.62}$$

The first two equations of (6.62) can be used to calculate divergence at $t + \Delta t$:

$$\delta(t + \Delta t) = [(U_T)_x + (V_T)_y] - \Delta t \left\{ [\Phi(t + \Delta t)]_{xx} + [\Phi(t + \Delta t)]_{yy} \right\}; \qquad (6.63)$$

by substituting this value into the third equation of system (6.61), we get:

$$\Phi(t + \Delta t) = \Phi_T - m^2 \Phi^* \Delta t \left[(U_T)_x + (V_T)_y \right] + m^2 \Phi^* \Delta t^2 \left\{ [\Phi(t + \Delta t)]_{xx} + [\Phi(t + \Delta t)]_{yy} \right\}. \qquad (6.64)$$

This equation implicitly giving the value of $\Phi(t + \Delta t)$ is a Helmholtz equation just like equation (6.48) and can therefore be solved by the same methods. Having $\Phi(t + \Delta t)$, we can finally obtain $U(t + \Delta t)$ and $V(t + \Delta t)$ by using the first two equations of (6.61).

The semi-implicit semi-Lagrangian scheme just set out will be stable insofar as the various constraints met in Paragraphs 5.4.4.1 and 5.4.4.3 and in this paragraph are properly satisfied:

- the condition relating to the explicit processing of the Coriolis term, (5.68);
- the condition ensuring convergence of the point I search algorithm, (6.60);
- the condition as to the choice of grid points for interpolation (5.88).

These conditions are nonetheless not very restrictive given the scales handled in the numerical prediction models, which allow us to use a markedly greater time step with the semi-Lagrangian semi-implicit scheme than with the simple semi-implicit scheme. We have seen that the quality of advection processing is related to the interpolation scheme used, the cost of which is greater when greater accuracy is sought; however, the possibility of increasing the time step justifies the extra computing costs due to semi-Lagrangian processing, for which a number of variants have been imagined.

6.2.5.2 Variants of semi-Lagrangian processing

The implicit processing of Coriolis terms was addressed when studying the linear model in Paragraph 5.4.4.2; these terms are calculated as the arithmetic means of their values taken at point O at time $t - \Delta t$ and at point G at time $t + \Delta t$. Consequently, the values of $U(t + \Delta t)$ and $V(t + \Delta t)$ are obtained from the values of the variables at t and $t - \Delta t$ and of the geopotential $\Phi(t + \Delta t)$ by solving the linear system of two equations with two unknowns formed by the two equations of momentum. The C grid, however, is not really suitable for this type of processing, as the variables U and V are not expressed at the same point; we shall therefore give preference in this case to an A or B grid.

The drawback with the semi-Lagrangian method is the cost of the various interpolations; accordingly many variants have been proposed to simplify or do away with costly computations. One simplification proposed by Kaas (1987), and refined by Tanguay et al. (1992), consists in evaluating the residual explicit terms at time t at intermediate point I by calculating the arithmetic means of their values at point G and at point O.

A variant of the semi-Lagrangian scheme was proposed by Ritchie (1986). It consisted in determining the grid point O' as close as possible to the origin point O, and then processing the fictitious displacement from O' to G in a Lagrangian way (Figure 6.5).

Starting from the vector relation $\overrightarrow{OG} = \overrightarrow{O'G} + \overrightarrow{OO'}$, the total derivative can be calculated as:

$$\frac{d}{dt} \equiv \frac{\bar{d}}{dt} + m^2 \left(U' \frac{\partial}{\partial x} + V' \frac{\partial}{\partial x} \right),$$

where $\bar{d}/dt$ designates the total derivative calculated for the displacement $\overrightarrow{O'G}$, and the 'velocities' $m^2 U'$ and $m^2 V'$ are the components of the vector $\overrightarrow{OO'}/2\Delta t$. Lagrangian processing is performed for a fictitious displacement $\overrightarrow{O'G}$ associated with symmetric implicit processing of the linear terms over points O' and G, while the residual advection term is evaluated, as for the other nonlinear terms, explicitly at point I' at time t. The interpolation at the origin point O disappears and the interpolation of the nonlinear terms at point I' can be simplified by calculating the arithmetic means of the values taken at points O' and G. The linear analysis that can be done on this scheme by following the procedure in Chapter 5 shows that this scheme conserves the wave amplitude, which is the whole point of it.

The principle of the two time level scheme for moving directly to the variables at $t + \Delta t$ from their values at $t - \Delta t$ was evoked when discussing the linearized model with implicit processing of the Coriolis terms in Paragraph 5.4.4.2. It is still necessary, though, to calculate the values at the intermediate time t, both to calculate the trajectories of the particles and to evaluate the residual nonlinear terms at point I (either by spatial interpolation or by the simple arithmetic mean of the values taken by these terms at point O and at point G).

The variables can be determined at time t by time extrapolation involving the values of variables determined at previous times (Temperton and Staniforth, 1987). The two time level scheme using, in this context, $2\Delta t$, the variables are assumed to be known at times $t - \Delta t$ and $t - 3\Delta t$; the value at time t, extrapolated linearly in time, is therefore written:

$$\left[\mathbf{X}(t) \right]_I = \frac{1}{2} \left(3 \left[\mathbf{X}(t - \Delta t) \right]_I - \left[\mathbf{X}(t - 3\Delta t) \right]_I \right).$$

For a two time level scheme using a time step Δt, we find for the value extrapolated at time $t + \Delta t/2$ the value given by Temperton and Staniforth:

$$\left[\mathbf{X}(t + \Delta t / 2) \right]_I = \frac{1}{2} \left(3 \left[\mathbf{X}(t) \right]_I - \left[\mathbf{X}(t - \Delta t) \right]_I \right).$$

As this formulation can lead to instabilities, another formulation involving the values of variables at points O and G has been proposed by Hortal (2002):

$$\left[\mathbf{X}(t + \Delta t / 2) \right]_I = \frac{1}{2} \left(2 \left[\mathbf{X}(t) \right]_O + \left[\mathbf{X}(t) \right]_G - \left[\mathbf{X}(t - \Delta t) \right]_O \right).$$

The two time level algorithms are highly advantageous since they allow us to advance in time twice as fast for the same accuracy and do not lead to any spurious computational solutions.

6.3 Spectral processing of the shallow water model

6.3.1 Formulation of the equations

Spectral processing is used for processing the shallow water model on the sphere. We use equations (2.14) with zero orography ($\Phi_s = 0$), by employing the natural geographic coordinate system λ and $\mu = \sin\varphi$ and the reduced wind components $U = au\cos\varphi$ and $V = av\cos\varphi$:

$$
\left.
\begin{aligned}
\frac{\partial U}{\partial t} &= (\zeta + f)V - \frac{\partial}{\partial\lambda}(K + \Phi), \\[4pt]
\frac{\partial V}{\partial t} &= -(\zeta + f)U - (1-\mu^2)\frac{\partial}{\partial\mu}(K + \Phi), \\[4pt]
\frac{\partial \Phi}{\partial t} &= -\nabla\cdot(\Phi\mathbf{V}), \\[4pt]
K &= \frac{U^2 + V^2}{2a^2(1-\mu^2)}, \text{ kinetic energy.}
\end{aligned}
\right\}
\tag{6.65}
$$

The relative vorticity ζ and divergence η are given by the expressions:

$$
\zeta = \frac{1}{a^2(1-\mu^2)}\left[\frac{\partial V}{\partial\lambda} - (1-\mu^2)\frac{\partial U}{\partial\mu}\right] \quad \text{and} \quad \eta = \frac{1}{a^2(1-\mu^2)}\left[\frac{\partial U}{\partial\lambda} + (1-\mu^2)\frac{\partial V}{\partial\mu}\right].
\tag{6.66}
$$

The wind components can be calculated via the streamfunction ψ and the velocity potential χ:

$$
\left.
\begin{aligned}
U &= \frac{\partial\chi}{\partial\lambda} - (1-\mu^2)\frac{\partial\psi}{\partial\mu}, \\[4pt]
V &= \frac{\partial\psi}{\partial\lambda} + (1-\mu^2)\frac{\partial\chi}{\partial\mu}.
\end{aligned}
\right\}
\tag{6.67}
$$

The streamfunction ψ and the velocity potential χ are then obtained from the vorticity ζ and from the divergence η by the Poisson equations:

$$
\zeta = \nabla^2\psi \quad \text{and} \quad \eta = \nabla^2\chi.
\tag{6.68}
$$

By calling ξ the absolute vorticity $\zeta + f$, the evolution equations for vorticity, divergence, and geopotential are written:

$$\frac{\partial \zeta}{\partial t} = -\frac{1}{a^2(1-\mu^2)}\left\{\frac{\partial}{\partial \lambda}(\xi U)+(1-\mu^2)\frac{\partial}{\partial \mu}(\xi V)\right\} = A(\lambda,\mu),$$

$$\frac{\partial \eta}{\partial t} = \frac{1}{a^2(1-\mu^2)}\left\{\frac{\partial}{\partial \lambda}(\xi V)-(1-\mu^2)\frac{\partial}{\partial \mu}(\xi U)\right\} - \nabla^2(\Phi+K) = B(\lambda,\mu)-\nabla^2(\Phi+K),$$

$$\frac{\partial \Phi}{\partial t} = -\frac{1}{a^2(1-\mu^2)}\left\{\frac{\partial}{\partial \lambda}(\Phi U)+(1-\mu^2)\frac{\partial}{\partial \mu}(\Phi V)\right\} = C(\lambda,\mu).$$

$$(6.69)$$

The quantities $A(\lambda,\mu)$, $B(\lambda,\mu)$, and $C(\lambda,\mu)$, containing quadratic terms, are calculated on the Gaussian grid associated with the truncation chosen for the model, as seen in Subsection 4.3.7. The spectral coefficients A_n^m, B_n^m, and C_n^m as well as Φ_n^m and K_n^m are obtained from successive Fourier and then Legendre transforms. The calculation is detailed here for the quantity A_n^m.

It should be emphasized that in this part, the index m henceforth refers to the *zonal wavenumber* and is not to be confused with the scale factor involved in the formulation of conformal projection equations.

The Fourier transform is written:

$$A_m(\mu) = \frac{1}{2\pi}\int_0^{2\pi} A(\lambda,\mu)e^{-im\lambda}d\lambda,$$

$$(6.70)$$

and the Legendre transform:

$$A_n^m = \frac{1}{2}\int_{-1}^{+1} A_m(\mu)P_n^m(\mu)d\mu.$$

$$(6.71)$$

By replacing $A(\lambda,\mu)$ by its expression given in (6.69), we get:

$$A_m(\mu) = -\frac{1}{a^2(1-\mu^2)}\frac{1}{2\pi}\int_0^{2\pi}\left[\frac{\partial}{\partial \lambda}(U\xi)+(1-\mu^2)\frac{\partial}{\partial \mu}(V\xi)\right]e^{-im\lambda}d\lambda$$

$$= -\frac{1}{a^2(1-\mu^2)}\left[im(U\xi)_m+(1-\mu^2)\frac{\partial}{\partial \mu}(V\xi)_m\right],$$

$$(6.72)$$

where $(X)_m$ denotes the Fourier coefficient of quantity X.

By performing the Legendre transform next, we get:

$$A_n^m = -\frac{1}{2a^2}\int_{-1}^{+1} im(U\xi)_m P_n^m(\mu)\frac{d\mu}{1-\mu^2} - \frac{1}{2a^2}\int_{-1}^{+1}\frac{\partial}{\partial \mu}\left[(V\xi)_m\right]P_n^m(\mu)d\mu.$$

$$(6.73)$$

The integral of the second term of the right-hand side can be partially integrated and becomes:

$$\int_{-1}^{+1}\left[\frac{\partial}{\partial \mu}\left[(V\xi)_m\right]P_n^m(\mu)\right]d\mu = \left[(V\xi)_m P_n^m(\mu)\right]_{-1}^{+1} - \int_{-1}^{+1}\left[(V\xi)_m\frac{dP_n^m(\mu)}{d\mu}\right]d\mu.$$

$$(6.74)$$

The first term of the right-hand side cancels, since the value of V (like that of U) is zero at the poles. The integral of the right-hand side is calculated by using Gaussian quadrature, by performing the discrete sum on the K latitudes μ_k with the weights w_k. We thus get for A_n^m, and in the same way for B_n^m and C_n^m, expressions (6.75):

$$\left.\begin{aligned}
A_n^m &= -\frac{1}{2}\sum_{k=1}^{K}\left[im(U\xi)_m P_n^m(\mu_k)-(V\xi)_m(1-\mu_k^2)\frac{dP_n^m(\mu_k)}{d\mu}\right]\frac{w_k}{a^2(1-\mu_k^2)}, \\
B_n^m &= \frac{1}{2}\sum_{k=1}^{K}\left[im(V\xi)_m P_n^m(\mu_k)+(U\xi)_m(1-\mu_k^2)\frac{dP_n^m(\mu_k)}{d\mu}\right]\frac{w_k}{a^2(1-\mu_k^2)}, \\
C_n^m &= -\frac{1}{2}\sum_{k=1}^{K}\left[im(U\Phi)_m P_n^m(\mu_k)-(V\Phi)_m(1-\mu_k^2)\frac{dP_n^m(\mu_k)}{d\mu}\right]\frac{w_k}{a^2(1-\mu_k^2)}.
\end{aligned}\right\} \quad (6.75)$$

The coefficients Φ_n^m and K_n^m are also obtained by performing a Fourier transform followed by a Legendre transform. The evolution equations for the spectral coefficients are therefore written:

$$\left.\begin{aligned}
\frac{\partial \zeta_n^m}{\partial t} &= \frac{1}{4\pi}\iint_S AY_n^{-m}d\lambda d\mu = A_n^m, \\
\frac{\partial \eta_n^m}{\partial t} &= \frac{1}{4\pi}\iint_S\left[B-\nabla^2(\Phi+K)\right]Y_n^{-m}d\lambda d\mu = B_n^m+\frac{n(n+1)}{a^2}(K_n^m+\Phi_n^m), \\
\frac{\partial \Phi_n^m}{\partial t} &= \frac{1}{4\pi}\iint_S CY_n^{-m}d\lambda d\mu = C_n^m.
\end{aligned}\right\} \quad (6.76)$$

We can then apply the Euler scheme for the first time step and then the centred explicit (leapfrog) scheme for the subsequent time steps, which leads to the time extrapolation formulas:

$$\left.\begin{aligned}
\zeta_n^m(t+\Delta t) &= \zeta_n^m(t-\Delta t)+2\Delta t A_n^m(t), \\
\eta_n^m(t+\Delta t) &= \eta_n^m(t-\Delta t)+2\Delta t\left[B_n^m(t)+\frac{n(n+1)}{a^2}\left(K_n^m(t)+\Phi_n^m(t)\right)\right], \\
\Phi_n^m(t+\Delta t) &= \Phi_n^m(t-\Delta t)+2\Delta t C_n^m(t).
\end{aligned}\right\} \quad (6.77)$$

From the values $\zeta_n^m(t+\Delta t)$ and $\eta_n^m(t+\Delta t)$, we calculate the values of the stream-function and of the velocity potential at time $t+\Delta t$ by solving the Poisson equations $\zeta=\nabla^2\psi$ and $\eta=\nabla^2\chi$, which directly yield:

$$\psi_n^m(t+\Delta t)=\left[-\frac{n(n+1)}{a^2}\right]^{-1}\zeta_n^m(t+\Delta t) \quad \text{and} \quad \chi_n^m(t+\Delta t)=\left[-\frac{n(n+1)}{a^2}\right]^{-1}\eta_n^m(t+\Delta t). \quad (6.78)$$

The inverse Legendre transform on the Gaussian grid allows us to obtain the Fourier coefficients below from the spectral coefficients ζ_n^m, η_n^m, Φ_n^m, ψ_n^m, and χ_n^m:

$$\left.\zeta_m(\mu) = \sum_{n=m}^{M} \zeta_n^m P_n^m(\mu), \qquad \psi_m(\mu) = \sum_{n=m}^{M} \psi_n^m P_n^m(\mu), \right.$$

$$\left.\eta_m(\mu) = \sum_{n=m}^{M} \chi_n^m P_n^m(\mu), \qquad \chi_m(\mu) = \sum_{n=m}^{M} \chi_n^m P_n^m(\mu), \right\}$$ (6.79)

$$\left.\Phi_m(\mu) = \sum_{n=m}^{M} \Phi_n^m P_n^m(\mu), \right.$$

as well as:

$$\left.\left[(1-\mu^2)\frac{d\psi}{d\mu}\right]_m = \sum_{n=m+1}^{M} \psi_n^m (1-\mu^2)\frac{dP_n^m(\mu)}{d\mu}, \right\}$$

$$\left[(1-\mu^2)\frac{d\chi}{d\mu}\right]_m = \sum_{n=m}^{M} \chi_n^m (1-\mu^2)\frac{dP_n^m(\mu)}{d\mu}. \right\}$$ (6.80)

The latter quantities can be used to calculate the Fourier coefficients for the variables U and V:

$$\left.U_m(\mu) = im\chi_m(\mu) - \left[(1-\mu^2)\frac{d\psi}{d\mu}\right]_m, \right\}$$

$$V_m(\mu) = im\psi_m(\mu) + \left[(1-\mu^2)\frac{d\chi}{d\mu}\right]_m. \right\}$$ (6.81)

The final stage consists in performing the inverse Fourier transform which, from the values of the Fourier coefficients $U(\mu)$, $V(\mu)$, $\zeta(\mu)$, $\eta(\mu)$, and $\Phi(\mu)$ on the Gaussian grid latitudes, gives the values of the variables on the Gaussian grid in physical space: $U(\lambda, \mu)$, $V(\lambda, \mu)$, $\zeta(\lambda, \mu)$, $\eta(\lambda, \mu)$, and $\Phi(\lambda, \mu)$ at time $t + \Delta t$.

6.3.2 Semi-implicit processing

Examination of the evolution equations for vorticity and divergence shows that only the latter equation contains adaptation terms. Semi-implicit processing therefore concerns the continuity equation and the divergence equation only, with the vorticity equation continuing to be processed explicitly. By adopting the same notations as used in Subsection 6.2.4, the two equations of the semi-implicit system are therefore written:

$$\left.\overline{\eta}_t^t = B(t) - \nabla^2 K(t) - \overline{\nabla^2 \Phi}^{2t}, \right\}$$

$$\overline{\Phi}_t^t = C(t) + \Phi^* \eta(t) - \Phi^* \overline{\eta}^{2t}. \right\}$$ (6.82)

After performing the Fourier and Legendre transforms, we get the evolution equations for the spectral coefficients, which are written:

$$\left.\eta_n^m(t + \Delta t) = (B_T)_n^m + \frac{n(n+1)\Delta t}{a^2} \Phi_n^m(t + \Delta t), \right\}$$

$$\Phi_n^m(t + \Delta t) = (C_T)_n^m - \Phi^* \Delta t \, \eta_n^m(t + \Delta t), \right\}$$ (6.83)

with:

$$(B_T)_n^m = \eta_n^m(t-\Delta t) + 2\Delta t\left\{B_n^m(t) + \frac{n(n+1)}{a^2}\left[K_n^m(t) + \frac{1}{2}\left(\Phi_n^m(t-\Delta t)\right)\right]\right\},$$

$$(C_T)_n^m = \Phi_n^m(t-\Delta t) + 2\Delta t\left\{C_n^m(t) + \Phi^*\left[\eta_n^m(t) - \frac{1}{2}\left(\Phi_n^m(t-\Delta t)\right)\right]\right\}. \qquad (6.84)$$

Quantities $(B_T)_n^m$ and $(C_T)_n^m$ are the spectral coefficients of quantities B_T and C_T, which only involve the quantities calculated from variables at times t and $t - \Delta t$.

Substituting the value of $\eta_n^m(t + \Delta t)$ from the first equation into the second equation finally provides the equation:

$$\Phi_n^m(t+\Delta t) = (C_T)_n^m - \Phi^*\Delta t\,(B_T)_n^m - \Phi^*\frac{n(n+1)\Delta t^2}{a^2}\Phi_n^m(t+\Delta t). \qquad (6.85)$$

This Helmholtz equation for $\Phi_n^m(t + \Delta t)$ in spectral space can be particularly easily solved and yields the solution:

$$\Phi_n^m(t+\Delta t) = \frac{(C_T)_n^m - \Phi^*\Delta t.(B_T)_n^m}{\left[1 + \Phi^*\dfrac{n(n+1)\Delta t^2}{a^2}\right]}. \qquad (6.86)$$

The divergence coefficient $\eta_n^m(t + \Delta t)$ can be deduced immediately from $\Phi_n^m(t + \Delta t)$ by using the first equation of (6.83), while the vorticity coefficient $\zeta_n^m(t + \Delta t)$ is obtained explicitly by using the first equation of (6.77).

6.3.3 Semi-Lagrangian processing

For semi-Lagrangian processing, it is more judicious to replace the formulation using the vorticity and divergence by a formulation using the components of the wind (Ritchie, 1988), which allows the advection terms to re-appear, these being essential for implementing such a scheme. Keeping the same notations as before, the equations are written:

$$\frac{\partial U}{\partial t} = -\frac{1}{a^2(1-\mu^2)}\left[U\frac{\partial U}{\partial \lambda} + (1-\mu^2)V\frac{\partial U}{\partial \mu}\right] + fV - \frac{\partial \Phi}{\partial \lambda},$$

$$\frac{\partial V}{\partial t} = -\frac{1}{a^2(1-\mu^2)}\left[U\frac{\partial V}{\partial \lambda} + (1-\mu^2)V\frac{\partial V}{\partial \mu}\right] - fU - 2\mu K - (1-\mu^2)\frac{\partial \Phi}{\partial \mu},$$

$$\frac{\partial \Phi}{\partial t} = -\frac{1}{a^2(1-\mu^2)}\left[U\frac{\partial \Phi}{\partial \lambda} + (1-\mu^2)V\frac{\partial \Phi}{\partial \mu}\right] - \Phi\eta, \qquad (6.87)$$

$$\text{with } K = \frac{U^2+V^2}{2a^2(1-\mu^2)}, \quad \text{and } \eta = \frac{1}{a^2(1-\mu^2)}\left[\frac{\partial U}{\partial \lambda} + (1-\mu^2)\frac{\partial V}{\partial \mu}\right].$$

This formulation brings out derivatives with respect to μ, that can be calculated from derivatives with respect to λ using the vorticity and divergence, since:

$$
\left.\begin{aligned}
(1-\mu^2)\frac{\partial U}{\partial \mu} &= \frac{\partial V}{\partial \lambda} - a^2(1-\mu^2)\zeta \,, \\
(1-\mu^2)\frac{\partial V}{\partial \mu} &= -\frac{\partial U}{\partial \lambda} + a^2(1-\mu^2)\eta \,.
\end{aligned}\right\}
\tag{6.88}
$$

However, the explicit treatment of the metric term $2\mu K$ leads to instability in the vicinity of the poles and so it is better to replace the first two equations of (6.77) by a vector equation for the vector $\mathbf{V}_H$; this allows us to handle the problem of advection on the sphere directly before returning to the components U and V. The equation for horizontal momentum is then written:

$$
\frac{\partial \mathbf{V}_H}{\partial t} + (\mathbf{V}_H \cdot \nabla_H)\mathbf{V}_H = -f\mathbf{k} \times \mathbf{V}_H - \nabla_H \Phi \,.
\tag{6.89}
$$

Discretization for semi-implicit semi-Lagrangian processing is then written:

$$
\frac{[\mathbf{V}_H(t+\Delta t)]_G - R^{OG}[\mathbf{V}_H(t-\Delta t)]_O}{2\Delta t} = -R^{IG}[f\mathbf{k} \times \mathbf{V}_H(t)]_I - \frac{[\nabla\Phi(t+\Delta t)]_G + R^{OG}[\nabla\Phi(t-\Delta t)]_O}{2},
\tag{6.90}
$$

notations R^{OG} and R^{IG} designating the operators for rotation of a horizontal vector on the sphere, one from point O (origin of the trajectory) to grid point G and one from point I (mid-point).

These operators are obtained by writing that in the course of a rotation a vector conserves the angle it makes with the unit vector tangential to its trajectory, assimilated to a great circle. This operator, which reflects a rotation in the tangent plane, is of the form:

$$
R^{PG} = \begin{pmatrix} p & q \\ -q & p \end{pmatrix},
$$

where the quantities p and q are given by:

$$
p = \frac{\cos\varphi^G \cos\varphi^P + (1+\sin\varphi^G \sin\varphi^P)\cos(\lambda^G - \lambda^P)}{1+\cos\Theta},
$$

$$
q = \frac{(\sin\varphi^G + \sin\varphi^P)\cos(\lambda^G - \lambda^P)}{1+\cos\Theta}.
$$

The quantities φ^G, λ^G and φ^P, λ^P denote the latitudes and longitudes of points G and P (the latter designating either the origin point O, or the median point I) while Θ denotes the corresponding angle of rotation along the great circle (Figure 6.6).

The final equation of system (6.77) for the geopotential is written more simply:

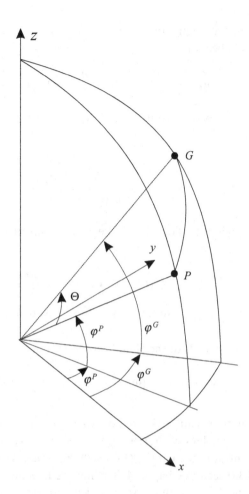

Figure 6.6   Trajectory on the sphere.

$$\frac{[\Phi(t+\Delta t)]_G - \Phi(t-\Delta t)]_O}{2\Delta t} = -\left[(\Phi(t)-\Phi^*)\eta(t)\right]_I - \Phi^*\frac{[\eta(t+\Delta t)]_G + [\eta(t-\Delta t)]_O}{2}. \quad (6.91)$$

The resulting system for the values at grid point G and at time $t + \Delta t$ is written:

$$\left.\begin{aligned}
U(t+\Delta t) &= U_T - \Delta t\,\frac{\partial}{\partial\lambda}\Phi(t+\Delta t),\\[4pt]
V(t+\Delta t) &= V_T - \Delta t(1-\mu^2)\frac{\partial}{\partial\mu}\Phi(t+\Delta t),\\[4pt]
\Phi(t+\Delta t) &= \Phi_T - \Delta t\Phi^*\eta(t+\Delta t).
\end{aligned}\right\} \quad (6.92)$$

The intermediate values U_T, V_T, and Φ_T are obtained from equations (6.90) and (6.91) on the Gaussian grid and concern only quantities known at points I and O at times t and $t - \Delta t$.

The first two equations of (6.92) allow us to obtain an equation for divergence and the system becomes:

$$\eta(t+\Delta t) = \frac{1}{a^2(1-\mu^2)}\left[\frac{\partial U_T}{\partial \lambda}+(1-\mu^2)\frac{\partial V_T}{\partial \mu}\right]-\Delta t \nabla^2 \Phi(t+\Delta t), \left.\begin{array}{c}\\\\\end{array}\right\}$$

$$\Phi(t+\Delta t) = \Phi_T - \Delta t \Phi^* \eta(t+\Delta t). \qquad\qquad\qquad\qquad\qquad\qquad \right\} \qquad (6.93)$$

By writing:

$$D_T = \frac{1}{a^2(1-\mu^2)}\left[\frac{\partial U_T}{\partial \lambda}+(1-\mu^2)\frac{\partial V_T}{\partial \mu}\right],$$

after applying the Fourier and Legendre transforms, we obtain in the spectral space the system:

$$\eta_n^m(t+\Delta t) = (D_T)_n^m + \Delta t \frac{n(n+1)}{a^2}\Phi_n^m(t+\Delta t), \left.\begin{array}{c}\\\\\end{array}\right\}$$

$$\Phi_n^m(t+\Delta t) = (\Phi_T)_n^m - \Delta t \Phi^* \eta_n^m(t+\Delta t), \qquad\qquad\qquad \right\} \qquad (6.94)$$

which finally leads to the Helmholtz equation:

$$\Phi_n^m(t+\Delta t) = (\Phi_T)_n^m - \Delta t \Phi^*(D_T)_n^m - \Delta t^2 \Phi^* \frac{n(n+1)}{a^2}\Phi_n^m(t+\Delta t). \qquad (6.95)$$

This equation is quite like that obtained in (6.85) and so is solved in the same way.

Once the values of $\Phi_n^m(t+\Delta t)$ have been determined, we also calculate the spectral coefficients for the quantity $(1-\mu^2)\partial\Phi/\partial\mu$, which is obtained by application of the relation (4.38) seen in Chapter 4. The inverse Legendre and Fourier transforms allow us to find the values of $\Phi(t+\Delta t)$ and of $(1-\mu^2)\partial\Phi/\partial\mu$ on the Gaussian grid. We take the opportunity of using the Fourier coefficients to calculate the coefficients of $\partial\Phi/\partial\lambda$ so this quantity is on the grid too.

The several variants of the semi-implicit semi-Lagrangian scheme addressed in Paragraph 6.2.5.2 can also be adapted to processing on the sphere. It is important to observe that the Coriolis term can be integrated into the Lagrangian processing (allowing us to reach a two time level scheme without extrapolating it) by returning to the original vector formulation of the equation for horizontal momentum, which is written:

$$\frac{d}{dt}\left[\mathbf{V}_H + \Omega \times \mathbf{r}\right] = -\nabla_H \Phi,$$

Ω being the vector for rotation of the Earth and $\mathbf{r}$ the vector for the radius from the centre of the Earth. Moreover, this way of working, proposed by Rochas (1990) and detailed by Temperton (1997), is generally essential to reach a two time level integration scheme while conveniently performing the Lagrangian processing. Care must be taken to keep the relevant trajectory systematically orthogonal to the vertical: this leads to more complicated processing when working on the sphere but greatly simplifies the equations when working on a plane conformal projection.

6.4 Practical use of the shallow water model

The zero divergence barotropic model was used initially to represent the evolution of the geopotential at 500 hPa, which was supposed to be representative of a *mean atmosphere*.

It may seem natural too to apply the shallow water model (primitive equation) barotropic model to a fluid layer some 5500 m thick and whose free surface corresponds to the geopotential at 500 hPa. Such a choice implies then that the mean geopotential Φ^* has a value in the vicinity of 55 000 J/kg, which amounts to setting mean divergence at a much lower value than is actually observed at this level. Consequently, with forecasts made in this way, long-wave retrogression may be shown up, in much the same way as was observed by using the zero divergence model, but to a lesser extent.

Long-wave stabilization consists here in setting an average divergence value close to the observed value, by choosing an average value of the geopotential Φ^* of the order of 15 000 J/kg. This modification is simple to implement, as we need only to integrate the shallow water model using a mean geopotential corresponding to an effective depth of 1.5 km, as follows:

$$\Phi = \Phi_{500\,\text{hPa}} - \Phi_0 \quad \text{with} \ \ \Phi_0 = 40\,000 \ \text{J/kg}.$$

This modification then allows us to use the shallow water model to simulate the evolution of the 500 hPa surface in a relatively realistic way.

It is obvious that the simplified representations of the atmosphere covered in this chapter cannot claim to faithfully represent the evolution of the real three-dimensional atmosphere. To obtain such realistic representations, we need to take into account the vertical dimension of the atmosphere to construct *baroclinic* models for which the atmosphere must be treated as a set of stacked layers. This is the subject of Chapters 7 and 8.

Baroclinic model equations

7.1 Introduction

The shallow water model made it possible to examine the way to use various algorithms and to specify their respective properties (horizontal discretization, time integration). Making a more realistic model to describe the evolution of a baroclinic atmosphere requires allowance for the vertical dimension of the atmosphere and the use of suitable discretization. We commonly speak then of *baroclinic models*. The earliest baroclinic models used for operational forecasting were built on systems of *filtered equations* characterized by introducing a balance equation between the mass field and the wind field, thereby excluding the possibility of inertia-gravity wave propagation, but allowing the use of a comparatively large time step. However, the increased power of computers in the 1970s was soon to allow the primitive equations to be used to construct models for operational forecasting.

The primitive equations were used with various types of vertical coordinate. Once the drawbacks of the pressure coordinate had been identified, the normalized pressure coordinate (commonly called sigma and noted σ) was proposed for formulating the condition at the lower boundary of the atmosphere in a simple way. Introducing this coordinate, however, was not without its drawbacks for the highest layers, which were more apt to be processed by the pressure coordinate. A new *hybrid* coordinate (a combination of σ and p) has provided a fairly satisfactory answer to the problem and has now found its place in many models using either the primitive (hydrostatic) equations or the (nonhydrostatic) Euler equations. These different coordinates are presented in this chapter and the integral invariants of the system are studied in the context of a general formulation encompassing both the primitive equations and the Euler equations.

7.2 Introducing a general vertical coordinate

7.2.1 The transformation formulas

We take a general vertical coordinate s, which is assumed to be a monotonic function of altitude z, and therefore suitable for uniquely identifying a point on the vertical. With this general vertical coordinate, altitude becomes a dependent variable, $z = z(x, y, s, t)$, and the new form of equations is obtained by using the derivation formulas in the new coordinate system.

The spatial derivatives at constant s are given by:

$$\left(\frac{\partial}{\partial l}\right)_{s \text{ cst}} = \left(\frac{\partial}{\partial l}\right)_{z \text{ cst}} + \left(\frac{\partial z}{\partial l}\right)_{s \text{ cst}} \frac{\partial}{\partial z}, \tag{7.1}$$

and the derivatives in the vertical by:

$$\frac{\partial}{\partial s} = \frac{\partial z}{\partial s} \frac{\partial}{\partial z}, \tag{7.2}$$

the notation $\partial z / \partial s$ always being understood as for constant x, y, and t.

7.2.2 The total derivative expression

The expression of the total derivative as a function of the partial derivatives is now written:

$$\frac{d}{dt} = \frac{\partial}{\partial t} + u \left(\frac{\partial}{\partial x}\right)_{s \text{ cst}} + v \left(\frac{\partial}{\partial y}\right)_{s \text{ cst}} + \dot{s} \frac{\partial}{\partial s}, \tag{7.3}$$

where the variable $\dot{s}$ denotes the generalized vertical velocity ds/dt. The variables u and v are functions of x, y, s, t and the horizontal derivatives are then calculated for a constant s.

7.3 Application to the primitive equations

The transformation formulas (7.1) and (7.2) together with the expression for the total derivative given by (7.3) are now used to get the expression of the primitive equations (equations (2.1) to (2.5) of Chapter 2) by using the vertical coordinate s.

7.3.1 The hydrostatic equation

Given relation (7.2) and the equation of state, the hydrostatic equation is now written:

$$\frac{\partial \Phi}{\partial s} = -\frac{RT}{p} \left(\frac{\partial p}{\partial s}\right). \tag{7.4}$$

7.3.2 The pressure force term

Formula (7.1) allows us to write the pressure force in equation (2.1) as follows:

$$-\frac{1}{\rho}\nabla_z p = -\frac{1}{\rho}\left(\nabla_s p - \frac{\partial p}{\partial z}\nabla_s z\right).$$

By using the equation of state and the hydrostatic relation, this becomes:

$$-\frac{1}{\rho}\nabla_z p = -\left(\frac{RT}{p}\nabla_s p + \nabla_s \Phi\right), \tag{7.5}$$

with $\Phi = gz$ denoting the geopotential.

7.3.3 The continuity equation

The continuity equation (2.3) can still be written:

$$\frac{d}{dt}(\ln \rho) + \nabla_z \cdot \mathbf{V} + \frac{\partial w}{\partial z} = 0. \tag{7.6}$$

By applying the transformation formulas (7.1) and (7.2), the divergence terms of equation (7.6) are written:

$$\nabla_z \cdot \mathbf{V} + \frac{\partial w}{\partial z} = \nabla_s \cdot \mathbf{V} - \frac{\partial s}{\partial z}\frac{\partial \mathbf{V}}{\partial s}\nabla_s z + \frac{\partial w}{\partial s}\frac{\partial s}{\partial z}. \tag{7.7}$$

Since $w = dz/dt$, the derivation formula (7.3) can be used to expand $\partial w/\partial s$ and yields:

$$\frac{\partial w}{\partial s} = \frac{\partial}{\partial t}\left(\frac{\partial z}{\partial s}\right) + \mathbf{V}\cdot\nabla_s \frac{\partial z}{\partial s} + \frac{\partial \mathbf{V}}{\partial s}\cdot\nabla_s z + \frac{\partial \dot{s}}{\partial s}\frac{\partial z}{\partial s} + \dot{s}\frac{\partial}{\partial s}\left(\frac{\partial z}{\partial s}\right). \tag{7.8}$$

By substituting this expression in (7.7) we get:

$$\nabla_z \cdot \mathbf{V} + \frac{\partial w}{\partial z} = \nabla_s \cdot \mathbf{V} + \frac{\partial s}{\partial z}\left(\frac{\partial}{\partial t} + \mathbf{V}\cdot\nabla_s + \dot{s}\frac{\partial}{\partial s}\right)\frac{\partial z}{\partial s} + \frac{\partial \dot{s}}{\partial s}. \tag{7.9}$$

Since $\partial s/\partial z = (\partial z/\partial s)^{-1}$, the second term of the right-hand side of (7.9) can be simply written $d\,[\ln(\partial z/\partial s)]/dt$, such that the continuity equation (7.6) then takes the form.

$$\frac{d}{dt}\left[\ln\left(\rho\frac{\partial z}{\partial s}\right)\right] + \nabla_s \cdot \mathbf{V} + \frac{\partial \dot{s}}{\partial s} = 0, \tag{7.10}$$

which, in view of the hydrostatic relation, can again be written:

$$\frac{d}{dt}\left[\ln\left(\frac{\partial p}{\partial s}\right)\right] + \nabla_s \cdot \mathbf{V} + \frac{\partial \dot{s}}{\partial s} = 0. \tag{7.11}$$

By expanding the total derivative appearing in (7.11), we get:

$$\left(\frac{\partial p}{\partial s}\right)^{-1}\left[\frac{\partial}{\partial t}\left(\frac{\partial p}{\partial s}\right)+\mathbf{V}\cdot\nabla_s\,\frac{\partial p}{\partial s}\;+\;\dot{s}\frac{\partial}{\partial s}\left(\frac{\partial p}{\partial s}\right)\right]+\nabla_s\cdot\mathbf{V}+\frac{\partial\dot{s}}{\partial s}=0,$$

which, after rearrangement, finally gives for the continuity equation:

$$\frac{\partial}{\partial t}\left(\frac{\partial p}{\partial s}\right)=-\nabla_s\cdot\left(\frac{\partial p}{\partial s}\mathbf{V}\right)-\frac{\partial}{\partial s}\left(\dot{s}\frac{\partial p}{\partial s}\right). \qquad (7.12)$$

The transition from the altitude vertical coordinate z to the general vertical coordinate s does not alter the prognostic type of the continuity equation (which brings out the time derivative of $\partial p/\partial s$) while the use of pressure as the vertical coordinate leads to a purely diagnostic relation for the continuity equation (as this time derivative identically becomes zero).

The choice of such a general vertical coordinate s allows us to define a domain of vertical integration corresponding to the physical problem: a rigid lower boundary of the domain ($s = s_b$), corresponding to ground level (where $\Phi = \Phi_s$, geopotential at the level of the model topography) and an elastic upper boundary of the domain ($s = s_t$), corresponding to a constant pressure value $p = p_t$ (where p_t may be zero). Moreover, the zero flux conditions (generalized vertical velocity $\dot{s} = 0$) at the bottom and top of the working domain make it possible to ensure the conservative properties of the equations.

7.3.4 The surface pressure tendency equation

For the sake of concise notation, from now on we abstain from noting the subscript s for all the horizontal derivatives calculated with constant s.

Integrating equation (7.12) in the vertical from top ($s = s_t$), to the bottom ($s = s_b$), of the working domain gives:

$$\frac{\partial p_s}{\partial t}-\frac{\partial p_t}{\partial t}=-\int_{s_t}^{s_b}\nabla\cdot\left(\frac{\partial p}{\partial s}\mathbf{V}\right)ds-\left[\left(\dot{s}\frac{\partial p}{\partial s}\right)_{s_b}-\left(\dot{s}\frac{\partial p}{\partial s}\right)_{s_t}\right], \qquad (7.13)$$

where $p_s = p(s_b)$ is the surface pressure and $p_t = p(s_t)$, the pressure at the top.

By choosing as the upper boundary of the working domain a constant surface pressure p_t and by adopting zero flux conditions at the bottom and top of the atmosphere so considered ($\dot{s} = 0$, for $s = s_b$ and $s = s_t$), we get:

$$\frac{\partial p_s}{\partial t}=-\int_{s_t}^{s_b}\nabla\cdot\left(\frac{\partial p}{\partial s}\mathbf{V}\right)ds. \qquad (7.14)$$

This equation shows that the surface pressure tendency is equal, except for the sign, to the mean value of divergence of the vector $(\partial p/\partial s)\mathbf{V}$ (in other words, to the mean value in the vertical of the horizontal momentum divergence). It shows the importance of precisely evaluating this quantity when integrating the primitive equation model, since it yields the value of the surface pressure tendency.

7.3.5 The vertical velocity equation

Integrating equation (7.12) in the vertical from the top ($s = s_t$, assuming it is a constant pressure level) to a level s corresponding to pressure p gives:

$$\left(\dot{s} \frac{\partial p}{\partial s} \right) = - \int_{s_t}^{s} \nabla \cdot \left(\frac{\partial p}{\partial s'} \mathbf{V} \right) ds' - \frac{\partial p}{\partial t}. \tag{7.15}$$

In this equation, the derivative $\partial p / \partial t$ is taken for constant x, y, and s.

7.4 Various vertical coordinates

7.4.1 The drawbacks of the pressure coordinate

The use of the pressure coordinate, introduced by Eliassen (1949), leads to a particularly simple formulation of the continuity equation. It then becomes possible to construct a baroclinic model by considering the atmosphere to be a stack of layers separated by isobaric surfaces. A difficulty arises, though, when trying to properly prescribe the condition at the lower boundary of the atmosphere, since the isobaric surfaces intersect the orography. New vertical coordinates depending not only on pressure but also on surface pressure have been proposed to solve this problem.

7.4.2 The sigma coordinate

We now consider a vertical coordinate, depending only on pressure p and surface pressure p_s, which varies monotonically as a function of p (and therefore also as a function of z, because of the hydrostatic assumption).

The normalized pressure coordinate σ, introduced by Phillips (1957), is defined as:

$$\sigma = \frac{p - p_t}{p_s - p_t}, \tag{7.16}$$

where p is the pressure, p_s the pressure at the surface, which is the lower boundary of the domain, and p_t the pressure at the upper boundary of the domain. This coordinate is termed the *normalized pressure coordinate* since it varies from $\sigma = 0$, at the top ($p = p_t$), to $\sigma = 1$, at the bottom ($p = p_s$). The surface pressure p_s is calculated at the level of the topography $z_s(x, y)$ associated with the model, the characteristics of which depend on the horizontal resolution adopted.

The atmosphere between the upper level ($p = p_t$) and the surface (at the level of the model's topography) can be represented by a stack of layers bounded by iso-σ

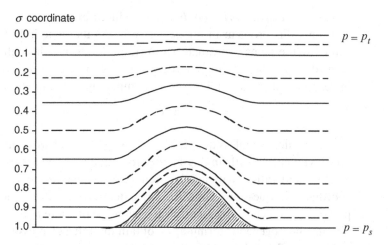

σ coordinate

Figure 7.1 Shape of layers defined by the iso-sigma surfaces: the solid lines indicate the layer boundaries, while the dashed lines indicate the representative levels of the layers.

surfaces, hugging the orography (Figure 7.1). The quantity $\dot\sigma = d\sigma/dt$ represents the generalized vertical velocity associated with this vertical coordinate.

In order to conserve the various integral invariants of the system (total mass, total energy, total angular momentum, etc.), we assume zero flux conditions at the top and bottom of the working domain by choosing:

$$\dot\sigma = 0 \ \text{ at the surface } \sigma = 0,$$
$$\dot\sigma = 0 \ \text{ at the surface } \sigma = 1.$$

Where the top of the domain corresponds to the zero pressure surface ($p_t = 0$), the σ coordinate is quite simply written:

$$\sigma = \frac{p}{p_s}, \text{ which has as its consequence: } \frac{\partial p}{\partial s} \equiv \frac{\partial p}{\partial \sigma} = p_s.$$

7.4.3 The progressive hybrid coordinate

The sigma coordinate is perfectly suitable for formulating the lower boundary condition of the atmosphere. However, the horizontal variations in altitude of the iso-σ surfaces tend to reflect those of the topography, even towards the upper boundary of the atmosphere. In this case, determining the initial values of the variables of the model at the upper levels may require hazardous extrapolations above the high topography, which entails a risk of imbalance between the two terms of the geopotential gradient (given by the right-hand side of relation 7.5), which have high absolute values but opposing signs.

The *progressive hybrid coordinate* introduced by Simmons and Burridge (1981) circumvents this problem in an elegant way. As its name indicates, this coordinate is similar to the sigma coordinate for the lower levels but, through a seamless transition, coincides with the pressure coordinate for the highest levels. This new vertical coordinate, noted simply s, can be expressed implicitly by the relation:

$$p(s) = f(s)p_0 + g(s)p_s,\tag{7.17}$$

where the dimensionless quantities $f(s)$ and $g(s)$, which are functions of the variable s (always noted explicitly in this chapter so as not to confuse them with the Coriolis parameter f and with gravity acceleration g), determine the characteristics of the hybrid coordinate in question. Thus, for any value s, after setting the value of the surface pressure p_s and choosing an arbitrary reference pressure p_0 (e.g. 1013.25 hPa) so as to obtain a homogeneous formula, we obtain the value of pressure p.

Relation (7.17) shows that the pure pressure coordinate corresponds to $f(s) = s/p_0$ and $g(s) = 0$, whereas the pure sigma coordinate corresponds to $f(s) = 0$ and $g(s) = s$. In addition, so as to obtain the normalized vertical coordinate, we take as boundary values:

$$\text{at the top, for } p = p_t, \quad s(p_t, p_s) = 0,$$
$$\text{at the bottom, for } p = p_s, \quad s(p_s, p_s) = 1.$$

Formula (7.17) therefore yields the relations:

$$p_t = f(0)p_0 + g(0)p_s,$$
$$p_s = f(1)p_0 + g(1)p_s,$$

which set the values of $f(s)$ and $g(s)$ for $s = 0$ and $s = 1$:

$$f(0) = \frac{p_t}{p_0} \quad \text{and} \quad g(0) = 0,$$
$$f(1) = 0 \quad \text{and} \quad g(1) = 1.$$

The pressure p_t at the top of the atmosphere can also take the value $p_t = 0$.

These elements allows us to outline the shape the functions $f(s)$ and $g(s)$ must be to obtain the desired result (Figure 7.2a), that is, to have the s coordinate coincide with the p coordinate at the upper levels and with the σ coordinate at the lower levels. However, it is necessary to ensure that the choice made does indeed give, for p, an increasing function of s for any possible value of surface pressure, p_s.

To define the vertical discretization of the model, a practical way to determine the values of $f(s)$ and $g(s)$ is to take *a priori* the shape of the profiles $p_1(s)$ and $p_2(s)$ we wish to obtain for two extreme values of surface pressure $p_{s\min}$ and $p_{s\max}$ (Figure 7.2b):

$$p_1(s_k) = f(s_k)p_0 + g(s_k)p_{s\min},$$
$$p_2(s_k) = f(s_k)p_0 + g(s_k)p_{s\max}.$$

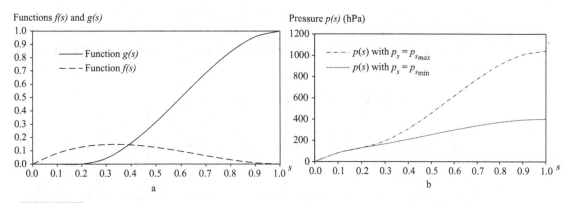

Figure 7.2 (a) Shape of functions $f(s)$ and $g(s)$, here with $g(s) = 0$ for $0 < s < 0.2$. (b) profiles $p(s)$ for $p_s = p_{s\,min}$ and for $p_s = p_{s\,max}$.

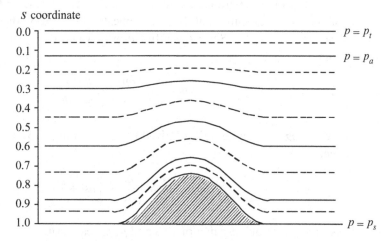

Figure 7.3 Shape of layers defined by the iso-s surface; plotted using the same conventions as for Figure 7.1.

These two relations, computed for each of the values s_k corresponding to the surfaces chosen to delimit the layers of the model, yield the numerical values of $f(s_k)$ and $g(s_k)$. Analytical relations can be obtained for $f(s)$ and $g(s)$ by making an adjustment with polynomial functions and ensuring that they still lead to correct profiles for $p_1(s)$ and $p_2(s)$.

The new plot of the distribution of layers obtained with this progressive hybrid coordinate in Figure 7.3 shows the specific feature of the upper layers that are bounded by isobaric surfaces. The difference with the plot in Figure 7.1 appears comparatively slight *a priori* because the vertical scale is linear in pressure; a linear scale along the altitude z would show that the amplitude of deformation due to topography always remains large by using the σ coordinate, whereas it declines progressively with the progressive hybrid coordinate s, whenever the function $g(s)$ tends towards zero when s tends towards zero.

By virtue of relation (7.5) applied both to the progressive hybrid coordinate and to the pressure coordinate, we can write:

$$\nabla_s \Phi = \nabla_p \Phi - \frac{RT}{p} g(s) \nabla p_s, \text{ with } g(s) = \frac{\partial p}{\partial p_s}.$$

As the atmosphere is almost barotropic, the isobaric geopotential gradient is very weak; by using the σ coordinate, we have $g(\sigma) = \sigma$, which implies a steep iso-σ gradient, since $\nabla_s \Phi \approx -RT \nabla \ln p_s \approx \nabla \Phi_s$: the iso-$\sigma$ surface heights follow the topography at all altitudes. By using a hybrid s coordinate such that $g(s)$ becomes zero for high levels, then $\nabla_s \Phi = \nabla_p \Phi \approx 0$, which eliminates the drawback encountered previously.

Compared with the equations written in the pressure coordinates, the formulation of the equations using the general coordinate s is more complicated: the expression of the isobaric pressure gradient is now the sum of two terms and the thermodynamic equation becomes somewhat more complicated because of the expression for vertical velocity $\omega = dp/dt$. Since p is a function of s and of p_s, the vertical velocity ω is written:

$$\omega = \frac{dp}{dt} = \frac{\partial p}{\partial s} \dot{s} + \frac{\partial p}{\partial p_s} \frac{dp_s}{dt}. \tag{7.18}$$

By replacing $\dot{s} \frac{\partial p}{\partial s}$ by its value derived from equation (7.15), and by expanding $\frac{dp_s}{dt}$, we get:

$$\omega = -\int_{s_t}^{s} \nabla \cdot \left(\frac{\partial p}{\partial s'} \mathbf{V} \right) ds' + \mathbf{V} \cdot \nabla p. \tag{7.19}$$

Pressure p being only a function of s and p_s, we can write:

$$\frac{\partial p}{\partial t} = \frac{\partial p}{\partial p_s} \frac{\partial p_s}{\partial t} = g(s) \frac{\partial p_s}{\partial t}. \tag{7.20}$$

Given the expression of the surface pressure tendency (7.14), this relation can be used to express the equation giving the vertical velocity (7.15) in a diagnostic form (explained a little further on in this section).

By using this vertical coordinate s, we can therefore write the primitive equations for a rotating atmosphere without exchange with its surroundings (adiabatic and frictionless) as follows:

• the equation of momentum:

$$\frac{\partial \mathbf{V}}{\partial t} = -\mathbf{k} \times (\zeta + f) \mathbf{V} - \dot{s} \frac{\partial \mathbf{V}}{\partial s} - \nabla K - (\nabla \Phi + \frac{RT}{p} \nabla p), \tag{7.21}$$

with $\zeta = \mathbf{k} \cdot (\nabla \times \mathbf{V})$ and $K = \|\mathbf{V}\|^2/2$;

- the thermodynamic equation:

$$\frac{\partial T}{\partial t} = -\mathbf{V}\cdot\nabla T - \dot{s}\frac{\partial T}{\partial s} - \frac{RT}{C_p}\left[\frac{1}{p}\int_{s_t}^{s} \nabla\cdot\left(\frac{\partial p}{\partial s'}\mathbf{V}\right)ds' - \frac{1}{p}(\mathbf{V}\cdot\nabla p)\right]; \qquad (7.22)$$

- the surface pressure tendency equation:

$$\frac{\partial p_s}{\partial t} = -\int_{s_t}^{s_b} \nabla\cdot\left(\frac{\partial p}{\partial s}\mathbf{V}\right)ds. \qquad (7.23)$$

Relations (7.21) to (7.23) form a set of *prognostic* equations for the model's state variables: the horizontal wind $\mathbf{V}$ (two components), thermodynamic temperature T, and surface pressure p_s. These equations give the expression of their time derivatives (tendencies). The first two variables $\mathbf{V}$ and T depend on the two horizontal coordinates and one vertical coordinate, while the variable p_s depends on the two horizontal coordinates only.

The right-hand sides of equations (7.21), (7.22), and (7.23) involve the geopotential Φ and the generalized vertical velocity $\dot{s}$. Two *diagnostic* equations are therefore added to the set of *prognostic* equations, allowing the latter quantities to be expressed as a function of the state variables:

- the hydrostatic equation (in integrated form):

$$\Phi = \Phi_s - R\int_{s_b}^{s} \frac{T}{p}\frac{\partial p}{\partial s'}ds', \Phi_s \text{ designating the surface geopotential;} \qquad (7.24)$$

- the vertical velocity equation:

$$\left(\dot{s}\frac{\partial p}{\partial s}\right) = -\int_{s_t}^{s} \nabla\cdot\left(\frac{\partial p}{\partial s'}\mathbf{V}\right)ds' + \frac{\partial p}{\partial p_s}\int_{s_t}^{s_b} \nabla\cdot\left(\frac{\partial p}{\partial s}\mathbf{V}\right)ds. \qquad (7.25)$$

The set of prognostic and diagnostic equations forms a complete system for the model's state variables.

The evolution of the specific humidity due to the model dynamics (in the case of an atmosphere without exchange with its surroundings and without phase change of water vapour) comes down to a straightforward advection equation:

$$\frac{\partial q}{\partial t} = -\mathbf{V}\cdot\nabla q - \dot{s}\frac{\partial q}{\partial s}, \qquad (7.26)$$

which merely reflects the conservation of specific humidity q in the course of motion. More generally, any evolution equation for another conservative quantity (or assumed to be such by first approximation) is of the same type as (7.26).

7.5 Generalization to nonhydrostatic equations

7.5.1 The role of 'hydrostatic pressure'

Computing power has made it possible to increase the spatial resolution of models steadily and to handle ever smaller scale phenomena. With meshes of the order of a few kilometres, the hydrostatic hypothesis ceases to hold up and the Euler equations must be used. In recent years, the development of models has accelerated thanks to advances made by algorithms allowing us to increase the time step without affecting stability. Tanguay, Robert, and Laprise (1990) showed, using the vertical coordinate z (altitude), that semi-implicit and semi-Lagrangian techniques allowed a nonhydrostatic model to be integrated effectively with the Euler equations. Examination of the continuity equation in its most general form, done by Laprise (1992), showed that the use of a mass type coordinate in the vertical made it possible to obtain a diagnostic form of the continuity equation (analogous to that obtained in pressure coordinates for the primitive equations).

The continuity equation for any vertical coordinate s follows directly from equation (7.10) and is obtained by expansion similar to that at the beginning of Subsection 7.3.3; it is written:

$$\frac{\partial}{\partial t}\left(\rho\frac{\partial z}{\partial s}\right) = -\nabla_s \cdot \left(\rho\frac{\partial z}{\partial s}\mathbf{V}\right) - \frac{\partial}{\partial s}\left(\rho\frac{\partial z}{\partial s}\dot{s}\right). \tag{7.27}$$

By choosing a coordinate s such that the quantity $\rho\partial z/\partial s$ is a constant, the continuity equation then takes on a diagnostic form. The coordinate π that satisfies the hydrostatic relation:

$$\frac{\partial \pi}{\partial z} = -\rho g = -\frac{p}{RT}g, \tag{7.28}$$

fully meets this condition, the value of the constant then being $-1/g$. If the upper boundary of the working domain is pushed back to infinity where $p_t = 0$, we can choose $\pi_t = 0$ and π then becomes the *hydrostatic pressure* defined by:

$$\pi(z) = \int_\infty^z \rho g\,dz - \int_\infty^z \frac{p}{RT}g\,dz. \tag{7.29}$$

There is no reason this hydrostatic pressure π should be equal to the *true pressure*, p, which is a prognostic variable (and justifies introducing the notation π). As the condition $\pi_t = 0$ at infinity is commonly used, we henceforth term *hydrostatic pressure* (for convenience, but improperly) the coordinate π defined by relation (7.28), independently of the upper boundary conditions chosen when integrating that relation.

A *normalized hydrostatic pressure* coordinate may also be defined, together with its extension to a progressive hybrid coordinate. We thus obtain a formulation that is very close to that of the primitive equations studied earlier. This similarity is not fortuitous: the coordinate π for the Euler equations, and the coordinate p for the primitive equations, are both *mass type* coordinates (allowing us to define a certain mass of atmosphere).

7.5.2 The normalized 'hydrostatic pressure' hybrid coordinate

The Euler equations (2.7), introduced in Subsection 2.2.2, use altitude z as the vertical coordinate. They may be re-written by using as the vertical coordinate the new progressive hybrid coordinate proposed by Laprise (1992).

$$\pi(s) = f(s)\pi_0 + g(s)\pi_s,$$

π_0 being a reference hydrostatic pressure (e.g. 1013.25 hPa) and π_s the hydrostatic pressure at the surface.

The left-hand sides of equations (2.7) are not formally modified, but in their expression as a function of time and space, partial derivatives must take account of the new system of coordinates.

The horizontal derivatives at constant z can be expressed as a function of the horizontal derivatives at constant s by using formula (7.1). We then obtain for the pressure gradient:

$$\nabla_z p = \nabla_s p - \frac{\partial p}{\partial z}\nabla_s z. \tag{7.30}$$

Given that:

$$\frac{\partial p}{\partial z} = \frac{\partial p}{\partial s}\frac{\partial s}{\partial \pi}\frac{\partial \pi}{\partial z}, \tag{7.31}$$

the pressure force in the equation for horizontal momentum is written:

$$\frac{RT}{p}\nabla_z p = \frac{RT}{p}\nabla_s p + \frac{\partial p}{\partial s}\left(\frac{\partial \pi}{\partial s}\right)^{-1}\nabla_s \Phi. \tag{7.32}$$

Given the relation (7.28) defining hydrostatic pressure π, the right-hand side of the evolution equation of vertical velocity is transformed as below:

$$-\frac{RT}{p}\frac{\partial p}{\partial z} - g = -g\left[1 - \frac{\partial p}{\partial s}\left(\frac{\partial \pi}{\partial s}\right)^{-1}\right]. \tag{7.33}$$

The right-hand side of the thermodynamic equation for temperature T is not formally modified, but the total derivative of pressure must be expressed in the new system of coordinates.

The hydrostatic equation takes the form:

$$\frac{\partial \Phi}{\partial s} = \frac{\partial \Phi}{\partial \pi}\frac{\partial \pi}{\partial s} = -\frac{RT}{p}\frac{\partial \pi}{\partial s}. \qquad (7.34)$$

The horizontal and vertical derivatives, which are used for expressing divergence in the evolution equation of true pressure p, must be modified by applying the transformation formulas used for calculating the gradient in relation (7.30).

We get:

$$\nabla_z \cdot \mathbf{V} = \nabla_s \cdot \mathbf{V} - \left(\frac{\partial \mathbf{V}}{\partial z}\right)\cdot\nabla_s z. \qquad (7.35)$$

Given equation (7.34), the previous relation may also be written, by introducing the geopotential $\Phi = gz$, in the form:

$$\nabla_z \cdot \mathbf{V} = \nabla_s \cdot \mathbf{V} + \frac{p}{RT}\frac{\partial s}{\partial \pi}\left(\frac{\partial \mathbf{V}}{\partial s}\right)\cdot\nabla_s \Phi. \qquad (7.36)$$

The derivative with respect to z of the vertical velocity $w = dz/dt$ is written:

$$\frac{\partial w}{\partial z} = \frac{\partial w}{\partial s}\frac{\partial s}{\partial \pi}\frac{\partial \pi}{\partial z} = -g\frac{p}{RT}\frac{\partial s}{\partial \pi}\frac{\partial w}{\partial s}, \qquad (7.37)$$

this expression being used to obtain the new form of the evolution equation of true pressure.

The evolution equation of true pressure thus obtained may be replaced by the equation resulting from combining the evolution equation of true pressure with the thermodynamic equation:

$$\frac{d\Phi}{dt} = gw. \qquad (7.38)$$

This equation (7.38) corresponds exactly to the definition of vertical velocity $w = dz/dt$.

In view of the definition of hydrostatic pressure π:

$$\rho\frac{\partial z}{\partial s} = \rho\frac{\partial z}{\partial \pi}\frac{\partial \pi}{\partial s} = -\frac{1}{g}\frac{\partial \pi}{\partial s},$$

the continuity equation (7.27) becomes:

$$\frac{\partial}{\partial t}\left(\frac{\partial \pi}{\partial s}\right) = -\nabla_s\cdot\left(\frac{\partial \pi}{\partial s}\mathbf{V}\right) - \frac{\partial}{\partial s}\left(\frac{\partial \pi}{\partial s}\dot{s}\right). \qquad (7.39)$$

By taking as the upper boundary of the working domain a surface where hydrostatic pressure is constant ($\pi = \pi_t$), integrating the continuity equation (7.39) in the vertical, from $s = s_t$ to $s = s_b$, at which boundaries the generalized vertical velocity $\dot{s}$ is assumed to be zero, provides an evolution equation for surface hydrostatic pressure $\pi_s = \pi(s_b)$:

$$\frac{\partial \pi_s}{\partial t} = -\int_{s_t}^{s_b} \nabla_s \cdot \left(\frac{\partial \pi}{\partial s} \mathbf{V} \right) ds. \tag{7.40}$$

This equation for the tendency of 'hydrostatic surface pressure' π_s corresponds to the equation for the surface pressure tendency p_s (7.23), for the primitive equations.

The generalized vertical velocity is obtained by integrating the continuity equation (7.39) of $s = s_t$ up to the level s in question, giving:

$$\left(\dot{s} \frac{\partial \pi}{\partial s} \right) = -\int_{s_t}^{s} \nabla_s \cdot \left(\frac{\partial \pi}{\partial s'} \mathbf{V} \right) ds' + \frac{\partial \pi}{\partial \pi_s} \int_{s_t}^{s_b} \nabla_s \cdot \left(\frac{\partial \pi}{\partial s} \mathbf{V} \right) ds, \tag{7.41}$$

corresponding to the equation (7.25) for the primitive equations.

Lastly, the continuity equation (7.39) can again be written:

$$\dot{\pi} = \frac{d\pi}{dt} = \mathbf{V}.\nabla \pi - \int_{s_t}^{s} \nabla \cdot \left(\frac{\partial \pi}{\partial s'} \mathbf{V} \right) ds'. \tag{7.42}$$

With the Euler equations, the total derivative of π is given by a diagnostic relation, whereas that of p is given by a prognostic equation.

7.5.3 A comprehensive synthetic formulation of the equations

The (nonhydrostatic) Euler equations can be written in such a way that the (hydrostatic) primitive equations can be deduced from them simply. This formulation, which introduces the binary indicator γ (with γ equal to 1 or 0 to switch from nonhydrostatic to hydrostatic formulation), is presented here; to alleviate the notation, the subscript s noting the derivation with constant s has been omitted.

• Evolution equation for horizontal wind:

$$\frac{\partial \mathbf{V}}{\partial t} = -\mathbf{k} \times (\zeta + f)\mathbf{V} - \dot{s} \frac{\partial \pi}{\partial s} \frac{\partial \mathbf{V}}{\partial \pi} - \nabla K - \left[\frac{\partial p}{\partial s} \left(\frac{\partial \pi}{\partial s} \right)^{-1} \nabla \Phi + RT \nabla \ln p \right], \tag{7.43}$$

with $\zeta = \mathbf{k} \cdot (\nabla \times \mathbf{V})$ and $K = \|\mathbf{V}\|^2/2$.

• Evolution equation for vertical velocity w:

$$\gamma \frac{\partial w}{\partial t} = -g \left[1 - \frac{\partial p}{\partial s} \left(\frac{\partial \pi}{\partial s} \right)^{-1} \right]. \tag{7.44}$$

• Evolution equation for temperature:

$$\frac{\partial T}{\partial t} = -\mathbf{V}.\nabla T - \dot{s} \frac{\partial \pi}{\partial s} \frac{\partial T}{\partial \pi} + \frac{RT}{C_p} \frac{d \ln p}{dt}. \tag{7.45}$$

• Evolution equation for true pressure:

$$\frac{\partial \ln p}{\partial t} = -\mathbf{V}\cdot\nabla \ln p - \dot{s}\frac{\partial \pi}{\partial s}\frac{\partial \ln p}{\partial \pi} - \frac{C_p}{C_v}\left[\nabla\cdot\mathbf{V} + \frac{p}{RT}\left(\frac{\partial \pi}{\partial s}\right)^{-1}\left(\frac{\partial \mathbf{V}}{\partial s}\right)\cdot\nabla\Phi - g\frac{p}{RT}\left(\frac{\partial \pi}{\partial s}\right)^{-1}\frac{\partial w}{\partial s}\right]. \quad (7.46)$$

• Continuity equation:

$$\frac{\partial}{\partial t}\left(\frac{\partial \pi}{\partial s}\right) = -\nabla\cdot\left(\frac{\partial \pi}{\partial s}\mathbf{V}\right) - \frac{\partial}{\partial s}\left(\frac{\partial \pi}{\partial s}\dot{s}\right); \quad (7.47)$$

from this equation, we can deduce both the equation for hydrostatic surface pressure:

$$\frac{\partial \pi_s}{\partial t} = -\int_{s_t}^{s_b} \nabla\cdot\left(\frac{\partial \pi}{\partial s}\mathbf{V}\right) ds, \quad (7.48)$$

and the generalized vertical velocity equation:

$$\left(\dot{s}\frac{\partial \pi}{\partial s}\right) = -\int_{s_t}^{s} \nabla\cdot\left(\frac{\partial \pi}{\partial s'}\mathbf{V}\right) ds' + \frac{\partial \pi}{\partial \pi_s}\int_{s_t}^{s_b} \nabla\cdot\left(\frac{\partial \pi}{\partial s}\mathbf{V}\right) ds. \quad (7.49)$$

• Integrated hydrostatic equation, giving the geopotential:

$$\Phi = \Phi_s - \int_{s_b}^{s} \frac{RT}{p}\frac{\partial \pi}{\partial s'} ds', \quad (7.50)$$

Φ_s being the surface geopotential.

The system contains six prognostic equations for the state variables: the three components of the wind, u, v, and w, the thermodynamic temperature, T, and true pressure, p, as well as hydrostatic surface pressure π_s. Two further diagnostic equations are added to the system, giving the geopotential, Φ, and the generalized vertical velocity, $\dot{s}$.

This synthetic formulation can be used to find the (hydrostatic) primitive equations very easily, these being obtained by making $\gamma = 0$. The fact that $\gamma = 0$ implies $\partial\pi/\partial s = \partial p/\partial s$, which means that π is equal to p to within an additive constant, which can be assumed to be invariant along the horizontal without loss of generality. The variable π can therefore be replaced by the variable p, since it is only involved through its derivatives.

So, by replacing p by π, equation (7.43) for horizontal momentum gives us equation (7.21) again; the equation for vertical velocity w (7.44) quite simply disappears. The thermodynamic equation (7.45) gives equation (7.22) again. Equation (7.46) giving the evolution of the true pressure p (which had been obtained by combining the thermodynamic equation and the continuity equation) has become redundant. The continuity equation (7.47), which by vertical integration leads to the equation for evolution of surface hydrostatic pressure π_s (7.48) and the diagnostic equation for generalized vertical velocity $\dot{s}$ (7.49), can be used to obtain the evolution equation for surface pressure

p_s (7.23) and the diagnostic relation (7.25), respectively. Lastly, the diagnostic equation for geopotential (7.50) gives us the integrated hydrostatic relation (6.42) again. It is to be noticed that equation (7.42) yields an expression for vertical velocity ω corresponding to that given by (7.19).

The formulation of the hydrostatic system now includes just four prognostic equations for the following historical variables: the two components of the horizontal wind, u and v, thermodynamic temperature, T, and (surface) pressure, p_s. To this system are added the two diagnostic relations giving the geopotential Φ and generalized vertical velocity $\dot{s}$.

7.6 Conservation properties of the equations

7.6.1 The expression of global parameters

The equations used in the atmospheric models are based on laws of physics that are simply the reflection of the principles of conservation for certain quantities. Hence, it is logical to examine how the various terms of the equations combine to ensure the conservation of these quantities that are known as integral invariants. It will be important to use numerical schemes that enable us to satisfy the same properties of conservation. The various integral invariants are examined here for the Euler system of equations, using the vertical coordinate s relating to hydrostatic pressure π.

Integration is performed over the total bounded volume of the atmosphere V where the equations were defined. In the case of a *global model*, this is a spherical slab bounded by Earth's surface and the chosen upper boundary; in the case of a *limited area model*, the integral invariants are only conserved if the lateral boundary conditions ensure that the incoming and outgoing flows are equal.

With the chosen vertical coordinate, the element of mass $d\mathfrak{M}$, surface dS, and thickness dz is written (using the vertical coordinate π):

$$d\mathfrak{M} = \frac{1}{g}\left(\frac{\partial \pi}{\partial s}\right) ds \, dS;$$

the absence of a minus sign in front of this expression (contrary to what is implied by the definition of hydrostatic pressure π) is justified by the fact that the vertical integration with respect to s is from top to bottom (contrary to vertical integration with respect to z).

The expression for total mass $\mathfrak{M}$ is obtained by integrating over the three-dimensional working domain V:

$$\mathfrak{M} = \frac{1}{g}\iiint_V \left(\frac{\partial \pi}{\partial s}\right) ds \, dS. \tag{7.51}$$

The total angular momentum $\mathfrak{I}$, compatible with the thin layer approximation, as defined in Subsection 2.2.2, is written:

$$\mathfrak{I} = \frac{1}{g}\iiint_V a\cos\varphi\,(u + \Omega a\cos\varphi)\left(\frac{\partial\pi}{\partial s}\right)ds\,dS \tag{7.52}$$

The total kinetic energy $\mathfrak{K}$ for horizontal and vertical momentum is written:

$$\mathfrak{K} = \frac{1}{g}\iiint_V \left(K + \gamma\frac{w^2}{2}\right)\left(\frac{\partial\pi}{\partial s}\right)ds\,dS \tag{7.53}$$

The total internal energy $\mathfrak{Q}$ is written:

$$\mathfrak{Q} = \frac{1}{g}\iiint_V C_v T\left(\frac{\partial\pi}{\partial s}\right)ds\,dS. \tag{7.54}$$

The total energy, $\mathfrak{E}$, which is the sum of the kinetic energy, the internal energy, and the potential energy, is therefore given by the expression:

$$\mathfrak{E} = \frac{1}{g}\iiint_V \left(K + \gamma\frac{w^2}{2} + C_v T + \Phi\right)\left(\frac{\partial\pi}{\partial s}\right)ds\,dS. \tag{7.55}$$

7.6.2 Conservation of mass

The evolution equation for total mass $\mathfrak{M}$ is deduced immediately from the continuity equation:

$$\frac{\partial\mathfrak{M}}{\partial t} = -\frac{1}{g}\iiint_V \nabla\cdot\left(\frac{\partial\pi}{\partial s}\mathbf{V}\right)ds\,dS - \frac{1}{g}\iiint_V \frac{\partial}{\partial s}\left(\frac{\partial\pi}{\partial s}\dot{s}\right)ds\,dS. \tag{7.56}$$

The first term of the right-hand side of (7.56) cancels when integrating over the sphere or over a limited area with suitable boundary conditions (Green-Ostrogradsky theorem); the same goes for the second term when integrating in the vertical, given the boundary conditions ($\dot{s} = 0$) chosen at the top and bottom of the domain. The tendency of total mass $\mathfrak{M}$ is therefore zero, which implies that it is conserved.

7.6.3 Conservation of angular momentum

The evolution equation for angular momentum, obtained by taking the derivatives of both sides of equation (7.52), is written:

$$\frac{\partial\mathfrak{I}}{\partial t} = \frac{1}{g}\frac{\partial}{\partial t}\iiint_V \left[a\cos\varphi\,(u + \Omega a\cos\varphi)\frac{\partial\pi}{\partial s}\right]ds\,dS \tag{7.57}$$

This can be decomposed as:

$$\frac{\partial \mathfrak{J}}{\partial t} = \frac{1}{g} \iiint_V \left[a\cos\varphi \frac{\partial u}{\partial t} \frac{\partial \pi}{\partial s} \right] ds \, dS$$

$$+ \frac{1}{g} \iiint_V \left[a\cos\varphi \, (u + \Omega a\cos\varphi) \frac{\partial}{\partial t} \left(\frac{\partial \pi}{\partial s} \right) \right] ds \, dS. \tag{7.58}$$

By using the equation of momentum (7.43) and the continuity equation (7.47), we obtain after rearrangement and by using the definition of hydrostatic pressure π:

$$\frac{\partial \mathfrak{J}}{\partial t} = -\frac{1}{g} \iiint_V \left\{ \nabla \cdot \left[a\cos\varphi \, (u + \Omega a\cos\varphi) \left(\frac{\partial \pi}{\partial s} \mathbf{V} \right) \right] \right\} ds \, dS$$

$$- \frac{1}{g} \iiint_V \left\{ \frac{\partial}{\partial s} \left[a\cos\varphi \, (u + \Omega a\cos\varphi) \left(\frac{\partial \pi}{\partial s} \dot{s} \right) \right] \right\} ds \, dS \tag{7.59}$$

$$+ \frac{1}{g} \iiint_V \frac{\partial}{\partial s} \left[\Phi \left(\frac{\partial p}{\partial \lambda} \right) \right] ds \, dS.$$

The first term of the right-hand side of (7.59) cancels when integrated on the sphere or over a limited area with suitable boundary conditions; the same goes for the second term when integrated in the vertical, given the boundary conditions ($\dot{s} = 0$) chosen at the top and bottom of the domain. Lastly, the third term of the right-hand side of (7.59) is written, after vertical integration from s_t to s_b:

$$\frac{1}{g} \iiint_V \frac{\partial}{\partial s} \left[\Phi \left(\frac{\partial p}{\partial \lambda} \right) \right] ds \, dS = \frac{1}{g} \iint_S \left[\Phi_{s_t} \left(\frac{\partial p_{s_t}}{\partial \lambda} \right) \right] dS - \frac{1}{g} \iint_S \left[\Phi_s \left(\frac{\partial p_{s_b}}{\partial \lambda} \right) \right] dS. \tag{7.60}$$

The first of these two terms is zero if $p(s_t)$ is chosen to be constant. The second term is nonzero (except for the case of an idealized Earth with no topography) and represents the contribution of topography to the variations in the global angular momentum of the atmosphere. It is then the angular momentum of the Earth and the atmosphere together that is conserved.

The conservation properties of the angular momentum are strictly the same when we move from the Euler equations to the primitive equations.

7.6.4 The conservation of energy

The evolution equation for total energy breaks down as below:

$$\frac{\partial \mathfrak{E}}{\partial t} = \frac{1}{g} \iiint_V C_v \frac{\partial T}{\partial t} \left(\frac{\partial \pi}{\partial s} \right) ds \, dS + \frac{1}{g} \iiint_V \frac{\partial \Phi}{\partial t} \left(\frac{\partial \pi}{\partial s} \right) ds \, dS$$

$$+ \frac{1}{g} \iiint_V C_v T \frac{\partial}{\partial s} \left(\frac{\partial \pi}{\partial t} \right) ds \, dS + \frac{1}{g} \iiint_V \Phi \frac{\partial}{\partial s} \left(\frac{\partial \pi}{\partial t} \right) ds \, dS$$

$$+ \frac{1}{g} \iiint_V \mathbf{V} \cdot \frac{\partial \mathbf{V}}{\partial t} \left(\frac{\partial \pi}{\partial s} \right) ds \, dS + \frac{\gamma}{g} \iiint_V w \frac{\partial w}{\partial t} \left(\frac{\partial \pi}{\partial s} \right) ds \, dS \tag{7.61}$$

$$+ \frac{1}{g} \iiint_V \frac{\|\mathbf{V}\|^2}{2} \frac{\partial}{\partial s} \left(\frac{\partial \pi}{\partial t} \right) ds \, dS + \frac{\gamma}{g} \iiint_V \frac{w^2}{2} \frac{\partial}{\partial s} \left(\frac{\partial \pi}{\partial t} \right) ds \, dS.$$

By using the evolution equations for temperature (7.45), horizontal wind (7.43), and vertical velocity w (7.44), the continuity equation (7.45), and the Mayer relation ($C_p = C_v + R$), we get the expression (after some rather laborious algebra):

$$\frac{\partial \mathfrak{E}}{\partial t} = -\frac{1}{g} \iiint_V \nabla \cdot \left[\left(K + \gamma \frac{w^2}{2} + C_p T + \Phi \right) \frac{\partial \pi}{\partial s} \mathbf{V} \right] ds\, dS$$
$$-\frac{1}{g} \iiint_V \frac{\partial}{\partial s} \left[\left(K + \gamma \frac{w^2}{2} + C_p T + \Phi \right) \frac{\partial \pi}{\partial s} \dot{s} \right] ds\, dS \quad (7.62)$$
$$+\frac{1}{g} \iiint_V \frac{\partial}{\partial s} \left[p \left(\frac{\partial \Phi}{\partial t} \right) \right] ds\, dS.$$

In doing the calculations to get to expression (7.62), a number of terms cancel out from the equations used. The scalar product of the vector $\mathbf{V}$ by the rotation term $-\mathbf{k} \times (\zeta + f)\mathbf{V}$ in the momentum equation (7.43) makes a zero contribution for the kinetic energy of the horizontal wind. Moreover, other terms from the development of tendencies in equation (7.61) appear with opposing signs and cancel exactly.

The first term of the right-hand side of (7.62) cancels when integrated on the sphere or over a limited domain with suitable boundary conditions; the same goes for the second term when integrated in the vertical, given the boundary conditions ($\dot{s} = 0$) chosen at the top and bottom of the domain. Lastly, the third term of the right-hand side of (7.62) gives, after vertical integration from s_t to s_b:

$$\frac{1}{g} \iint_S \left[\left(p \frac{\partial \Phi}{\partial t} \right)_{s=s_b} - \left(p \frac{\partial \Phi}{\partial t} \right)_{s=s_t} \right] dS = +\frac{1}{g} \iint_S p_s \frac{\partial \Phi_s}{\partial t} dS - \frac{1}{g} \iint_S p_t \frac{\partial \Phi_t}{\partial t} dS. \quad (7.63)$$

The first of these two terms is zero since the surface geopotential Φ_s is constant. The second also cancels if $p_t = 0$, assuming that the temperature remains bounded (the quantity $p_t \partial \Phi_t / \partial t$ tends towards 0 when p tends towards 0). By adopting a top of the domain with zero pressure, the tendency of total energy $\mathfrak{E}$ is therefore zero, which implies that total energy is indeed conserved in these conditions. When pressure p_t is nonzero, the quantity $\mathfrak{E}$ is not conserved, as allowance must then be made for the work done by the pressure force corresponding to fluctuations in altitude of the upper boundary of the working domain defined by $p = p_t$.

Returning to the expression for total energy (7.55), it will be noticed that, after integration by parts and applying the definition of hydrostatic pressure, the integral with respect to s is written:

$$\int_{s_b}^{s_t} \Phi \frac{\partial \pi}{\partial s} ds = \left[\Phi_t \pi_t - \Phi_s \pi_s \right] + \int_{s_b}^{s_t} \frac{\pi}{p} RT \frac{\partial \pi}{\partial s} ds,$$

or alternatively:

$$\int_{s_b}^{s_t} \Phi \frac{\partial \pi}{\partial s} ds = \pi_t \left(\Phi_t - \Phi_s \right) + \int_{s_b}^{s_t} \left(\Phi_s + \frac{\pi}{p} RT \right) \frac{\partial \pi}{\partial s} ds. \quad (7.64)$$

Moving from the Euler equations to the primitive equations, with the hydrostatic assumption, p is identified with π and the preceding relation is written:

$$\int_{s_b}^{s_t} \Phi \frac{\partial p}{\partial s} ds = p_t \left(\Phi_t - \Phi_s \right) + \int_{s_b}^{s_t} \left(\Phi_s + RT \right) \frac{\partial p}{\partial s} ds.$$

When $p_t = 0$ (top of the domain pushed back to infinity), we have the Margules relation (Lorenz, 1967):

$$\int_{s_b}^{s_t} \Phi \frac{\partial p}{\partial s} ds = \int_{s_b}^{s_t} \left(\Phi_s + RT \right) \frac{\partial p}{\partial s} ds. \tag{7.65}$$

It follows that the expression for the total energy conserved by the primitive equations can be written, given the Mayer relation and because $\gamma = 0$:

$$\mathfrak{E} = \frac{1}{g} \iiint_V \left(K + C_p T + \Phi_s \right) \left(\frac{\partial p}{\partial s} \right) ds \, dS. \tag{7.66}$$

As was the case with the Euler equations, where the pressure at the top p_t is nonzero, this quantity is not conserved, as allowance must also be made for the term introducing the variable thickness of the atmosphere.

7.7 Conclusion

The progressive hybrid general coordinate just defined allows us to solve in a relatively satisfactory way the problems posed by the formulation of lower and upper boundary conditions of the domain of integration. The zero flux conditions at the top and bottom of the working domain mean in particular that the integral invariants are conserved.

However, the zero flux condition at the top of the domain is not without its drawbacks: in particular it is responsible for the reflection of vertically propagating gravity waves generated by orography. Herzog (1995) proposed a *radiative* formulation for the upper boundary condition (where the waves can escape freely) applied to the systems of equations using the progressive hybrid coordinate based on earlier work by Klemp and Durran (1983) and by Bougeault (1983).

The choice of any general coordinate implies, of course, that we can determine the values of state variables of the model over a number of levels defined for fixed values of that coordinate. In the past, the initial state of the model obtained from available observations by *objective analysis* was defined rather on isobaric levels and pre-processing involving interpolations or extrapolations was necessary to obtain variables on the model's levels. There are now methods for determining the variables on the model's levels directly from observations, so avoiding this operation that degrades information. However, from fields on the model levels, post-processing is essential to restore the model variables on isobaric levels, height levels, and more generally on any level defined by using another coordinate system. It should be pointed out that extrapolation *below relief* (e.g. for determining the sea level pressure) sometimes requires hazardous assumptions and so must be interpreted with great caution.

Some baroclinic models

8.1 Introduction

The problems relating to the horizontal discretization of models and the implementation of the semi-implicit algorithm for shallow water models have been studied in the foregoing chapters. After introducing a very general vertical coordinate in Chapter 7, we now examine the problems of vertical discretization of the primitive equations in an endeavour to obtain a solution which can best conserve the integral invariants.

We then present the implementation of explicit and semi-implicit time integration algorithms applied successively to the case of a primitive equation model by using the grid point method and the spectral method for the horizontal representation of fields.

8.2 The context of discretization

8.2.1 The equations

The primitive equations describing the evolution of the atmosphere in the absence of exchanges with its surroundings (frictionless adiabatic system) are written by using the progressive hybrid vertical coordinate studied in Chapter 7:

$$p(s, p_s) = f(s)p_0 + g(s)p_s.$$

The evolution equation of the horizontal wind vector $\mathbf{V}$ is written:

$$\frac{\partial \mathbf{V}}{\partial t} = -\mathbf{k} \times (\zeta + f)\mathbf{V} - \dot{s}\frac{\partial \mathbf{V}}{\partial s} - \nabla K - \left(\nabla\Phi + \frac{RT}{p}\nabla p\right), \tag{8.1}$$

with $\zeta = \mathbf{k}. (\nabla \times \mathbf{V})$ and $K = \|\mathbf{V}\|^2/2$.

The evolution equation for temperature (thermodynamic equation) is written:

$$\frac{\partial T}{\partial t} = -\mathbf{V}\cdot\nabla T - \dot{s}\frac{\partial T}{\partial s} - \frac{RT}{C_p}\left[\frac{1}{p}\int_{s_t}^{s} \nabla\cdot\left(\frac{\partial p}{\partial s'}\mathbf{V}\right) ds' - \frac{1}{p}(\mathbf{V}\cdot\nabla p)\right]. \tag{8.2}$$

The surface pressure tendency equation is written:

$$\frac{\partial p_s}{\partial t} = -\int_{s_t}^{s_b} \nabla \cdot \left(\frac{\partial p}{\partial s} \mathbf{V} \right) ds. \tag{8.3}$$

In addition to these *prognostic* equations, we need the two *diagnostic* equations for computing the generalized vertical velocity $\dot{s}$ and the geopotential Φ.

The equation for the generalized vertical velocity is written:

$$\dot{s} \frac{\partial p}{\partial s} = -\int_{s_t}^{s} \nabla \cdot \left(\frac{\partial p}{\partial s'} \mathbf{V} \right) ds' + \frac{\partial p}{\partial p_s} \int_{s_t}^{s_b} \nabla \cdot \left(\frac{\partial p}{\partial s} \mathbf{V} \right) ds. \tag{8.4}$$

The hydrostatic equation is written:

$$\Phi = \Phi_s - R \int_{s_b}^{s} \frac{T}{p} \frac{\partial p}{\partial s'} ds', \tag{8.5}$$

Φ_s designating the surface geopotential.

8.2.2 The layers, levels, and positions of variables

Vertical discretization consists in decomposing the atmosphere into a stack of layers. For a given column of the atmosphere, the model variables characterize the state of these layers and are only a mean value. However, these variables can be attributed to a level that is representative of the layer. We then speak of representative *levels* and of *layers* bounded by *interlayer surfaces*. In Figure 8.1 the interlayer surfaces are represented by solid lines whereas the levels characteristic of the layers, located somewhere between the interlayer surfaces, are shown by dotted lines (zero topography is assumed, so as to simplify the drawing).

Two interlayer surfaces play a specific role: the lower boundary and the upper boundary of the working domain. The lower boundary corresponds to the surface $s = s_b$, which is in contact with the model's topography, and the upper boundary $s = s_t$ is here taken to coincide with the zero pressure surface ($p_t = p(s_t) = 0$). This choice ensures the integral invariants are conserved (as shown in Section 7.6) and we can now properly call these boundary surfaces the base and top of the atmosphere represented here.

The prognostic variables (horizontal wind and temperature), calculated on the levels, are noted $\mathbf{V}_k$ and T_k, with the subscript k varying from 1 (highest level) to N (lowest level). There are therefore $\tilde{N} + 1$ interlayer surfaces with the subscript $\tilde{k}$, ranging from $\tilde{0}$ to $\tilde{N}$ (the indices with a *tilde* identifying these surfaces). Accordingly, the top of the atmosphere corresponds to the surface index $\tilde{k} = \tilde{0}$ while the base of the atmosphere corresponds to surface index $\tilde{k} = \tilde{N}$.

The divergence of momentum $\nabla \cdot [(\partial p / \partial s) \mathbf{V}]$ is naturally enough computed on the representative levels of the layers.

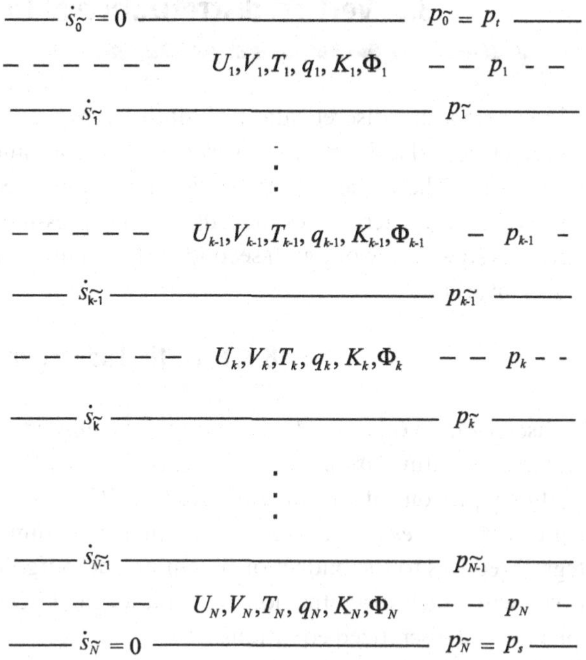

$$\dot{s}_{\tilde{0}} = 0 \hspace{3cm} p_{\tilde{0}} = p_t$$

$$U_1, V_1, T_1, q_1, K_1, \Phi_1 \hspace{1cm} p_1$$

$$\dot{s}_{\tilde{1}} \hspace{4cm} p_{\tilde{1}}$$

$$\vdots$$

$$U_{k-1}, V_{k-1}, T_{k-1}, q_{k-1}, K_{k-1}, \Phi_{k-1} \hspace{0.5cm} p_{k-1}$$

$$\dot{s}_{\widetilde{k-1}} \hspace{4cm} p_{\widetilde{k-1}}$$

$$U_k, V_k, T_k, q_k, K_k, \Phi_k \hspace{1cm} p_k$$

$$\dot{s}_{\tilde{k}} \hspace{4cm} p_{\tilde{k}}$$

$$\vdots$$

$$\dot{s}_{\widetilde{N-1}} \hspace{4cm} p_{\widetilde{N-1}}$$

$$U_N, V_N, T_N, q_N, K_N, \Phi_N \hspace{1cm} p_N$$

$$\dot{s}_{\tilde{N}} = 0 \hspace{3cm} p_{\tilde{N}} = p_s$$

Figure 8.1   Vertical discretizaton: the layers and levels.

The surface pressure p_s, which is a variable that is independent of the coordinate s, is calculated on the surface $\tilde{k} = \tilde{N}$, where the surface geopotential $\Phi_{\tilde{N}}$ Φ_s, which is a time-independent parameter, is also defined.

The generalized vertical velocity $\dot{s}$ is calculated on the interlayer surfaces; this allows us to take account easily of the zero flux boundary conditions, by setting $\dot{s} = 0$ at the base of the atmosphere $(s = s_b)$ and at the top of the atmosphere $(s = s_t)$. Given the position of the divergence term for the horizontal wind on the levels and of the continuity equation expression, this choice also allows us to evaluate the vertical derivative of the quantity $\dot{s}\ (\partial p/\partial s)$ in central differences.

$$\left[\frac{\partial}{\partial s}\left(\dot{s}\frac{\partial p}{\partial s}\right)\right]_k = \frac{1}{s_{\tilde{k}} - s_{\widetilde{k-1}}}\left[\left(\dot{s}\frac{\partial p}{\partial s}\right)_{\tilde{k}} - \left(\dot{s}\frac{\partial p}{\partial s}\right)_{\widetilde{k-1}}\right]. \tag{8.6}$$

Likewise, given the position of the temperatures and of the hydrostatic equation expression, the geopotential $\tilde{\Phi}$ is calculated first on the interlayer surfaces, a choice that allows us to calculate the term $\partial \Phi / \partial (\ln p)$ in central differences:

$$\left(\frac{\Phi_{\tilde{k}} - \Phi_{\widetilde{k-1}}}{\ln p_{\tilde{k}} - \ln p_{\widetilde{k-1}}}\right)_k = -RT_k. \tag{8.7}$$

8.3 Vertical discretization of the equations

We look here at the discretization of all the terms of the equations involving the vertical dimension, that is, both the vertical advections and the integrals with respect to the variable s. The vertical discretization proposed by Simmons and Burridge (1981), which is presented here, gives precedence to expressing derivatives in suitable central differences so we can work with second-order accuracy and ensure conservation of the integral invariants.

8.3.1 Vertical advection

The discretization of vertical advection terms must ensure that mass, total energy, and angular momentum are conserved. Thus the *vertical divergences* that appear in the evolution equations of the integral invariants (7.51) to (7.55) must be expressed in the form of *discrete vertical divergences*, so that their sum along the vertical (discretized integral) reduces to the budget of incoming and outgoing flows, because terms cancel out stepwise. Such discretization thus allows the Green-Ostrogradsky relation to be applied to the discretized equations.

For the conservation of mass, angular momentum, and energy to be preserved, the following relation must hold:

$$\left(\dot{s}\frac{\partial X}{\partial s}\right)_k \delta p_k + X_k\left[\varsigma_{\tilde{k}} - \varsigma_{\widetilde{k-1}}\right] = \left[X_{\tilde{k}}\;\varsigma_{\tilde{k}} - X_{\widetilde{k-1}}\;\varsigma_{\widetilde{k-1}}\right], \tag{8.8}$$

where $\dot{\varsigma} = \dot{s}\,(\partial p/\partial s)$ characterizes the generalized vertical velocity (do not mistake this Greek letter, *final sigma*, for the Greek letter *zeta* which is generally used to designate vorticity); X_k is a generic prognostic variable at level k; $X_{\tilde{k}}$ and $X_{\widetilde{k-1}}$ are the values of this same variable evaluated on the interlayer surfaces $\tilde{k}$ and $\widetilde{k-1}$, and δp_k, the thickness of the layer in pressure: $\delta p_k = p_{\tilde{k}} - p_{\widetilde{k-1}}$.

By choosing for $X_{\tilde{k}}$ a linear combination of X_k and X_{k+1}: $X_{\tilde{k}} = \lambda X_{k+1} + (1-\lambda)X_k$, condition (8.8) leads the advection term, calculated for level k, to be written:

$$\left(\dot{s}\frac{\partial X}{\partial s}\right)_k = \frac{1}{\delta p_k}\left[\lambda\varsigma_{\tilde{k}}\left(X_{k+1} - X_k\right) + (1-\lambda)\,\varsigma_{\widetilde{k-1}}\left(X_k - X_{k-1}\right)\right]. \tag{8.9}$$

To preserve the conservation of total energy, which involves a quadratic quantity (kinetic energy), we must also ensure that the expression:

$$2X_k\frac{1}{\delta p_k}\left[\lambda\varsigma_{\tilde{k}}\left(X_{k+1} - X_k\right) + (1-\lambda)\,\varsigma_{\widetilde{k-1}}\left(X_k - X_{k-1}\right)\right]\delta p_k + X_k^2\left(\varsigma_{\tilde{k}} - \varsigma_{\widetilde{k-1}}\right) \tag{8.10}$$

can be written in the form of a discrete vertical divergence.

This expression expands as:

$$2\lambda\dot{\varsigma}_{\widetilde{k}}\ X_k X_{k+1} - 2(1-\lambda)\dot{\varsigma}_{\widetilde{k-1}}\ X_{k-1} X_k + (1-2\lambda)X_k^2(\dot{\varsigma}_{\widetilde{k}} + \dot{\varsigma}_{\widetilde{k-1}}).$$

By choosing $\lambda = 0.5$, the X_k^2 term disappears and expression (8.10) is reduced to:

$$\dot{\varsigma}_{\widetilde{k}}\ X_k X_{k+1} - \dot{\varsigma}_{\widetilde{k-1}}\ X_{k-1} X_k, \tag{8.11}$$

which is indeed a discrete vertical divergence cancelling term by term when summed over k.

8.3.2 Surface pressure evolution equation

From equation (8.3), the discretized evolution equation for surface pressure is obtained very simply by replacing the integral by a finite sum:

$$\frac{\partial p_s}{\partial t} = -\sum_{k=1}^{N} \nabla\cdot(\mathbf{V}_k \delta p_k). \tag{8.12}$$

The tendency of surface pressure at the bottom of an atmospheric column is therefore equal, except for the sign, to the sum of divergence of the momentum of all the layers making up the column.

8.3.3 Diagnostic equation for generalized vertical velocity

The diagnostic equation for generalized vertical velocity (8.4) is naturally written in discrete form:

$$\dot{\varsigma}_{\widetilde{k}} = \left(\dot{s}\frac{\partial p}{\partial s}\right)_{\widetilde{k}} = -\sum_{i=1}^{k} \nabla\cdot(\mathbf{V}_i \delta p_i) + \left(\frac{\partial p}{\partial p_s}\right)_{\widetilde{k}}\sum_{i=1}^{N} \nabla\cdot(\mathbf{V}_i \delta p_i), \tag{8.13}$$

with $(\partial p/\partial p_s)_{\widetilde{k}} = g(s_{\widetilde{k}})$, $g(s)$, being the function characterizing dependence on surface pressure p_s in the definition of the s coordinate.

8.3.4 Diagnostic equation for geopotential

Since temperature is a characteristic of the layer, it is logical to consider the atmosphere locally as a stack of isothermal layers. The hydrostatic relation, applied to layer k of temperature T_k, is written:

$$\Phi_{\widetilde{k}} - \Phi_{\widetilde{k-1}} = -RT_k\ \ln\ (p_{\widetilde{k}}/p_{\widetilde{k-1}}). \tag{8.14}$$

Determining the geopotential Φ_k on the model levels, where pressure is p_k, remains essential for calculating the pressure force in the momentum equation. The pressure p_k may be chosen fairly arbitrarily (for primitive equations at any rate), insofar as $p_{\widetilde{k-1}} < p_k < p_{\widetilde{k}}$; the final choice of the value of p_k will be specified a little further on.

The geopotential Φ_k is obtained by applying the hydrostatic relation to the underlying half-layer and to the overlying half-layer:

$$\Phi_k = \Phi_{\tilde{k}} + RT_k \ \ln\ (p_{\tilde{k}}/\, p_k), \text{ and } \Phi_{\widetilde{k-1}} = \Phi_k \ + RT_k \ \ln\ (p_k/\,p_{\widetilde{k-1}}) \qquad (8.15)$$

expressions that remain consistent with relation (8.14).

By using the notations:

$$\alpha_k = \ln\ \frac{p_{\tilde{k}}}{p_k}, \ \beta_k = \ln\frac{p_k}{p_{\widetilde{k-1}}} \ \text{ and } \ \gamma_k = \alpha_k + \beta_k = \ln\frac{p_{\tilde{k}}}{p_{\widetilde{k-1}}}, \qquad (8.16)$$

the geopotential Φ_k on a level k is given by the relation:

$$\Phi_k = \Phi_s + \sum_{i=k+1}^{N} \gamma_i RT_i + \alpha_k RT_k, \qquad (8.17)$$

which is the integral form of the hydrostatic relation when numerically integrated.

8.3.5 Pressure force term

To conserve the total angular momentum, we must be able to express the product of the pressure force by the thickness in pressure:

$$-\left[\nabla\Phi + RT\frac{\nabla p}{p}\right]_k \delta p_k, \qquad (8.18)$$

in the form of a discrete vertical divergence, allowing terms to be cancelled stepwise when summed in the vertical. Given the hydrostatic relation, this expression expands as:

$$-\left[\nabla\Phi + RT\frac{\nabla p}{p}\right]_k \delta p_k = -\nabla\Phi_k \ \delta p_k + \left(\frac{\partial\Phi}{\partial p}\nabla p\right)_k \delta p_k,$$

which can also be written:

$$-\left[\nabla\Phi + RT\frac{\nabla p}{p}\right]_k \delta p_k = -\nabla\Phi_k \delta p_k + \left[\frac{\partial}{\partial p}(\Phi\nabla p)\right]_k \delta p_k - \Phi_k\nabla(\delta p_k).$$

By expressing the derivative $\partial(\Phi\nabla p)/\partial p$ at level k using central differences, the expression (8.18) takes the form:

$$-\left[\nabla\Phi + RT\frac{\nabla p}{p}\right]_k \delta p_k = -\nabla(\Phi_k\delta p_k) + (\Phi_{\tilde{k}}\nabla p_{\tilde{k}} \ - \ \Phi_{\widetilde{k-1}}\nabla p_{\widetilde{k-1}}), \qquad (8.19)$$

which involves a discrete vertical divergence in the second term of the right-hand side allowing term by term cancellation when summed in the vertical.

The discretization given in (8.19) can be used to calculate:

$$\left[RT\frac{\nabla p}{p} \right]_k = \frac{1}{\delta p_k}\left[-\Phi_{\widetilde{k}}\nabla p_{\widetilde{k}} + \Phi_{\widetilde{k-1}}\nabla p_{\widetilde{k-1}} + \Phi_k\nabla(\delta p_k) \right].$$ (8.20)

By using relations (8.15) and by introducing the coefficients defined in (8.16), the pressure force at level k takes the form:

$$-\left[\nabla\Phi + RT\frac{\nabla p}{p} \right]_k = -\left[\nabla\Phi_k + \frac{RT_k}{\delta p_k}(\alpha_k\nabla p_{\widetilde{k}} + \beta_k\nabla p_{\widetilde{k-1}}) \right].$$ (8.21)

By expressing the pressure gradient ∇p as a function of the surface pressure gradient ∇p_s, we get:

$$-\left[\nabla\Phi + RT\frac{\nabla p}{p} \right]_k = -\left\{ \nabla\Phi_k + \frac{RT_k}{\delta p_k}\left[\alpha_k g(s_{\widetilde{k}}) + \beta_k g(s_{\widetilde{k-1}}) \right]\nabla p_s \right\}.$$ (8.22)

By introducing the condensed notation:

$$\varepsilon_k = [\alpha_k g(s_{\widetilde{k}}) + \beta_k g(s_{\widetilde{k-1}})]/\delta p_k,$$ (8.23)

the discretized expression of pressure force is finally written:

$$-\left[\nabla\Phi + RT\frac{\nabla p}{p} \right]_k = -\left[\nabla\Phi_k + RT_k\varepsilon_k\nabla p_s \right].$$ (8.24)

8.3.6 Energy conversion term

The term of conversion between potential energy and kinetic energy that appears in the thermodynamic equation is written:

$$\frac{R}{C_p}\frac{T}{p}\omega = -\frac{RT}{C_p}\left[\frac{1}{p}\int_{s_t}^{s} \nabla\cdot\left(\frac{\partial p}{\partial s'}\mathbf{V} \right) ds' \right] + \frac{R}{C_p}\frac{T}{p}(\mathbf{V}\cdot\nabla p).$$ (8.25)

Notice that when the total energy is calculated, the quantity:

$$\left(C_p\frac{\partial T}{\partial t} + \mathbf{V}\cdot\frac{\partial \mathbf{V}}{\partial t} \right)\frac{\partial p}{\partial s}ds$$

brings out two identical expressions with opposite signs coming for one from the term $-(RT/p)\nabla p$ in the momentum equation and for the other from the term $C_p^{-1}(RT/p)\mathbf{V}\times\nabla p$ in the thermodynamic equation. They must therefore be given strictly identical expressions to ensure the conservation of total energy by the discretized system. Given the expression for the pressure isobaric gradient in (8.24), the second term of the right-hand side of relation (8.25) must therefore necessarily be written:

$$\left[\frac{R}{C_p}\frac{T}{p}(\mathbf{V}\cdot\nabla p)\right]_k = \frac{RT_k}{C_p}\varepsilon_k\,\mathbf{V}_k\cdot\nabla p_s. \tag{8.26}$$

To ensure total energy is conserved, it is necessary to bring out the first term of the right-hand side of relation (8.25) in the form of a discrete vertical divergence. Applying the hydrostatic relation gives:

$$\left[-\frac{RT}{p}\int_{s_t}^{s}\nabla\cdot\left(\frac{\partial p}{\partial s'}\mathbf{V}\right)ds'\right]\frac{\partial p}{\partial s}ds = \frac{\partial\Phi}{\partial s}\left[\int_{s_t}^{s}\nabla\cdot\left(\frac{\partial p}{\partial s'}\mathbf{V}\right)ds'\right]ds,$$

an expression that may take the form:

$$\left[-\frac{RT}{p}\int_{s_t}^{s}\nabla\cdot\left(\frac{\partial p}{\partial s'}\mathbf{V}\right)ds'\right]\frac{\partial p}{\partial s}ds = \frac{\partial}{\partial s}\left[\Phi\int_{s_t}^{s}\nabla\cdot\left(\frac{\partial p}{\partial s'}\mathbf{V}\right)ds'\right]ds - \Phi\nabla\cdot\left(\frac{\partial p}{\partial s}\mathbf{V}\right)ds, \tag{8.27}$$

which clearly brings out a vertical divergence.

The corresponding discretized expression is therefore:

$$\left[-\frac{RT}{p}\int_{s_t}^{s}\nabla\cdot\left(\frac{\partial p}{\partial s'}\mathbf{V}\right)ds'\right]_k\delta p_k = \left[\Phi_{\tilde{k}}\sum_{i=1}^{k}\nabla\cdot(\mathbf{V}_i\delta p_i) - \Phi_{\widetilde{k-1}}\sum_{i=1}^{k-1}\nabla\cdot(\mathbf{V}_i\delta p_i)\right] - \Phi_k\nabla\cdot(\mathbf{V}_k\delta p_k), \tag{8.28}$$

which may be written:

$$\left[-\frac{RT}{p}\int_{s_t}^{s}\nabla\cdot\left(\frac{\partial p}{\partial s'}\mathbf{V}\right)ds'\right]_k\delta p_k = \left[\Phi_{\tilde{k}}\sum_{i=1}^{k}\nabla\cdot(\mathbf{V}_i\delta p_i) - \Phi_{\widetilde{k-1}}\sum_{i=1}^{k-1}\nabla\cdot(\mathbf{V}_i\delta p_i)\right]$$
$$-\left[\Phi_k\sum_{i=1}^{k}\nabla\cdot(\mathbf{V}_i\delta p_i) - \Phi_k\sum_{i=1}^{k-1}\nabla\cdot(\mathbf{V}_i\delta p_i)\right].$$

By gathering the sums with the same limits, we get:

$$\left[-\frac{RT}{p}\int_{s_t}^{s}\nabla\cdot\left(\frac{\partial p}{\partial s'}\mathbf{V}\right)ds'\right]_k\delta p_k = \left[(\Phi_{\tilde{k}}-\Phi_k)\sum_{i=1}^{k}\nabla\cdot(\mathbf{V}_i\,\delta p_i) + (\Phi_k-\Phi_{\widetilde{k-1}})\sum_{i=1}^{k-1}\nabla\cdot(\mathbf{V}_i\delta p_i)\right]. \tag{8.29}$$

Applying the discretized hydrostatic relation as per relations (8.15) and introducing coefficients defined in (8.16) we can write:

$$\left[-\frac{RT}{p}\int_{s_t}^{s}\nabla\cdot\left(\frac{\partial p}{\partial s'}\mathbf{V}\right)ds'\right]_k = -\frac{RT_k}{\delta p_k}\left[\alpha_k\sum_{i=1}^{k}\nabla\cdot(\mathbf{V}_i\,\delta p_i) + \beta_k\sum_{i=1}^{k-1}\nabla\cdot(\mathbf{V}_i\delta p_i)\right]. \tag{8.30}$$

The term of conversion between potential energy and kinetic energy in the thermo-dynamic equations is therefore written:

$$\left[\frac{RT}{C_p} \frac{\omega}{p} \right]_k = -\frac{RT_k}{C_p} \frac{1}{\delta p_k} \left[\gamma_k \sum_{i=1}^{k-1} \nabla \cdot (\mathbf{V}_i \delta p_i) + \alpha_k \nabla \cdot (\mathbf{V}_k \delta p_k) \right] + \frac{RT_k}{C_p} \varepsilon_k \mathbf{V}_k \cdot \nabla p_s. \quad (8.31)$$

It is important to note that, in the general case, the quantities δp_k, α_k, β_k, and γ_k are not constants, but depend at each instant on the value of the pressure p at level s, which is related to the surface pressure p_s; in the case of the most general s coordinate, these quantities must be recalculated for each time step of the model.

8.3.7 Location of the pressure levels

The way the various terms involved in the equations are calculated means that the conservation properties of the integral invariants can be preserved. In those computations, the prognostic variables are calculated on the levels, whereas the diagnostic variables are calculated on the interlayer surfaces. So far no assumption whatsoever has been made about the exact position of the level within the layer, which is logical enough insofar as the prognostic variables represent mean values for the layer. The problem of the location of the pressure level within the layer only arises when determining the initial conditions. Whichever method is used to ascertain the initial values at the grid points from a background (climatic values, recent forecast) and a set of observations, it is essential to know the exact position of the variables to be initialized if we are to be able to calculate their distance from the observations. At the end of the forecast, it is necessary too to know the position of the variables on the vertical so as to make the interpolations (or extrapolations) allowing us to retrieve the variables predicted by the model on the isobaric or constant height levels so users may correctly interpret the corresponding fields.

For the sake of consistency, we can write:

$$\left[\frac{RT}{p} \nabla p \right]_k = \frac{RT_k \nabla p_k}{p_k}. \quad (8.32)$$

Given the discretization adopted for $\left[\dfrac{RT}{p} \nabla p \right]_k$ in (8.19), we then have:

$$\ln p_k = \frac{1}{\delta p_k} \left[p_{\tilde{k}} \ln p_{\tilde{k}} - p_{\widetilde{k-1}} \ln p_{\widetilde{k-1}} \right] - 1. \quad (8.33)$$

This quantity, of the type $[\delta(p \ln p)]/\delta p - 1$, may be considered an approximate expression of $\ln p$; it is not particularly problematic to evaluate, even when the pressure p tends towards zero, since in that event, the quantity $p \ln p$ tends towards zero too.

This determination of the pressure level p_k, that arises here from relation (8.32), remains arbitrary, though, and other choices may be made so long as we are working with the primitive equations; however, with the Euler equations, considerations as to conservation and stability may make this determination more restrictive (Bubnova et al., 1995).

8.3.8 Alternative solutions for vertical discretization

The layout of variables along the vertical corresponds to the *Lorenz* arrangement proposed to ensure the conservation of energy (Lorenz, 1960). This choice is not without its drawbacks, though: as shown by Hollingsworth (1995), the relation defining geopotential on the model's *levels* arising from this arrangement may prompt the development of a *non-physical* temperature wave in the vertical. The geopotential perturbations Φ_k' corresponding to temperature perturbations T_k' (relative to a basic state given by Φ_k and T_k) verify the relation:

$$\Phi_{k-1}' - \Phi_k' = RT_k' \ln \frac{p_{\tilde{k}}}{p_k} + RT_{k-1}' \ln \frac{p_k}{p_{\widetilde{k-1}}}.$$

If the perturbation Φ_N' is zero, it can be seen that the occurrence of temperature perturbations T_k' of successive opposite signs is compatible with zero perturbation of the geopotential Φ_k' throughout the atmospheric column, which is obviously impossible in the real world.

Another arrangement of variables, the *Charney-Phillips* arrangement, was proposed by Charney and Phillips (1953) to satisfy the constraint of conservation of the quasi-geostrophic potential vorticity in a quasi-geostrophic model. This arrangement differs from Lorenz's in that the temperatures are calculated on *interlayer surfaces*. With the variables arranged in this way, the geopotential perturbations $\Phi_{\tilde{k}}'$ corresponding to the temperature perturbations T_k' then verify the relation:

$$\Phi_{k-1}' - \Phi_k' = RT_{\widetilde{k-1}}' \ln \frac{p_k}{p_{k-1}},$$

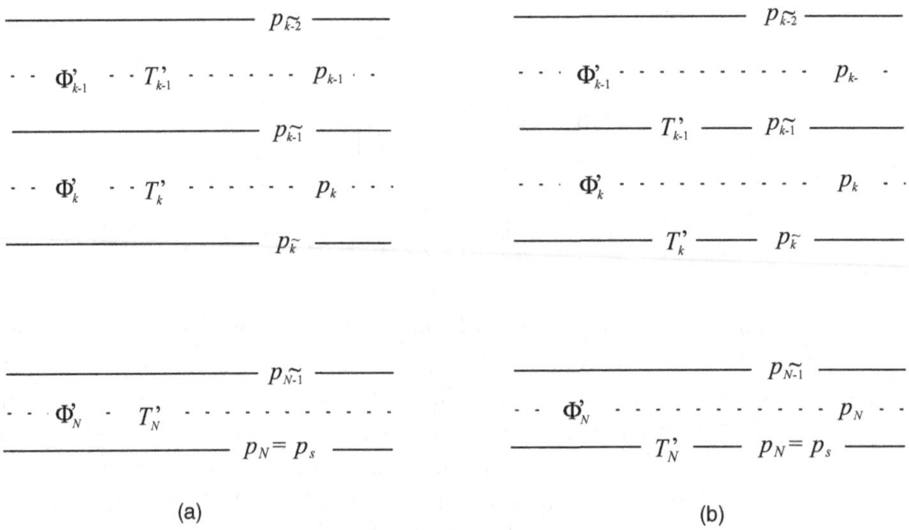

Figure 8.2   Vertical grid arrangements: (a) Lorenz; (b) Charney-Phillips.

showing that only zero temperature perturbations T_k' can yield zero perturbations of the geopotential Φ_k'.

Figure 8.2 shows the location of the geopotential perturbations Φ' corresponding to temperature perturbations T' for both types of arrangement of variables.

The Charney-Phillips arrangement has certain advantages when it comes to geostrophic adjustment, as shown by Arakawa and Moorthi (1988) and by Leslie and Purser (1992). However, it has the effect of doubling the number of trajectories to be calculated to apply the semi-Lagrangian method. A detailed study of the behaviour of numerical simulations with various arrangements of variables on the vertical was made by Cullen and James (1994).

8.4 A sigma coordinate and finite difference model

Here we present an example of discretization of a primitive equation baroclinic model using an Arakawa C grid and Cartesian horizontal coordinates defined by choosing a conformal projection of the sphere on the plane (polar stereographic projection, Lambert conical projection, or Mercator cylindrical projection), characterized by the map scale factor m. This technique was used for the limited area forecast model PERIDOT, used operationally by Météo-France in the 1980s (Imbard et al., 1987).

As pointed out in Chapter 2, it is convenient to characterize the horizontal wind by the reduced wind components $U = u/m$ and $V = v/m$. On the vertical, we use the pure sigma coordinate with the top of the atmosphere corresponding to the zero pressure surface. It is also convenient to use the logarithm of surface pressure $Z = \ln p_s$ as the prognostic variable rather than the actual surface pressure.

8.4.1 Simplifications with the pure sigma coordinate

The pure sigma coordinate is defined by:

$$p = \sigma p_s, \text{ which implies: } f(\sigma) \equiv 0 \text{ and } g(\sigma) = \sigma.$$

With this type of vertical coordinate, the coefficients α_k, β_k, and γ_k defined in (8.16) simplify to:

$$\alpha_k = \ln \frac{\sigma_{\tilde{k}}}{\sigma_k}, \quad \beta_k = \ln \frac{\sigma_k}{\sigma_{\widetilde{k-1}}}, \text{ and } \gamma_k = \alpha_k + \beta_k = \ln \frac{\sigma_{\tilde{k}}}{\sigma_{\widetilde{k-1}}}. \tag{8.34}$$

These expressions are then constants that depend only on the chosen location for the levels and interlayer surfaces.

The use of the pure sigma coordinate also simplifies the expression of the coefficient ε_k given in (8.23), which takes the form:

$$\varepsilon_k = \frac{1}{p_s \delta \sigma_k}(\sigma_{\tilde{k}} \ln \sigma_{\tilde{k}} - \sigma_{\widetilde{k-1}} \ln \sigma_{\widetilde{k-1}}) - \frac{\ln \sigma_k}{p_s}.$$

By choosing the location of the levels σ_k as indicated in (8.33), this expression reduces to the very simple form:

$$\varepsilon_k = 1 / p_s, \tag{8.35}$$

which is independent of the level in question.

8.4.2 The location of variables on the C grid

As for the location of the different variables on the C grid, comparison of the surface pressure p_s tendency equation (obtained by vertically integrating the continuity equation) with the equation for the evolution of the geopotential of the free surface Φ in the case of the shallow water model (which is merely the continuity equation) shows that these two quantities formally play analogous roles. Temperatures T_k and surface pressure p_s should therefore be calculated at the same locations as the geopotential Φ_k on the C grid. The same goes for the divergence of momentum, for kinetic energy, and for the map scale factor m (which is dependent on latitude alone). As in the shallow water model, the potential absolute vorticity $\xi^* = (\zeta + f)/p_s$ is calculated at the central point of the grid (Figure 8.3).

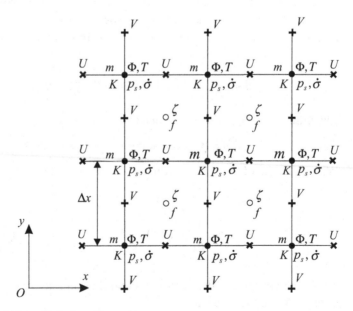

Figure 8.3   Location of variables on the Arakawa C type grid.

8.4.3 The discretized equations

The divergence of momentum for a layer of thickness δp_k is written:

$$\nabla \cdot (\mathbf{V_k} \delta p_k) = [\nabla \cdot (p_s \mathbf{V_k})] \delta \sigma_k.$$

By analogy with the formulation used for the continuity equation in the shallow water model, this expression is discretized as:

$$m^2 \, [(\overline{p}_s^x U_k)_x + (\overline{p}_s^y V_k)_y] \, \delta \sigma_k = m^2 \, \widetilde{D}_k \, \delta \sigma_k, \text{ with: } \widetilde{D}_k = [(\overline{p}_s^x U_k)_x + (\overline{p}_s^y V_k)_y];$$

the expression $\widetilde{D}_k$, which is in the form of a *discrete horizontal divergence*, is calculated on the levels of the model at the same points as the temperature T_k and the map scale factor m.

The equation for generalized vertical velocity (8.4) is discretized in the form:

$$\dot{\varsigma}_{\widetilde{k}} = (p_s \, \dot{\sigma})_{\widetilde{k}} = m^2 \left(\sigma \sum_{i=1}^{N} \widetilde{D}_i \, \delta \sigma_i - \sum_{i=1}^{k} \widetilde{D}_i \, \delta \sigma_i \right). \tag{8.36}$$

The integrated hydrostatic equation (8.5) is discretized as already indicated in (8.17) and is used for calculating the geopotential on the levels k:

$$\Phi_k = \Phi_s + \sum_{i=k+1}^{N} \gamma_i R T_i + \alpha_k R T_k, \tag{8.37}$$

with coefficients α_k and γ_k taking the values indicated in (8.34).

The evolution equation for the variable $Z = \ln p_s$ is deduced immediately from the surface pressure evolution equation (8.3) discretized as in (8.12):

$$\frac{\partial Z}{\partial t} = -\frac{m^2}{p_s} \sum_{k=1}^{N} \widetilde{D}_k \, \delta \sigma_k. \tag{8.38}$$

As was done in the shallow water model, the energy K_k and potential absolute vorticity ξ_k^* are discretized as follows at the locations shown in Figure 8.3:

$$K_k = m^2 \frac{\overline{U_k^2}^x + \overline{V_k^2}^y}{2}, \tag{8.39}$$

$$\xi_k^* = \frac{\overline{m^2}^{xy} \left[(V_k)_x - (U_k)_y \right] + f}{\overline{p}_s^{xy}}. \tag{8.40}$$

Vertical advection for temperature is easily calculated at the same points as the temperature T_k, by using the expression given in (8.9) with $\lambda = 0, 5$ so as to ensure the total energy is conserved; this term for the vertical advection of temperature is noted in short form:

$$\mathcal{A}_T(T_k) = \left(\dot{\sigma}\frac{\partial T}{\partial \sigma}\right)_k = \frac{1}{p_s}\frac{1}{2\delta\sigma_k}\left[\ \dot{\varsigma}_{\tilde{k}}\ (T_{k+1} - T_k) + \ \dot{\varsigma}_{\widetilde{k-1}}\ (T_k - T_{k-1})\right]. \tag{8.41}$$

Vertical advection for variables U_k and V_k requires horizontal means of vertical velocity $\dot{\varsigma}$ to be calculated, since this latter quantity is located at the same points as the variable T_k. The vertical advections for the wind components are then written in short form:

$$\mathcal{A}_U(U_k) = \left(\dot{\sigma}\frac{\partial U}{\partial \sigma}\right)_k = \frac{1}{\overline{p_s}^x}\frac{1}{2\delta\sigma_k}\left[\ \overline{\dot{\varsigma}_{\tilde{k}}}^{\,x}\ (U_{k+1} - U_k) + \ \overline{\dot{\varsigma}_{\widetilde{k-1}}}^{\,x}\ (U_k - U_{k-1})\right], \tag{8.42}$$

$$\mathcal{A}_V(V_k) = \left(\dot{\sigma}\frac{\partial V}{\partial \sigma}\right)_k = \frac{1}{\overline{p_s}^y}\frac{1}{2\delta\sigma_k}\left[\ \overline{\dot{\varsigma}_{\tilde{k}}}^{\,y}\ (V_{k+1} - V_k) + \ \overline{\dot{\varsigma}_{\widetilde{k-1}}}^{\,y}\ (V_k - V_{k-1})\right], \tag{8.43}$$

and are calculated at points U_k and V_k, respectively.

As in the shallow water model, the pressure force is discretized as simply as can be done given the position of the different variables on the C grid. Taking up the results established in Subsection 8.3.5 and using the expression of ε_k given in (8.35), we then obtain for the two components of the geopotential isobaric gradient:

$$-[(\Phi_k)_x + R\overline{T_k}^x Z_x] \text{ and } - [(\Phi_k)_y + R\overline{T_k}^y Z_y], \tag{8.44}$$

with the geopotential Φ_k being calculated from the temperatures T_k using (8.37).

By following the principles adopted in Subsection 6.2.2 for discretization of the shallow water model on the C grid, the evolution equations of the two wind components for the baroclinic model are written:

$$\frac{\partial U_k}{\partial t} = +\overline{\xi_k^*}^{\,y}\ \overline{\overline{p_s^y V_k}}^{\,xy} - \mathcal{A}_U(U_k) - (K + \Phi)_x - R\overline{T_k}^x Z_x, \tag{8.45}$$

$$\frac{\partial V_k}{\partial t} = -\overline{\xi_k^*}^{\,x}\ \overline{\overline{p_s^x U_k}}^{\,xy} - \mathcal{A}_V(V_k) - (K + \Phi)_y - R\overline{T_k}^y Z_y. \tag{8.46}$$

In view of the choices made in Subsections 8.3.4 and 8.3.5 for vertical discretization, the evolution equation for temperature is written:

$$\frac{\partial T_k}{\partial t} = -\frac{m^2}{p_s}\left[\ \overline{\overline{p_s^x U_k(T_k)_x}}^{\,x} + \overline{\overline{p_s^y V_k(T_k)_y}}^{\,y}\ \right] - \mathcal{A}_T(T_k) - \frac{RT_k}{C_p p_s}\frac{m^2}{\delta\sigma_k}\left[\ \gamma_k\sum_{i=1}^{k-1}\tilde{D}_i\,\delta\sigma_i + \alpha_k\tilde{D}_k\,\delta\sigma_k\ \right]$$
$$+\frac{R}{C_p}\frac{m^2}{p_s}\left[\ \overline{\overline{p_s^x\ U_k\overline{T_k}^x Z_x}}^{\,x} + \overline{\overline{p_s^y\ V_k\overline{T_k}^y Z_y}}^{\,y}\ \right]. \tag{8.47}$$

Relations (8.38), (8.45), (8.46), and (8.47) together form the set of discretized prognostic equations; they give the expression of the tendencies for the logarithm of surface pressure $Z = \ln p_s$, the reduced components of the horizontal wind U_k and V_k, and for the temperature T_k. To calculate these tendencies, we must also add the two diagnostic equations (8.36) and (8.37), that allow us to calculate the geopotential Φ_k as a function of the temperature T_k and the vertical velocity $\dot{\varsigma}_k = (p_s \, \dot{\sigma})_k$ depending on the divergence of the momentum $m^2 \, \tilde{D}_k \delta \sigma_k$.

The discretization adopted according to the principles set out in Subsection 6.2.2 for the horizontal and in Section 8.3 for the vertical allows us to conserve, as far as can be done, the various integral invariants of the system of primitive equations.

The total mass of the atmosphere is written:

$$\mathfrak{M} = \frac{1}{g} \sum_S \sum \frac{p_s}{m^2} \, \delta x \, \delta y,$$

the double summation being extended to the entire surface S of the working domain, which is the union of the elementary surface elements $\delta x \delta y$.

As the discretization brings out the *discrete horizontal divergence* $\tilde{D}_k$ in the right-hand side of the evolution equation of the variable Z, (8.38) can be used to verify that the expression:

$$\frac{\partial \mathfrak{M}}{\partial t} = \frac{1}{g} \sum_S \sum \left[\frac{1}{m^2} p_s \frac{\partial Z}{\partial t} \right] \delta x \, \delta y$$

cancels with suitable boundary conditions (zero flux budget) when summed over all the points where the variable p_s is defined.

The absolute potential enstrophy is written:

$$\mathfrak{H} = \frac{1}{g} \sum_S \sum \left\{ \frac{\overline{p_s}^{xy}}{\overline{m^2}^{xy}} \sum_{k=1}^{N} \left[\frac{(\xi_k^*)^2}{2} \right] \delta \sigma_k \right\} \delta x \, \delta y.$$

The discretization of the equations of motion (8.45) and (8.46), copied from that adopted for the shallow water model in Subsection 6.2.2, can be used to verify that the expression:

$$\frac{\partial \mathfrak{H}}{\partial t} = \frac{1}{g} \sum_S \sum \left\{ \frac{1}{\overline{m^2}^{xy}} \sum_{k=1}^{N} \left[\xi_k^* \frac{\partial \xi_k^*}{\partial t} \overline{p_s}^{xy} + \frac{(\xi_k^*)^2}{2} \frac{\partial \overline{p_s}^{xy}}{\partial t} \right] \right\} \delta x \, \delta y.$$

cancels with suitable boundary conditions (zero flux budget), summing being done over all the points where the vorticity is defined (and the expression $\overline{p_s}^{xy}$).

The total energy is written:

$$\mathfrak{E} = \frac{1}{g} \sum_S \sum \left[\frac{p_s}{m^2} \sum_{k=1}^{N} (\Phi_s + C_p T_k + K_k) \, \delta \sigma_k \right] \delta x \, \delta y.$$

The discretization adopted for the pressure force in the momentum equations (8.45) and (8.46) and for the conversion term in the thermodynamic equation (8.47) allows us to verify that the expression:

$$\frac{\partial \mathfrak{E}}{\partial t} = -\frac{1}{g}\sum_S \sum \left[\frac{1}{m^2}\frac{\partial p_s}{\partial t}\sum_{k=1}^{N}(\Phi_s + C_p T_k + K_k)\,\delta\sigma_k \right]\delta x\,\delta y$$

$$+\frac{1}{g}\sum_S\sum\left[\sum_{k=1}^{N}\left(C_p\frac{p_s}{m^2}\frac{\partial T_k}{\partial t} + \overline{\overline{p_s^x U_k}\frac{\partial U_k}{\partial t}}^x + \overline{\overline{p_s^y V_k}\frac{\partial V_k}{\partial t}}^y \right)\delta\sigma_k\right]\delta x\,\delta y$$

reduces, provided suitable boundary conditions are used (zero flux budget), to the residual:

$$\frac{\partial \mathfrak{E}}{\partial t} = \frac{1}{g}\sum_S\sum\left[\sum_{k=1}^{N}\left(\overline{\overline{p_s^x U_k}\;\overline{\xi_k^*}^y\;\overline{p_s^y V_k}}^{xy}{}^x - \overline{\overline{p_s^y V_k}\;\overline{\xi_k^*}^x\;\overline{p_s^x U_k}}^{xy}{}^y \right)\delta\sigma_k\right]\delta x\,\delta y.$$

This does not cancel out exactly (unlike its analytical counterpart). Energy is therefore not entirely conserved, but this drawback arises from the choice of discretization of the equations of momentum that give precedence to the conservation of enstrophy, as for the shallow water model studied in Section 6.2.2.

8.4.4 Explicit time integration of the model

The model equations may be written very synthetically by introducing the column vectors **U**, **V**, and **T** whose components are the values of the parameters U_k, V_k, and T_k, with the subscript k ranging from 1 to N. The variable $Z = \ln p_s$ is a scalar quantity. Similarly, $\boldsymbol{\Phi}$ represents the column vector of the geopotential Φ^k, with the subscript k ranging from 1 to N. We note $\dot{\boldsymbol{\varsigma}}$ the column vector whose components are the values $\dot{\varsigma}_{\tilde{k}} = (p_s\dot{\sigma})_{\tilde{k}}$, with the subscript $\tilde{k}$ ranging from 1 to N. At the top of the atmosphere, at the zero pressure surface, $\dot{\varsigma}_{\tilde{0}} = 0$ and the relation (8.36) shows that we do indeed have $\dot{\varsigma}_{\tilde{N}} = 0$ at the bottom of the atmosphere.

The prognostic equations (8.38), (8.45), (8.46), and (8.47) take the form:

$$\left.\begin{aligned}
\frac{\partial Z}{\partial t} &= \mathcal{S}_Z(p_s, \mathbf{U}, \mathbf{V}),\\[4pt]
\frac{\partial \mathbf{U}}{\partial t} &= \mathcal{S}_U(p_s, \mathbf{U}, \mathbf{V}, \mathbf{T}, \dot{\boldsymbol{\varsigma}}, \boldsymbol{\Phi}),\\[4pt]
\frac{\partial \mathbf{V}}{\partial t} &= \mathcal{S}_V(p_s, \mathbf{U}, \mathbf{V}, \mathbf{T}, \dot{\boldsymbol{\varsigma}}, \boldsymbol{\Phi}),\\[4pt]
\frac{\partial \mathbf{T}}{\partial t} &= \mathcal{S}_T(p_s, \mathbf{U}, \mathbf{V}, \mathbf{T}, \dot{\boldsymbol{\varsigma}}).
\end{aligned}\right\} \tag{8.48}$$

The symbols $\mathcal{S}_z$, $\mathcal{S}_u$, $\mathcal{S}_v$, $\mathcal{S}_\tau$ designate the operators which, when applied to the specified variables, define the right-hand sides of the prognostic equations (8.38), (8.45), (8.46), and (8.47).

By introducing the column vector $\tilde{\mathbf{D}}$:

$$\tilde{\mathbf{D}} = [(\bar{p}_s^x \mathbf{U})_x + (\bar{p}_s^y \mathbf{V})_y],$$

(8.49)

and by calling $\mathbf{S}_k^T$ the line vector whose first k components are the thicknesses $\delta\sigma_i$ (with the subscript i ranging from 1 to k), the $N - k$ residual components being zero:

$$\mathbf{S}_k^T = [\delta\sigma_1 \ \ \delta\sigma_2 \ \ ... \ \ \delta\sigma_k \ \ 0 \ \ ... \ \ 0],$$

(8.50)

the two diagnostic equations (8.36) and (8.37), that allow the geopotential Φ to be calculated on the levels and the vertical velocity $\dot{\varsigma}_{\bar{k}}$ on the interlayer surfaces, may also be written as:

$$\dot{\boldsymbol{\varsigma}} = m^2 \, (\mathbf{S}_N^T \tilde{\mathbf{D}} - \mathbf{S}_k^T \tilde{\mathbf{D}}),$$

(8.51)

$$\boldsymbol{\Phi} = \Phi_s \mathbf{I} + \mathbf{BT},$$

(8.52)

the matrix $\mathbf{B}$ being the upper triangular matrix below:

$$\mathbf{B} = R \begin{bmatrix} \alpha_1 & \gamma_2 & \cdot & \gamma_k & \cdot & \gamma_N \\ 0 & \alpha_2 & \cdot & \gamma_k & \cdot & \gamma_N \\ \cdot & \cdot & \cdot & \cdot & \cdot & \cdot \\ 0 & 0 & \cdot & \alpha_k & \cdot & \gamma_N \\ \cdot & \cdot & \cdot & \cdot & \cdot & \cdot \\ 0 & 0 & \cdot & 0 & \cdot & \alpha_N \end{bmatrix}.$$

(8.53)

The model may be integrated very simply, as was explained for the shallow water model in Subsection 6.2.3, working in succession through the steps below:

- Beginning with the initial values (at time 0) of the prognostic variables $\mathbf{U}$, $\mathbf{V}$, $\mathbf{T}$, and p_s (therefore Z) at each point of the horizontal grid, we calculate the variable $\tilde{\mathbf{D}}$ defined in (8.49) then the vertical velocity $\dot{\boldsymbol{\varsigma}}$ using the diagnostic relation (8.51). The surface geopotential Φ_s and the variable $\mathbf{T}$ are used for calculating the geopotential $\boldsymbol{\Phi}$. We then have everything we need for calculating the right-hand sides of the equations in system (8.38).
- The right-hand sides of equations (8.38) being expressed at time 0, the discretization of the time derivatives using forward differences (Euler scheme) can be used to calculate the prognostic variables at time Δt, then the vertical velocity and geopotential at the same time Δt thanks to the diagnostic relations. We then have all of the variables at times 0 and Δt.
- The right-hand sides of equations (8.38) being expressed at time Δt, the discretization of the time derivatives using central differences (the *leapfrog* scheme) can be used to calculate the prognostic variables at time $2\Delta t$, then the vertical velocity and the geopotential at that same time $2\Delta t$ thanks to the diagnostic relations. We then have all of the variables at times 0, Δt, and $2\Delta t$.

- Now that we have the values of the variables at two successive times $t - \Delta t$ and t, we can continue to apply the previous procedure to calculate the values of the variables at time $t + \Delta t$. Stepwise, we thus calculate the values of the model's prognostic variables up to the required range.

The integration method is said to be *explicit*, since the values of the variables at time $t + \Delta t$ are obtained explicitly from the values of the variables at times $t - \Delta t$ and t.

As already explained in Subsection 6.2.3, the Euler scheme (forward differences) must be used for the first time step only. Accuracy for calculating the variables at time Δt is improved by using the technique described by Hoskins and Simmons (1975).

The explicit centred scheme is conditionally stable only. The CFL condition, for this model processed explicitly with the C grid, is analogous to the criterion laid down for the shallow water model in Subsection 6.2.3:

$$-1 \le 2\sqrt{2} \ (V^* + V_g) \ \Delta t / \Delta x \le +1, \tag{8.54}$$

V^* designating the maximum value of the synoptic wind and V_g the speed of propagation of the fastest gravity waves. The linear analysis of the linearized baroclinic model, which is not detailed here, shows that the speed of propagation of the fastest gravity wave (external wave) is of the order of magnitude of the speed of sound in the atmosphere, that is approximately 330 m/s; a maximum value of the synoptic wind of 50 m/s is a reasonable evaluation. For a mesh whose smallest dimension locally is $\Delta x = 100$ km, application of the CFL condition shows that we need a time step of 90 seconds (one and a half minutes) to ensure the numerical stability of the model.

When working on a limited area, it is easy to check that the discretization adopted for the right-hand sides of equations (8.38) does not allow them to be calculated on the boundaries of the horizontal working domain. It is therefore essential to define suitable *lateral boundary conditions* for the numerical model to be integrated.

A very crude approximation is to consider the boundary conditions are constant (the prognostic variables keep their initial values throughout the integration of the model). It is readily understandable that this approach can only be applied for short-term integrations so that the error generated on the boundaries of the domain does not have time to reach the central part of the domain on which the meteorological prediction is to be made. This method has been used, though, in the past for running quasi-hemispheric models for making less than 72-hour forecasts for Europe by placing the boundary of the horizontal domain in tropical regions, where the variability of meteorological fields is relatively low.

A procedure commonly used for working on a limited area is to determine the values of prognostic variables on the boundary of the domain by interpolating the results from forecasts made with a model with a larger horizontal mesh, operating over a much broader domain than the limited area in question, and whose lateral boundaries are far enough away from the area or are even nonexistent (a global domain). This *nested model* technique has proved very useful for economic small-scale limited-area forecasting. It is presented in a little more detail in Section 10.5 below.

8.4.5 Implementation of semi-implicit time integration

The semi-implicit method is applied to the baroclinic model along the same principles as those adopted for the shallow water model in Subsection 6.2.4.

As the implicit processing is to be done on the linear terms responsible for the propagation of gravity waves, the right-hand sides of the evolution equations (8.38) must first be put into a form that brings out these linear terms. As in the shallow water model, we must therefore consider both the pressure force term in the equation for the evolution of the horizontal wind $\mathbf{V}$ and the divergence terms arising from the formulation of the continuity equation in the evolution equation of $Z = \ln p_s$ as well as in the equation for the evolution of temperature T.

Contrary to what happens for the shallow water model, the pressure force at level k brings out a nonlinear term. However, by fixing a reference temperature $T_k^*(s)$ for each level, we can isolate a linear term:

$$\nabla \Phi_k + R T_k \nabla Z = \nabla \Phi_k + R T_k^* \nabla Z + R(T_k - T_k^*)\nabla Z.$$

By introducing the quantity $P_k = \Phi_k + R T_k^* Z$, the pressure force is written:

$$\nabla \Phi_k + R T_k \nabla Z = \nabla P_k + R(T_k - T_k^*)\nabla Z, \quad (8.55)$$

which clearly brings out the linear term ∇P_k.

As regards the divergence terms $\tilde{D}_k$ that occur in the evolution equation of Z and of temperature T, it can be seen that:

$$\tilde{D}_k = [(\overline{p}_s^x U_k)_x + (\overline{p}_s^y V_k)_y] \equiv p_s[(U_k)_x + (V_k)_y] + \overline{U_k(p_s)_x}^x + \overline{V_k(p_s)_y}^y. \quad (8.56)$$

This numerical identity means a linear divergence term can be brought out in the evolution equation of Z:

$$\frac{\partial Z}{\partial t} = -\frac{m^2}{p_s}\sum_{k=1}^{N}\left[\overline{U_k(p_s)_x}^x + \overline{V_k(p_s)_y}^y\right]\delta\sigma_k - m^2\sum_{k=1}^{N}D_k\delta\sigma_k, \quad (8.57)$$

$$\text{with } D_k = [(U_k)_x + (V_k)_y]. \quad (8.58)$$

As regards the equation for the evolution of temperature T, the linear divergence terms arise from the vertical advection term $\dot{s}\partial T_k^*(s)/\partial s$, and from the term of conversion between potential energy and kinetic energy $R T_k^*(s)\omega/(C_p p)$. By choosing for $T_k^*(s)$ a standard atmosphere type temperature profile, the implicit treatment of these two terms leads to an instability first reported by Simmons, Hoskins, and Burridge (1978) (and therefore termed *SHB instability*) and studied subsequently by Côté et al. (1993). The latter showed that the scheme becomes stable by choosing an isothermal vertical profile $T_k^*(s)$, a solution that has been adopted in numerous models. All the same, Bénard (2004) showed that the scheme remained stable even for non-isothermal $T_k^*(s)$ profiles so long as the advection term is treated explicitly. This is why we confine ourselves here to presenting just the implicit treatment of the conversion term that decomposes as below:

$$\frac{RT_k}{C_p}\frac{m^2}{\delta\sigma_k}\left[\gamma_k\sum_{i=1}^{k-1}\tilde{D}_i\,\delta\sigma_i+\alpha_k\,\tilde{D}_k\,\delta\sigma_k\right]=\frac{RT_k^*}{C_p p_s}\frac{m^2}{\delta\sigma_k}\left\{\gamma_k\sum_{i=1}^{k-1}\left[\overline{U_i(p_s)_x}^{\,x}+\overline{V_i(p_s)_y}^{\,y}\right]\delta\sigma_i\right\}$$

$$+\frac{RT_k^*}{C_p p_s}m^2\alpha_k\left[\overline{U_k(p_s)_x}^{\,x}+\overline{V_k(p_s)_y}^{\,y}\right]$$

$$+\frac{R(T_k-T_k^*)}{C_p p_s}\frac{m^2}{\delta\sigma_k}\left[\gamma_k\sum_{i=1}^{k-1}\tilde{D}_i\,\delta\sigma_i+\alpha_k\,\tilde{D}_k\,\delta\sigma_k\right]$$

$$+\frac{RT_k^*}{C_p}\frac{m^2}{\delta\sigma_k}\left[\gamma_k\sum_{i=1}^{k-1}D_i\delta\sigma_i+\alpha_k D_k\delta\sigma_k\right].$$

$$(8.59)$$

Calling **D** the column vector whose components are constituted by the D_k, with k ranging from 1 to N, the linear term (the fourth term) of the right-hand side of equation (8.59) is written in condensed fashion **AD**, with the lower triangular matrix **A** taking the form:

$$\mathbf{A}=\frac{R}{C_p}\begin{bmatrix}\alpha_1 T_1^* & 0 & . & 0 & . & 0\\ \dfrac{\gamma_2\delta\sigma_1}{\delta\sigma_2}T_2^* & \alpha_2 T_2^* & . & 0 & . & 0\\ . & & . & & . & .\\ \dfrac{\gamma_k\delta\sigma_1}{\delta\sigma_k}T_k^* & \dfrac{\gamma_k\delta\sigma_2}{\delta\sigma_k}T_k^* & . & \alpha_k T_k^* & . & 0\\ . & & . & & . & .\\ \dfrac{\gamma_N\delta\sigma_1}{\delta\sigma_N}T_N^* & \dfrac{\gamma_N\delta\sigma_2}{\delta\sigma_N}T_N^* & . & \dfrac{\gamma_N\delta\sigma_k}{\delta\sigma_N}T_N^* & . & \alpha_N T_N^*\end{bmatrix}.\qquad(8.60)$$

The hydrostatic equation (8.42) allows us to write **P**, the column vector whose components are the quantities P_k, with k ranging from 1 to N, in the form:

$$\mathbf{P}=\Phi_s\mathbf{I}+\mathbf{BT}+\mathbf{R}^*Z,\qquad(8.61)$$

$\mathbf{R}^*$ representing the column vector whose components are the quantities RT_k^*, with the subscript k ranging from 1 to N.

The system of equations (8.38) is discretized in a similar way as for the shallow water model, as detailed in Subsection 6.2.3: once the linear terms have been isolated as just described, they are processed in the centred implicit fashion. The implicit system is then written:

$$\left.\begin{array}{l}Z(t+\Delta t)=Z(t-\Delta t)+2\Delta t\mathcal{S}_z+2m^2\Delta t\,\mathbf{S}_N^{\mathrm{T}}\,\mathbf{D}(t)-m^2\Delta t\,\mathbf{S}_N^T\,[(\mathbf{D}(t-\Delta t)+\mathbf{D}(t+\Delta t)],\\[4pt] \mathbf{U}(t+\Delta t)=\mathbf{U}(t-\Delta t)+2\Delta t\mathcal{S}_U+2\Delta t\,\mathbf{P}(t)-\Delta t\,[(\mathbf{P}(t-\Delta t)+\mathbf{P}(t+\Delta t)]_x,\\[4pt] \mathbf{V}(t+\Delta t)=\mathbf{V}(t-\Delta t)+2\Delta t\mathcal{S}_V+2\Delta t\,\mathbf{P}(t)-\Delta t\,[(\mathbf{P}(t-\Delta t)+\mathbf{P}(t+\Delta t)]_y,\\[4pt] \mathbf{T}(t+\Delta t)=\mathbf{T}(t-\Delta t)+2\Delta t\mathcal{S}_T+2m^2\Delta t\,\mathbf{AD}(t)-m^2\Delta t\mathbf{A}\,[(\mathbf{D}(t-\Delta t)+\mathbf{D}(t+\Delta t)].\end{array}\right\}\quad(8.62)$$

By calling Z_T, $\mathbf{U}_T$, $\mathbf{V}_T$, and $\mathbf{T}_T$ the parts of the right-hand sides that are expressed using prognostic variables taken at times $t - \Delta t$ and t (*transitory* quantities) we get:

$$
\left.
\begin{aligned}
Z(t + \Delta t) &= Z_T - m^2 \Delta t \, \mathbf{S}_N^T \ \mathbf{D}(t + \Delta t), \\
\mathbf{U}(t + \Delta t) &= \mathbf{U}_T - \Delta t \ \mathbf{P}\big[(t + \Delta t)\big]_x, \\
\mathbf{V}(t + \Delta t) &= \mathbf{V}_T - \Delta t \ \mathbf{P}\big[(t + \Delta t)\big]_y, \\
\mathbf{T}(t + \Delta t) &= \mathbf{T}_T - m^2 \Delta t \mathbf{A} \ \mathbf{D}(t + \Delta t).
\end{aligned}
\right\}
\tag{8.63}
$$

By applying the finite difference operators $(\ )_x$ and $(\ )_y$ to both sides of the equations of momentum in system (8.62), we calculate:

$$
\mathbf{D}(t + \Delta t) = [\mathbf{U}(t + \Delta t)]_x + [\mathbf{V}(t + \Delta t)]_y,
\tag{8.64}
$$

which gives:

$$
\mathbf{D}(t + \Delta t) = \mathbf{D}_T - \Delta t \ \{[\mathbf{P}(t + \Delta t)]_{xx} + [\mathbf{P}(t + \Delta t)]_{yy}\},
\tag{8.65}
$$

with: $\mathbf{D}_T = [\mathbf{U}_T(t + \Delta t)]_x + [\mathbf{V}_T (t + \Delta t)]_y$.

By inserting this value of $\mathbf{D}(t + \Delta t)$ into the first and last equations of system (8.62), we get:

$$
\begin{aligned}
Z(t+\Delta t) &= Z_T - m^2 \Delta t \, \mathbf{S}_N^T \mathbf{D}_T + m^2 \Delta t^2 \, \mathbf{S}_N^T \ \Big\{ \ [\mathbf{P}(t+\Delta t)]_{xx} + [\mathbf{P}(t+\Delta t)]_{yy} \ \Big\}, \\
\mathbf{T}(t+\Delta t) &= \mathbf{T}_T - m^2 \Delta t \ \mathbf{A} \mathbf{D}_T + m^2 \Delta t \mathbf{A} \ \Big\{ \ [\mathbf{P}(t+\Delta t)]_{xx} + [\mathbf{P}(t+\Delta t)]_{yy} \ \Big\}.
\end{aligned}
$$

These two expressions allow us to calculate $\mathbf{P}(t + \Delta t)$:

$$
\begin{aligned}
\mathbf{P}(t+\Delta t) &= \Phi_s \mathbf{I} + \mathbf{B}\mathbf{T}_T + \mathbf{R} Z_T - m^2 \Delta t (\mathbf{B}\mathbf{A} + \mathbf{R}^* \mathbf{S}_N^T) \mathbf{D}_T \\
&\quad + m^2 \Delta t^2 (\mathbf{B}\mathbf{A} + \mathbf{R}^* \mathbf{S}_N^T) \ \Big\{ \ [\mathbf{P}(t+\Delta t)]_{xx} + [\mathbf{P}(t+\Delta t)]_{yy} \ \Big\},
\end{aligned}
$$

By writing $\mathbf{P}_T = \Phi_s \mathbf{I} + \mathbf{B}\mathbf{T}_T + \mathbf{R}^* Z_T$ and by calling $\mathbf{M}$ the square matrix $\mathbf{M} = \mathbf{B}\mathbf{A} + \mathbf{R}^* \mathbf{S}_N^T$, we finally obtain the equation giving $\mathbf{P}(t + \Delta t)$:

$$
\mathbf{P}(t+\Delta t) = \mathbf{P}_T - m^2 \Delta t \mathbf{M} \mathbf{D}_T + m^2 \Delta t \mathbf{M}\{[\mathbf{P}(t+\Delta t)]_{xx} + [\mathbf{P}(t+\Delta t)]_{yy}\}.
\tag{8.66}
$$

This matrix equation forms a coupled system of Helmholtz equations that can be solved simply by making a linear transformation on vectors $\mathbf{P}$ and $\mathbf{D}$ by which system (8.66) can be decoupled.

Let $\mathbf{Q}$ be the matrix whose column vectors are the eigenvectors of matrix $\mathbf{M}$, verifying the relation $\mathbf{Q}^{-1}\mathbf{M}\mathbf{Q} = \Lambda$, where Λ is the diagonal matrix of the eigenvalues of $\mathbf{M}$.

By making the transformation: $\mathbb{P} = \mathbf{Q}^{-1}\mathbf{P}$ and $\mathbb{D} = \mathbf{Q}^{-1}\mathbf{D}$, matrix equation (8.66) is written:

$$
\mathbf{Q}\mathbb{P} \ (t+\Delta t) = \mathbf{Q}\mathbb{P}_T - m^2 \Delta t \mathbf{M} \mathbf{Q} \mathbb{D}_T + m^2 \Delta t^2 \mathbf{M} \mathbf{Q} \ \Big\{ \ [\mathbb{P} \ (t+\Delta t)]_{xx} + [\mathbb{P} \ (t+\Delta t)]_{yy} \Big\}.
$$

Pre-multiplying the two sides of this equation by the inverse matrix $\mathbf{Q}^{-1}$ gives:

$$
\mathbb{P}(t+\Delta t) = \mathbb{P}_T - m^2 \Delta t \Lambda \mathbb{D}_T + m^2 \Delta t^2 \Lambda \ \Big\{ \ [\mathbb{P}(t+\Delta t)]_{xx} + [\mathbb{P}(t+\Delta t)]_{yy} \Big\}.
\tag{8.67}
$$

As matrix Λ is diagonal, the new equation reduces simply to N separate Helmholtz equations, that can be solved by tried and tested numerical methods (e.g. by using the over-relaxation method) to obtain the components of $\mathbb{P}(t + \Delta t)$.

The quantity $[\mathbb{P}(t + \Delta t)]_{xx} + [\mathbb{P}(t + \Delta t)]_{yy}$ is the classical expression of the Laplacian $\nabla^2 [\mathbb{P}(t + \Delta t)]$ with 2nd-order accuracy, calculated with the central point and the four nearest neighbours. This property arises from using the C grid and is not found with A or B grids.

Notice here that the presence of m^2, the square of the map scale factor, does not pose any special problem; it introduces only a spatial variability of the nonzero elements of the tridiagonal matrix of the linear systems corresponding to the discretized Helmholtz equations (8.67).

Once the quantity $\mathbb{P}(t + \Delta t)$ is obtained after solving these equations, calculating $\mathbf{P}(t + \Delta t) = \mathbf{Q}\mathbb{P}(t + \Delta t)$, substituted into the two equations of momentum in (8.63), allows us to obtain $\mathbf{U}(t + \Delta t)$ and $\mathbf{V}(t + \Delta t)$:

$$\mathbf{U}(t + \Delta t) = \mathbf{U}_T - \Delta t\, \mathbf{P}(t + \Delta t)_x,$$

$$\mathbf{V}(t + \Delta t) = \mathbf{V}_T - \Delta t\, \mathbf{P}(t + \Delta t)_y.$$

Relation (8.64) gives $\mathbf{D}(t + \Delta t)$, which we carry over into the two remaining equations of system (8.62) to calculate $Z(t + \Delta t)$ and $\mathbf{T}(t + \Delta t)$:

$$Z(t + \Delta t) = Z_T - m^2 \Delta t\, \mathbf{S}_N^{\mathsf{T}} \mathbf{D}(t + \Delta t),$$

$$\mathbf{T}(t + \Delta t) = \mathbf{T}_T - m^2 \Delta t\, \mathbf{A}\, \mathbf{D}(t + \Delta t).$$

$\mathbf{T}(t + \Delta t)$ may also be calculated differently, by using the fourth equation in (8.63):

$$\mathbf{T}(t + \Delta t) = \mathbf{B}^{-1}[\mathbf{P}(t + \Delta t) - \Phi_s\, \mathbf{I} - \mathbf{R}^*\, Z\,(t + \Delta t)].$$

In short, the operating procedure consists in first calculating for each of the variables Z, $\mathbf{U}$, $\mathbf{V}$, and $\mathbf{T}$ the transitory values Z_T, $\mathbf{U}_T$, $\mathbf{V}_T$, and $\mathbf{T}_T$ making up the right-hand sides of the implicit system (8.62), and then calculating the values of $\mathbf{P}_T$ and $\mathbf{D}_T$, that occur in the right-hand side of the system of Helmholtz equations. By applying to these quantities the linear transformation characterized by the matrix $\mathbf{Q}^{-1}$, we obtain the values of $\mathbb{P}_T$ and $\mathbb{D}_T$, from which we can calculate the right-hand side of the system of separate Helmholtz equations. Solving each of the Helmholtz equations by an appropriate method yields the value sought $\mathbb{P}(t + \Delta t)$. The linear transformation characterized by the matrix $\mathbf{Q}$ allows us to calculate $\mathbf{P}(t + \Delta t)$, and then the values of $\mathbf{U}(t + \Delta t)$, $\mathbf{V}(t + \Delta t)$, $\mathbf{D}(t + \Delta t)$, $Z(t + \Delta t)$, and $\mathbf{T}(t + \Delta t)$.

The linear transformation that is made to obtain the *diagonal system* of equations (8.67) is analogous to that used for calculating the *model's eigenmodes*, that is, the solutions of the linearized model. It is observed that applying this transformation provides, for the linear part, a system equivalent to a set of N shallow water models, the mean geopotentials of the free surface being given by each of the eigenvalues of matrix $\mathbf{M}$. Fully determining the model's eigenmodes also requires linearizing all of the advection terms, so as to isolate, for each of the equivalent models, the Rossby mode and the two inertia-gravity modes.

The system of linear equations corresponding to the Helmholtz equations may be solved by an iterative method of the over-relaxation type (Dady, 1969). To apply it, it is useful to start the iterative process from a background for the sought solution $\mathbb{P}(t + \Delta t)$ over the entire working domain including its boundaries. By choosing as background the result of the forecast made for the previous time step $P(t)$, we reduce the number of iterations required to converge compared with the choice of just any background. Moreover, the over-relaxation method converges faster when the matrix of the linear system is diagonal dominant. Close scrutiny of the implicit equations shows that this property is more pronounced when the mean geopotential of the free surface is lower (corresponding to greater mean divergence). For the usual reference profile values used for linearization (standard atmosphere, isothermal atmosphere), it proves to be the case that only the first two equations (for the external mode and first internal mode) take some time to converge. Compared with the explicit method, the implementation of the semi-implicit method is therefore far more costly in computing time for one time step. However, this drawback is offset by the possibility of using a longer time step.

The centred semi-implicit scheme formulated in the conditions set out at the beginning of this subsection is conditionally stable. The CFL criterion, for this model processed semi-implicitly with the C grid, is analogous to the criterion established for the shallow water model in Subsection 6.2.4:

$$-1 \leq \sqrt{2}\ V^*\ \Delta t\ /\ \Delta x \leq +1, \tag{8.68}$$

V^* denoting the maximum value of the synoptic wind. Taking the same numerical values as were used by way of example for determining the maximum time step in the explicitly processed model ($V^* = 50$ m/s and $\Delta x = 100$ km), applying the CFL criterion shows that a time step of 1200 seconds (20 minutes) is suitable for ensuring the numerical stability of the proposed model.

8.5 Formalization of the semi-implicit method

8.5.1 General formulation of the algorithm

The equations of a nonlinear model may be written in the very general form:

$$\frac{\partial \mathbf{X}}{\partial t} = \mathcal{S}(\mathbf{X}), \tag{8.69}$$

$\mathbf{X}$ being a vector representing the set of prognostic variables and $\mathcal{S}$ a nonlinear operator that, after defining a suitable basic state, may be put in the form: $\mathcal{S} = \mathcal{N} + \mathcal{L}$, $\mathcal{L}$ denoting a linear operator and $\mathcal{N}$ the residual, nonlinear operator.

Central difference discretization and semi-implicit processing applied to the operator $\mathcal{L}$ lead to the equation:

$$\mathbf{X}(t + \Delta t) = \mathbf{X}(t - \Delta t) + 2\Delta t \, \mathcal{N}\big[\mathbf{X}(t)\big] + \Delta t \, \mathcal{L}\big[\mathbf{X}(t - \Delta t) + \mathbf{X}(t + \Delta t)\big], \qquad (8.70)$$

which may also be written:

$$\mathbf{X}(t + \Delta t) = \mathbf{X}(t - \Delta t) + 2\Delta t \, \mathcal{S}\big[\mathbf{X}(t)\big] - 2\Delta t \, \mathcal{L}\big[\mathbf{X}(t)\big] + \Delta t \, \mathcal{L}\big[\mathbf{X}(t - \Delta t) + \mathbf{X}(t + \Delta t)\big].$$

By writing:

$$\mathbf{X}_E = \mathbf{X}(t - \Delta t) + 2\Delta t \, \mathcal{S}\big[\mathbf{X}(t)\big],$$

the value of $\mathbf{X}(t + \Delta t)$ obtained by applying the centred explicit scheme, the implicit equation is written:

$$\mathbf{X}(t + \Delta t) = \mathbf{X}_E + \Delta t \big\{ \mathcal{L}\big[\mathbf{X}(t - \Delta t)\big] - 2\mathcal{L}\big[\mathbf{X}(t)\big]\big\} + \Delta t \, \mathcal{L}\big[\mathbf{X}(t + \Delta t)\big],$$

that is:

$$\mathbf{X}(t + \Delta t) = \mathbf{X}_T + \Delta t \, \mathcal{L}\big[\mathbf{X}(t + \Delta t)\big], \qquad (8.71)$$

the value of $\mathbf{X}_T$ being given by:

$$\mathbf{X}_T = \mathbf{X}_E + \Delta t \big\{ \mathcal{L}\big[\mathbf{X}(t - \Delta t) - 2\mathbf{X}(t)\big]\big\}. \qquad (8.72)$$

This way of writing the implicit system shows that once a linearization characterized by the operator $\mathcal{L}$ has been adopted, it is a relatively straightforward matter to calculate the right-hand side $\mathbf{X}_T$ of the implicit equation. This is obtained by summing the value $\mathbf{X}_E$, obtained by applying the centred explicit scheme, and an additional linear term, that can be readily calculated once $\mathcal{L}$ has been determined.

8.5.2 Interpretation of the semi-implicit method

The semi-implicit method, which is used to determine $\mathbf{X}(t + \Delta t)$, is comparable to the Newton method used to solve a nonlinear equation of type $\mathcal{F}[\mathbf{X}(t + \Delta t)] = 0$, as shown by Bénard (2003). The solution can be obtained by successive iterations (written with the superscript in brackets) as below:

$$\mathbf{X}^{(k+1)}(t + \Delta t) = \mathbf{X}^{(k)}(t + \Delta t) - \Big[\nabla_{\mathbf{x}} \mathcal{F}\big(\mathbf{X}^{(k)}(t + \Delta t)\big)\Big]^{-1} \mathcal{F}\big(\mathbf{X}^{(k)}(t + \Delta t)\big), \qquad (8.73)$$

where $\nabla_{\mathbf{x}} \mathcal{F}[\mathbf{X}^{(k)}(t + \Delta t)]$ is the tangent linear operator calculated with the values of $\mathbf{X}(t + \Delta t)$ at iteration k. Implementing this algorithm of course requires the tangent linear operator to be invertible.

The Newton method for solving the fully implicit equation:

$$\mathbf{X}(t + \Delta t) - \mathbf{X}(t - \Delta t) - 2\Delta t \, \mathcal{S}\left(\frac{\mathbf{X}(t + \Delta t) + \mathbf{X}(t - \Delta t)}{2}\right) = 0 \qquad (8.74)$$

for the variable $\mathbf{X}(t + \Delta t)$ leads to the iterative algorithm:

$$\left[\mathbf{I} - \Delta t \nabla_\mathbf{x} \mathcal{S}^{(k)}\right]\mathbf{X}^{(k+1)}(t + \Delta t) = \left[\mathbf{I} - \Delta t \nabla_\mathbf{x} \mathcal{S}^{(k)}\right]\mathbf{X}^{(k)}(t + \Delta t)$$
$$- \left[\mathbf{X}^{(k)}(t + \Delta t) - \mathbf{X}(t - \Delta t) - 2\Delta t \, \mathcal{S}\left(\frac{\mathbf{X}^{(k)}(t + \Delta t) + \mathbf{X}(t - \Delta t)}{2}\right)\right],$$

where the notation:

$$\nabla_\mathbf{x} \mathcal{S}^{(k)} \equiv \nabla_\mathbf{x} \mathcal{S}\left[\frac{\mathbf{X}^{(k)}(t + \Delta t) + \mathbf{X}(t - \Delta t)}{2}\right]$$

denotes the tangent linear operator of $\mathcal{S}$.

After rearranging the various terms, this algorithm is written:

$$\left[\mathbf{I} - \Delta t \nabla_\mathbf{x} \mathcal{S}^{(k)}\right]\mathbf{X}^{(k+1)}(t + \Delta t) = \mathbf{X}(t - \Delta t) - \Delta t \nabla_\mathbf{x} \mathcal{S}^{(k)}\mathbf{X}^{(k)}(t + \Delta t)$$
$$+ 2\Delta t \, \mathcal{S}\left(\frac{\mathbf{X}^{(k)}(t + \Delta t) + \mathbf{X}(t - \Delta t)}{2}\right).$$

By choosing as the starting value $\mathbf{X}^{(0)}(t + \Delta t)$ for the linear extrapolation in time obtained from the values of $\mathbf{X}$ at times t and $t - \Delta t$:

$$\mathbf{X}^{(0)}(t + \Delta t) = 2\mathbf{X}(t) - \mathbf{X}(t - \Delta t),$$

which gives:

$$[\mathbf{X}^{(0)}(t + \Delta t) + \mathbf{X}(t - \Delta t)]/2 = \mathbf{X}(t),$$

the value of $\mathbf{X}(t + \Delta t)$ at the end of the first iteration is given by:

$$\left[\mathbf{I} - \Delta t \nabla_\mathbf{x} \mathcal{S}(\mathbf{X}(t))\right]\mathbf{X}(t + \Delta t) = \left[\mathbf{I} + \Delta t \nabla_\mathbf{x} \mathcal{S}(\mathbf{X}(t))\right]\mathbf{X}(t - \Delta t)$$
$$+ 2\Delta t\left[\mathcal{S}(\mathbf{X}(t)) - \nabla_\mathbf{x} \mathcal{S}(\mathbf{X}(t))\mathbf{X}(t)\right]. \qquad (8.75)$$

This formulation (8.75) is identified with the formulation (8.70) by taking:

$$\mathcal{L}(\mathbf{X}(t)) \equiv \nabla_\mathbf{x} \mathcal{S}(\mathbf{X}(t)) \text{ and } \mathcal{N}(\mathbf{X}(t)) \equiv \mathcal{S}(\mathbf{X}(t)) - \nabla_\mathbf{x} \mathcal{S}(\mathbf{X}(t))\mathbf{X}(t). \qquad (8.76)$$

This relation (8.76) clearly shows the close connection between the centred semi-implicit time integration scheme and the Newton's method, which, to be applied exactly, would require computing the tangent linear operator at each time step.

The classical semi-implicit method, as defined in the previous paragraph, is therefore a compromise for economically solving equation (8.74) by Newton's iterative method (8.73): we replace the tangent linear operator by a linear operator independent of the evolution (inverted once and for all at the beginning of integration of the model) and we make a single iteration after having chosen an approximate solution (linear temporal interpolation) as the starting solution. The choice of linear operator is far from intuitive, however, and a detailed linear analysis of the resulting scheme must be made (Bénard, 2004) to ensure it is stable.

Newton's method may also be applied by making more iterations so as to improve the accuracy of the semi-implicit scheme; in this perspective, time integration schemes requiring two iterations have been proposed by Côté et al. (1998) and by Cullen (2001) and their stability properties have been studied by Bénard (2003).

8.6 A variable resolution spectral model

Here we present an example of a primitive equation model on the sphere using the spectral method and the progressive hybrid vertical coordinate. This technique was implemented at Météo-France with a fixed spatial resolution for the operational weather forecasting model EMERAUDE, in a hemispheric version in 1985 (Coiffier et al., 1987a), and then in a global version in 1988. It was also used with variable horizontal resolution (Courtier and Geleyn, 1988), characterized by the map scale factor $\tilde{m}(\lambda, \mu)$ (noted with a tilde in this section to differentiate it from the zonal wavenumber m) for constructing the operational model ARPEGE/IFS (Déqué et al., 1994), developed jointly by Météo-France and the European Centre for Medium-Range Weather Forecasts (ECMWF) for the needs of weather forecasting and climate predictions in the early 1990s.

The equations on the transformed sphere using the pressure coordinate have been explained in Subsection 2.4.4. In the vertical, we adopt the progressive hybrid coordinate with a top of the atmosphere corresponding to the zero pressure level. As in the case of the grid point model, we also use the logarithm of surface pressure $Z = \ln p_s$ as the prognostic variable.

8.6.1 The equations

The prognostic equations for the reduced wind components $U = (au \cos \varphi)/\tilde{m}$, $V = (av \cos \varphi)/\tilde{m}$, temperature T, and surface pressure p_s are written:

$$
\left.
\begin{aligned}
\frac{\partial U}{\partial t} &= (\zeta + f)V - \dot{s}\frac{\partial p}{\partial s}\frac{\partial U}{\partial p} - \frac{\partial K}{\partial \lambda} - \left(\frac{\partial \Phi}{\partial \lambda} + \frac{RT}{p}\frac{\partial p}{\partial \lambda}\right), \\
\frac{\partial V}{\partial t} &= -(\zeta + f)U - \dot{s}\frac{\partial p}{\partial s}\frac{\partial V}{\partial p} - (1-\mu^2)\frac{\partial K}{\partial \lambda} - (1-\mu^2)\left(\frac{\partial \Phi}{\partial \mu} + \frac{RT}{p}\frac{\partial p}{\partial \mu}\right), \\
\frac{\partial T}{\partial t} &= -\mathbf{V}\cdot\nabla T - \dot{s}\frac{\partial T}{\partial s} - \frac{RT}{C_p}\left[\frac{1}{p}\int_{s_t}^{s}\nabla\cdot\left(\frac{\partial p}{\partial s'}\mathbf{V}\right)ds' - \frac{1}{p}(\mathbf{V}\cdot\nabla p)\right], \\
\frac{\partial p_s}{\partial t} &= -\int_{s_t}^{s_b}\nabla\cdot\left(\frac{\partial p}{\partial s}\mathbf{V}\right)ds.
\end{aligned}
\right\}
\quad (8.77)
$$

The diagnostic equations for vertical velocity and geopotential Φ are written:

$$\left(\dot{s}\frac{\partial p}{\partial s}\right)=-\int_{s_t}^{s}\nabla\cdot\left(\frac{\partial p}{\partial s'}\mathbf{V}\right)ds'+\frac{\partial p}{\partial p_s}\int_{s_t}^{s_b}\nabla\cdot\left(\frac{\partial p}{\partial s}\mathbf{V}\right)ds,$$

$$\left.\Phi=\Phi_s-R\int_{s_b}^{s}\frac{T}{p}\frac{\partial p}{\partial s'}ds',\right\}\tag{8.78}$$

Φ_s denoting surface geopotential.

Kinetic energy is given by:

$$K=\frac{\tilde{m}^2(U^2+V^2)}{2a^2(1-\mu^2)}.$$

Vorticity and divergence are given by:

$$\zeta=\frac{\tilde{m}^2}{a^2(1-\mu^2)}\left(\frac{\partial V}{\partial\lambda}-(1-\mu^2)\frac{\partial U}{\partial\mu}\right),\ \eta=\frac{\tilde{m}^2}{a^2(1-\mu^2)}\left(\frac{\partial U}{\partial\lambda}+(1-\mu^2)\frac{\partial V}{\partial\mu}\right).\tag{8.79}$$

The diagnostic relations by which the streamfunction ψ and velocity potential χ can be linked to vorticity and divergence are the Poisson equations below:

$$\zeta=\nabla^2\psi,\ \eta=\nabla^2\chi.\tag{8.80}$$

The reduced wind components are then obtained from the Helmholtz relations:

$$U=\frac{\partial\chi}{\partial\lambda}-(1-\mu^2)\frac{\partial\psi}{\partial\mu},$$

$$\left.V=(1-\mu^2)\frac{\partial\chi}{\partial\mu}+\frac{\partial\psi}{\partial\lambda}.\right\}\tag{8.81}$$

To facilitate the computations made in spectral space, it is practical to use instead of vorticity ζ and divergence η their respective reduced values $\zeta'=\zeta/\tilde{m}^2$ and $\eta'=\eta/\tilde{m}^2$. The Poisson equations (8.80) are then written:

$$\zeta'=\nabla'^2\psi,\ \eta'=\nabla'^2\chi,\tag{8.82}$$

with $\nabla'^2\equiv\dfrac{1}{a^2(1-\mu^2)}\left\{\dfrac{\partial^2}{\partial\lambda^2}+(1-\mu^2)\dfrac{\partial}{\partial\mu}\left[(1-\mu^2)\dfrac{\partial}{\partial\mu}\right]\right\}.$

8.6.2 Explicit time integration of the model

We use the vertical discretization described in Subsections 8.3.1 to 8.3.5, choosing therefore what is termed the *Lorenz* arrangement of variables.

The discretized momentum equations for the variables $\zeta' = \zeta/\tilde{m}^2$ and $\eta' = \eta/\tilde{m}^2$ are written:

$$\left. \begin{array}{l} \dfrac{\partial \zeta'_k}{\partial t} = \dfrac{1}{a^2(1-\mu^2)}\left[\dfrac{\partial (S_V)_k}{\partial \lambda} - (1-\mu^2)\dfrac{\partial (S_U)_k}{\partial \mu}\mu\right], \\[3mm] \dfrac{\partial \eta'_k}{\partial t} = \dfrac{1}{a^2(1-\mu^2)}\left[\dfrac{\partial (S_U)_k}{\partial \lambda} + (1-\mu^2)\dfrac{\partial (S_V)_k}{\partial \mu}\mu\right] - \nabla'^2 K_k - \nabla'^2 \Phi_k, \\[3mm] \zeta'_k = \dfrac{1}{a^2(1-\mu^2)}\left(\dfrac{\partial V_k}{\partial \lambda} - (1-\mu^2)\dfrac{\partial U_k}{\partial \mu}\right), \quad \eta'_k = \dfrac{1}{a^2(1-\mu^2)}\left(\dfrac{\partial U_k}{\partial \lambda} + (1-\mu^2)\dfrac{\partial V_k}{\partial \mu}\right), \end{array} \right\} \tag{8.83}$$

quantities $(S_U)_k$ and $(S_V)_k$ being given by:

$$\left. \begin{array}{l} (S_U)_k = (\tilde{m}^2 \zeta'_k + f)V_k - \dfrac{1}{2\delta p_k}\left[\dot{\varsigma}_{\tilde{k}}\left(U_{k+1} - U_k\right) + \dot{\varsigma}_{\widetilde{k-1}}\left(U_k - U_{k-1}\right)\right] \\[3mm] \qquad\quad - RT_k \varepsilon_k p_s \dfrac{\partial Z}{\partial \lambda}, \\[3mm] (S_V)_k = -(\tilde{m}^2 \zeta'_k + f)V_k - \dfrac{1}{2\delta p_k}\left[\dot{\varsigma}_{\tilde{k}}\left(V_{k+1} - V_k\right) + \dot{\varsigma}_{\widetilde{k-1}}\left(V_k - V_{k-1}\right)\right] \\[3mm] \qquad\quad - RT_k \varepsilon_k p_s (1-\mu^2)\dfrac{\partial Z}{\partial \mu}. \end{array} \right\} \tag{8.84}$$

The thermodynamic equation is discretized in the form:

$$\begin{aligned} \dfrac{\partial T_k}{\partial t} &= \dfrac{\tilde{m}^2}{a^2(1-\mu^2)}\left[U_k\dfrac{\partial T_k}{\partial \lambda} + V_k(1-\mu^2)\dfrac{\partial T_k}{\partial \mu}\right] - \dfrac{1}{2\delta p_k}\left[\dot{\varsigma}_{\tilde{k}}\left(T_{k+1} - T_k\right) + \dot{\varsigma}_{\widetilde{k-1}}\left(T_k - T_{k-1}\right)\right] \\[2mm] &\quad - \dfrac{RT_k}{C_p}\dfrac{1}{\delta p_k}\left[\gamma_k\sum_{i=1}^{k-1}\widetilde{D}_i + \alpha_k\widetilde{D}_k\right] + \dfrac{RT_k}{C_p}\varepsilon_k\dfrac{\tilde{m}^2 p_s}{a^2(1-\mu^2)}\left[U_k\dfrac{\partial Z}{\partial \lambda} + V_k(1-\mu^2)\dfrac{\partial Z}{\partial \mu}\right]. \end{aligned} \tag{8.85}$$

The equation for evolution of the logarithm of surface pressure is written:

$$\dfrac{\partial Z}{\partial t} = -\dfrac{1}{p_s}\sum_{k=1}^{N}\widetilde{D}_k, \text{ now with } \widetilde{D}_k = \nabla\cdot(\mathbf{V}_k\delta p_k); \tag{8.86}$$

because $\partial p/\partial p_s = g(s)$, this quantity (which is different from the one defined before for the σ coordinate model) is expanded as:

$$\widetilde{D}_k = \nabla\cdot(\mathbf{V}_k\delta p_k) = \tilde{m}^2\delta p_k\,\eta'_k + (g_{\tilde{k}} - g_{\widetilde{k-1}})\dfrac{\tilde{m}^2 p_s}{a^2(1-\mu^2)}\left[U_k\dfrac{\partial Z}{\partial \lambda} + V_k(1-\mu^2)\dfrac{\partial Z}{\partial \mu}\right],$$

The diagnostic equation for vertical velocity is written:

$$\dot{\varsigma}_{\tilde{k}} = \left(\dot{s} \frac{\partial p}{\partial s} \right)_{\tilde{k}} = -\sum_{i=1}^{k} \tilde{D}_i + g(s_k) \sum_{i=1}^{N} \tilde{D}_i. \tag{8.87}$$

And the hydrostatic equation is written:

$$\Phi_k = \Phi_s + \sum_{i=k+1}^{N} \gamma_i RT_i + \alpha_k RT_k, \tag{8.88}$$

with the quantities α_k, β_k, γ_k, and ε_k given by the relations:

$$\alpha_k = \ln \frac{p_{\tilde{k}}}{p_k}, \quad \beta_k = \ln \frac{p_k}{p_{\widetilde{k-1}}}, \quad \gamma_k = \alpha_k + \beta_k = \ln \frac{p_{\tilde{k}}}{p_{\widetilde{k-1}}} \quad \text{and} \quad \varepsilon_k = \frac{1}{\delta p_k} \left[\alpha_k . g(s_{\tilde{k}}) + \beta_k . g(s_{\widetilde{k-1}}) \right].$$

It is important to emphasize that, unlike what happens with the pure sigma coordinate, these quantities are not constants because they depend on the values of pressure at the *interlayer surfaces* $p_{\tilde{k}}$, which in turn depend on surface pressure p_s.

Adopting the vector notations analogous to those of Subsection 8.4.4 and calling ζ', η', ψ and χ the column vectors corresponding to reduced vorticity and divergence, streamfunction and velocity potential, respectively, the prognostic equations of the model are written in condensed form as below:

$$\left. \begin{aligned} \frac{\partial Z}{\partial t} &= \mathcal{S}_z(p_s, \mathbf{U}, \mathbf{V}), \\ \frac{\partial \zeta'}{\partial t} &= \mathcal{S}_\zeta(p_s, \mathbf{U}, \mathbf{V}, \mathbf{T}, \dot{\varsigma}), \\ \frac{\partial \eta'}{\partial t} &= \mathcal{S}_\eta(p_s, \mathbf{U}, \mathbf{V}, \mathbf{T}, \dot{\varsigma}) - \nabla'^2(\mathbf{K} + \mathbf{\Phi}), \\ \frac{\partial \mathbf{T}}{\partial t} &= \mathcal{S}_\mathbf{T}(p_s, \mathbf{U}, \mathbf{V}, \mathbf{T}, \dot{\varsigma}). \end{aligned} \right\} \tag{8.89}$$

The diagnostic relations for the reduced wind components are given by:

$$\left. \begin{aligned} \zeta' &= \nabla'^2 \psi, \quad \eta' = \nabla'^2 \chi, \\ \mathbf{U} &= \frac{\partial \chi}{\partial \lambda} - (1 - \mu^2) \frac{\partial \psi}{\partial \mu}, \quad \mathbf{V} = \frac{\partial \psi}{\partial \lambda} + (1 - \mu^2) \frac{\partial \chi}{\partial \mu}. \end{aligned} \right\} \tag{8.90}$$

This model is integrated explicitly by a procedure analogous to that used for the shallow water model in Subsection 6.3.1 and by carrying out the steps below:

• From the values of the prognostic variables known at time t on the transformation grid (Gaussian grid), we calculate the geopotential and vertical velocity, then the quantities $(S_U)_k$ and $(S_V)_k$ as well as the right-hand sides of the thermodynamic equation and the surface pressure tendency equation;

- The Fourier and Legendre transforms applied to the various quantities calculated at the grid points can be used to calculate the spectral coefficients of the right-hand sides of equations (8.89), that is, $\boldsymbol{\mathcal{S}}_Z$, $\boldsymbol{\mathcal{S}}_\zeta$, $\boldsymbol{\mathcal{S}}_\eta - \nabla'^2(\mathbf{K} + \boldsymbol{\Phi})$, and $\boldsymbol{\mathcal{S}}_T$;
- The evolution equations for the vectors representing the spectral coefficients are then written:

$$\left. \begin{aligned} \frac{\partial Z_n^m}{\partial t} &= \left(\boldsymbol{\mathcal{S}}_Z\right)_n^m, \\[2mm] \frac{\partial \boldsymbol{\zeta}\,'^m_n}{\partial t} &= \left(\boldsymbol{\mathcal{S}}_\zeta\right)_n^m, \\[2mm] \frac{\partial \boldsymbol{\eta}\,'^m_n}{\partial t} &= \left(\boldsymbol{\mathcal{S}}_\eta\right)_n^m + \frac{n(n+1)}{a^2}\mathbf{K}_n^m + \frac{n(n+1)}{a^2}\boldsymbol{\Phi}_n^m, \\[2mm] \frac{\partial \mathbf{T}_n^m}{\partial t} &= \left(\boldsymbol{\mathcal{S}}_T\right)_n^m. \end{aligned} \right\} \tag{8.91}$$

- Discretization of these time derivatives in forward differences at the first time step, then in central differences from the second time step onwards, allows us to obtain the values of quantities at time $t + \Delta t$;
- Having the values of the spectral coefficients of ζ' and η' at time $t + \Delta t$, it is easy to obtain the spectral coefficients of ψ and χ at this time step:

$$\psi_n^m(t+\Delta t) = -\frac{a^2}{n(n+1)}\zeta\,'^m_n(t+\Delta t) \text{ and } \chi_n^m(t+\Delta t) = \frac{a^2}{n(n+1)}\eta\,'^m_n(t+\Delta t). \tag{8.92}$$

- By performing the inverse Legendre transform on coefficients $\zeta\,'^m_n(t+\Delta t)$, $\eta\,'^m_n(t+\Delta t)$, $\psi_n^m(t+\Delta t)$, $\chi_n^m(t+\Delta t)$, $\mathbf{T}_n^m(t+\Delta t)$, and $Z_n^m(t+\Delta t)$, we obtain the Fourier coefficients $\zeta'_m(\mu)$, $\eta'_m(\mu)$, $\psi_m(\mu)$, $\chi_m(\mu)$, $\mathbf{T}_m(\mu)$, and $Z_m(\mu)$ at time $t + \Delta t$; Fourier coefficients of the reduced wind components $\mathbf{U}_m(\mu)$ and $\mathbf{V}_m(\mu)$ are deduced by using the relations:

$$\left. \begin{aligned} \mathbf{U}_m(\mu) &= \left[\frac{\partial \boldsymbol{\chi}}{\partial \lambda}\right]_m - \left[(1-\mu^2)\frac{d\psi}{d\mu}\right]_m = im\chi_m(\mu) - \sum_{n=m+1}^{M}\psi_n^m(1-\mu^2)\frac{dP_n^m(\mu)}{d\mu}, \\[2mm] \mathbf{V}_m(\mu) &= \left[\frac{\partial \boldsymbol{\psi}}{\partial \lambda}\right]_m + \left[(1-\mu^2)\frac{d\boldsymbol{\chi}}{d\mu}\right]_m = im\psi_m(\mu) + \sum_{n=m}^{M}\chi_n^m(1-\mu^2)\frac{dP_n^m(\mu)}{d\mu}. \end{aligned} \right\} \tag{8.93}$$

- The final step is to perform the inverse Fourier transform on the coefficients obtained so as to find the values of the variables $\zeta'(\lambda, \mu)$, $\eta'(\lambda, \mu)$, $\psi(\lambda, \mu)$, $\chi(\lambda, \mu)$, $\mathbf{T}(\lambda, \mu)$, $\mathbf{U}(\lambda, \mu)$, $\mathbf{V}(\lambda, \mu)$, and $Z(\lambda, \mu)$ on the transformation grid at time $t + \Delta t$.

This process (calculation on the transformation grid, transition to spectral coefficients, time step, return to the grid point values) is then repeated to perform the time integration until the desired range is reached.

The remarks on implementation of the centred explicit scheme that were noted in Subsection 8.4.4 for the grid point model hold for the spectral model. The stability criterion is slightly different, though, as it involves the truncation characteristics instead of the mesh size of the grid.

The linear analysis that can be made on the spectral model with triangular truncation of the maximum wavenumber M shows that the CFL stability condition is given by:

$$-1 \leq \left[\frac{V^* M}{a} + \frac{V_g \sqrt{N(N+1)}}{a} \right] \Delta t \leq +1, \tag{8.94}$$

with V^* designating the maximum value of the synoptic wind and V_g the value of the speed of propagation of the fastest gravity waves. By taking the same numerical values as used for the grid point models, we verify that, by choosing a maximum wavenumber $M = 100$, we ensure the model is stable by setting a time step of two and a half minutes (150 seconds). We thus achieve greater efficiency than with the equivalent grid point model using the C grid.

Using a variable resolution on the sphere with the presence of the map scale factor $\tilde{m}$ in the equations introduces additional constraints: like that mentioned in Subsection 2.4.4, relation (2.23) giving $\tilde{m}$ as a function of μ' (cosine of the colatitude on the transformed sphere) is linear and expressed therefore as a function of the surface harmonics Y_0^0 and Y_1^0. It follows that the second of the inequalities (4.50) seen in Subsection 4.3.7 (fixing the constraint that the Gaussian grid should satisfy to ensure accurate calculation of the nonlinear terms without aliasing), should be modified: $K < 3M + 2$. This new constraint therefore entails a slight increase in the number of points on the transformation grid.

8.6.3 Implementation of semi-implicit time integration

Notice first that the geopotential appears in the divergence equation only. Semi-implicit processing can therefore be applied on the divergence equation alone, while the vorticity equation continues to be processed explicitly.

When we work with the progressive hybrid coordinate, the coefficients α_k, β_k, γ_k, ε_k and thickness δp_k henceforth depend on the surface pressure p_s. To linearize the terms responsible for gravity wave propagation, we therefore need not only reference temperature values at each level T_k^*, but also a surface pressure reference value p_k^*. For these reference values, we can then determine the coefficients $\hat{\alpha}_k$, $\hat{\beta}_k$, $\hat{\gamma}_k$, $\hat{\varepsilon}_k$, and the thickness $\widehat{\delta p_k}$ independently of the prognostic variable p_s.

Introducing the map scale factor $\tilde{m}$ due to the variable solution also complicates the way of solving the implicit system. If the map scale factor does not vary too much over the working domain, then it may be written $\tilde{m} = \tilde{m}^* + \tilde{m}'$, with $\tilde{m}^*$ representing the mean value of $\tilde{m}$ over the working domain. The mean value $\tilde{m}^*$ is integrated into the linear terms dealt with implicitly while the residual terms containing $\tilde{m}'$ are processed explicitly. This is the assumption used in the remainder of this chapter to establish the equations corresponding to the semi-implicit processing of the system of equations.

By implicit treatment of the pressure force term and the divergence term in the surface pressure evolution equation and the temperature equation, we get:

$$
\left.
\begin{aligned}
Z_n^m(t + \Delta t) &= (Z_T)_n^m - \Delta t\, \mathbf{S}_N^T\, \tilde{m}^{*2}\, \boldsymbol{\eta}'_n^{\,m}(t + \Delta t), \\[4pt]
\boldsymbol{\eta}'_n^{\,m}(t + \Delta t) &= (\boldsymbol{\eta}'_T)_n^m + \Delta t\, \frac{n(n+1)}{a^2}\left[\widehat{\Phi}_n^{\,m}(t + \Delta t) + \mathbf{R}^* Z_n^m(t + \Delta t)\right], \\[4pt]
\mathbf{T}_n^m(t + \Delta t) &= (T_T)_n^m - \Delta t \mathbf{A}\, \tilde{m}^{*2}\, \boldsymbol{\eta}'_n^{\,m}(t + \Delta t),
\end{aligned}
\right\}
\tag{8.95}
$$

the transitory quantities $(Z_T)_n^m$, $(\boldsymbol{\eta}'_T)_n^m$, and $(\mathbf{T}_T)_n^m$ being written:

$$
\left.
\begin{aligned}
(Z_T)_n^m &= Z_n^m(t - \Delta t) + 2\Delta t (\boldsymbol{S}_Z)_n^m - \Delta t\, \mathbf{S}_N^T\, \tilde{m}^{*2}\left[\boldsymbol{\eta}'_n^{\,m}(t - \Delta t) - 2\boldsymbol{\eta}'_n^{\,m}(t)\right], \\[4pt]
(\boldsymbol{\eta}'_T)_n^m &= \boldsymbol{\eta}'_n^{\,m}(t - \Delta t) + 2\Delta t\left[(\boldsymbol{S}_\eta)_n^m + \frac{n(n+1)}{a^2}K_n^m(t) + \frac{n(n+1)}{a^2}\Phi_n^m(t)\right] \\[4pt]
&\quad + \Delta t\, \frac{n(n+1)}{a^2}\left\{\left[\widehat{\Phi}_n^{\,m}(t - \Delta t) - 2\widehat{\Phi}_n^{\,m}(t)\right] + \mathbf{R}^*\left[Z_n^m(t - \Delta t) - 2Z_n^m(t)\right]\right\}, \\[4pt]
(\mathbf{T}_T)_n^m &= \mathbf{T}_n^m(t - \Delta t) + 2\Delta t\,(\boldsymbol{S}_T)_n^m - \Delta t \mathbf{A}\, \tilde{m}^{*2}\left[\boldsymbol{\eta}'_n^{\,m}(t - \Delta t) - 2\boldsymbol{\eta}'_n^{\,m}(t)\right].
\end{aligned}
\right\}
\tag{8.96}
$$

The lower triangular matrix $\mathbf{A}$ takes the form:

$$
\mathbf{A} = \frac{R}{C_p}
\begin{bmatrix}
\widehat{\alpha}_1 T_1^* & 0 & . & 0 & . & 0 \\[6pt]
\dfrac{\widehat{\gamma}_2 \widehat{\delta p}_1}{\widehat{\delta p}_2} T_2^* & \widehat{\alpha}_2 T_2^* & . & 0 & . & 0 \\[6pt]
 & . & & . & & . \\[6pt]
\dfrac{\widehat{\gamma}_k \widehat{\delta p}_1}{\widehat{\delta p}_k} T_k^* & \dfrac{\widehat{\gamma}_k \widehat{\delta p}_2}{\widehat{\delta p}_k} T_k^* & . & \widehat{\alpha}_k T_k^* & . & 0 \\[6pt]
 & . & & . & & . \\[6pt]
\dfrac{\widehat{\gamma}_N \widehat{\delta p}_1}{\widehat{\delta p}_N} T_N^* & \dfrac{\widehat{\gamma}_N \widehat{\delta p}_2}{\widehat{\delta p}_N} T_2^* & . & \dfrac{\widehat{\gamma}_N \widehat{\delta p}_k}{\widehat{\delta p}_N} T_2^* & . & \widehat{\alpha}_N T_N^*
\end{bmatrix}.
\tag{8.97}
$$

The elements of column vector $\mathbf{S}_k$ are the values of thicknesses $\widehat{\delta p}_k$, with the subscript i ranging from 1 to k, the residual elements being zero:

$$
\mathbf{S}_k^T = \left[\widehat{\delta p}_1 \ \ \widehat{\delta p}_2 \ ... \ \widehat{\delta p}_k \ 0 \ ... \ 0\right].
\tag{8.98}
$$

The elements of column vector $\widehat{\Phi}_n^{\,m}$ are the linearized geopotential spectral coefficients on the levels k, the subscript k ranging from 1 to N:

$$
\widehat{\Phi}_n^{\,m} = (\Phi_s)_n^m \mathbf{I} + \mathbf{B} T_n^m,
\tag{8.99}
$$

the upper triangular matrix $\mathbf{B}$ being given by:

$$\mathbf{B} = R \begin{bmatrix} \hat{\alpha}_1 & \hat{\gamma}_2 & . & \hat{\gamma}_k & . & \hat{\gamma}_N \\ 0 & \hat{\alpha}_2 & . & \hat{\gamma}_k & . & \hat{\gamma}_N \\ . & . & . & . & . & . \\ 0 & 0 & . & \hat{\alpha}_k & . & \hat{\gamma}_N \\ . & . & . & . & . & . \\ 0 & 0 & . & 0 & . & \hat{\alpha}_N \end{bmatrix}. \tag{8.100}$$

The elements of vector $\mathbf{R}^*$ are the quantities RT_k^*, with the subscript k ranging from 1 to N:

$$\mathbf{R}^{*T} = Rp_s^* \, [\mathrm{T}_1^* \hat{\varepsilon}_1 \ \ \mathrm{T}_1^* \hat{\varepsilon}_2 \ . \ \mathrm{T}_k^* \hat{\varepsilon}_k \ . \ \mathrm{T}_N^* \hat{\varepsilon}_N]. \tag{8.101}$$

By inserting the value of $\hat{\Phi}_n^m(t + \Delta t)$, which is expressed as a function of the value of $\mathbf{T}_n^m(t + \Delta t)$, and the value of $Z_n^m(t + \Delta t)$ into the divergence equation $\boldsymbol{\eta}_n'^m(t + \Delta t)$ of system (8.95), after writing $\mathbf{M} = \mathbf{BA} + \mathbf{R}^* \mathbf{S}_N^T$, we obtain the matrix equation:

$$\left[\mathbf{I} + \tilde{m}^{*2} \, \Delta t^2 \, \frac{n(n+1)}{a^2} \mathbf{M} \right] \boldsymbol{\eta}_n'^m(t + \Delta t) = (\boldsymbol{\eta'}_T)_n^m - \Delta t \, \frac{n(n+1)}{a^2} \Big[\, \mathbf{R}^*(Z_T)_n^m + \mathbf{B}\,(\mathbf{T}_T)_n^m \, \Big], \tag{8.102}$$

This allows us to find the value of $\boldsymbol{\eta}_n'^m(t + \Delta t)$ by simply multiplying both sides of equation (8.102) by the matrix:

$$\left[\mathbf{I} + \tilde{m}^{*2} \, \Delta t^2 \, \frac{n(n+1)}{a^2} \mathbf{M} \right]^{-1},$$

which can be calculated in advance for all values of n.

We finally obtain:

$$\boldsymbol{\eta}_n'^m(t + \Delta t) = \left[\mathbf{I} + \tilde{m}^{*2} \, \Delta t^2 \, \frac{n(n+1)}{a^2} \mathbf{M} \right]^{-1} \left\{ (\boldsymbol{\eta'}_T)_n^m - \Delta t \, \frac{n(n+1)}{a^2} \Big[\mathbf{R}^*(Z_T)_n^m + \mathbf{B}(\mathbf{T}_T)_n^m \Big] \right\}, \tag{8.103}$$

The values of $Z_n^m(t + \Delta t)$ and $\mathbf{T}_n^m(t + \Delta t)$ can then be calculated by using the other equations of system (8.95). Then, just as in the explicit method, we solve the Poisson equations that yield the streamfunction $\psi_n^m(t + \Delta t)$ and the velocity potential $\chi_n^m(t + \Delta t)$.

It is particularly easy, then, to implement the semi-implicit algorithm in a spectral model, insofar as the map scale factor does not vary too much over the working domain. In this event, only the part consisting in processing the spectral coefficients fields is modified, and the system of Helmholtz equations is easily solved given how easy it is to calculate the Laplacian when working in the spectral coefficient space.

When the map scale factor varies greatly over the working domain, the method just described leads to an unstable scheme and the terms containing the map scale factor $\tilde{m}^2$ have to be processed implicitly. As this map scale factor is expressed by a polynomial in μ', its square is expressed by a polynomial in μ'^2. Consequently, given relation (4.35), equation (8.103) directly yielding $\eta_n'^m(t + \Delta t)$ is replaced by a pentadiagonal linear system for $\eta_{n-2}'^m(t + \Delta t)$, $\eta_{n-1}'^m(t + \Delta t)$, $\eta_n'^m(t + \Delta t)$, $\eta_{n+1}'^m(t + \Delta t)$, and $\eta_{n+2}'^m(t + \Delta t)$. This linear system is then solved by applying the Gaussian elimination method.

The linear analysis that can be performed on the spectral model with triangular truncation of maximum wavenumber M shows that the CFL condition is given by:

$$-1 \le \left[\frac{V^* M}{a} \right] \Delta t \le +1, \tag{8.104}$$

where V^* designates the maximum value of the synoptic wind. By taking the same numerical values as in the grid point model, it can be verified that by choosing a maximum wavenumber $M = 100$, we ensure the model is stable by taking a time step of 20 minutes (1200 seconds), as in the equivalent grid point model. However, we achieve greater efficiency than with the grid point model, because the surplus computation to complete a time step is far less.

It is important to emphasize that using the progressive hybrid coordinate requires, for isolating the linear terms and implementing the semi-implicit method, not only the choice of a reference temperature profile but also of a reference surface pressure. As said in Subsection 8.5.2, it is essential to ensure that the choice of linear operator leads to a stable scheme by performing a detailed linear analysis.

8.7 Lagrangian advection in baroclinic models

So as to work with a time step extending beyond the limits set by condition (8.68) for the grid point model and (8.104) for the spectral model, we can also process the advection terms in Lagrangian fashion.

In the case of a baroclinic model, advection and interpolation of variables must be handled three-dimensionally. Trajectory calculation is similar to that described for shallow water models in Subsection 6.2.5 for grid point models and in Subsection 6.3.3 for spectral models; however, it must take into account vertical displacement, making sure that the particle starting point does not exit the working domain. Interpolation of the variables at points located between the grid points must also be done three-dimensionally, which involves non-negligible additional computation compared with the two-dimensional case.

When the time step is increased, the use of the *semi-Lagrangian semi-implicit method* may lead, though, to generating wrong stationary solutions above mountains (Coiffier et al., 1987b). This problem may be overcome by averaging the forcing due to orography along the trajectory (Kaas, 1987), as shown by Tanguay et al. (1992). The phenomenon was analysed in detail in the context of a shallow water model with orography by Rivest et al. (1994), who showed that the problem arises from false numerical resonance of the part of the solution forced by orography; this numerical resonance may be eliminated by off-centring the semi-implicit scheme, but the time integration scheme then has only 1st-order accuracy. An alternative solution, that allows us to maintain 2nd-order accuracy in time, consists in processing the orographic forcing explicitly (Ritchie and Tanguay, 1996).

Despite the additional computation entailed by Lagrangian treatment of three-dimensional advection terms, the semi-Lagrangian semi-implicit method proves very advantageous for integrating operational forecasting models using primitive equations, for which great efficiency is sought. This method, which allows the time step to be increased, may also be extended to non-hydrostatic Euler equations. Various schemes have been proposed by Tanguay et al. (1990), Bubnová et al. (1995), as well as by Caya and Laprise (1999), and their stability properties have been studied by Bénard (2003).

9 Physical parameterizations

9.1 Introduction

The atmospheric models described in the previous chapters simulate the evolution of an atmosphere where no exchanges occur with its surroundings. Although inertia effects dominate the evolution of the atmosphere at the synoptic scale at mid-latitudes over short time periods (24 or 36 hours), when constructing realistic forecast models allowance must be made for the physical processes not handled by the adiabatic frictionless equations.

Exchanges of momentum, heat, and water vapour between the atmosphere and its surroundings (space and Earth) are the outcome of physical processes involving far smaller space and time scales than those taken into account by the dynamic model. As these processes cannot be simulated explicitly, we try to determine where and when they occur and to calculate their average effect for each of the model's prognostic variables (wind, temperature, water vapour). This is what is called *parameterizing* the effect of physical processes for *scales smaller than the scales resolved by the dynamic model* (or *sub-grid scales*). The parameterization of the effects of the sub-grid scale physical processes depends therefore on the space and time scales actually handled by the discretized model's equations.

Generally, the parameterization of sub-grid scale physical processes must make it possible to correctly simulate the various energy and hydrological cycles of the atmosphere and to evaluate the surface fluxes that determine the evolution of *current weather elements* (temperature, humidity, and wind close to the surface, rainfall, snowfall). Allowance for exchanges with the Earth requires the introduction of new variables such as the soil temperature and moisture, the evolution of which can be described by additional prognostic equations, and the various characteristics of the ground surface (two-dimensional fields) such as roughness length, proportion of vegetation, emissivity, albedo, and the like.

The parameterization of the sub-grid scale physical processes consists, then, in using mathematical relations to describe the way in which the model variables and these new variables interact under the influence of the various processes taken into account. This procedure leads to the calculation of terms, $\mathbf{F}$, Q, and M (algebraic inputs of momentum, energy, and water vapour due to physical processes) that occur in equations (2.1), (2.2), and (2.4) of Chapter 2 and are called, to simplify things, *physical tendencies*. These physical tendencies are themselves obtained by taking account of the contribution of all the partial physical tendencies due to each of the parameterized physical processes.

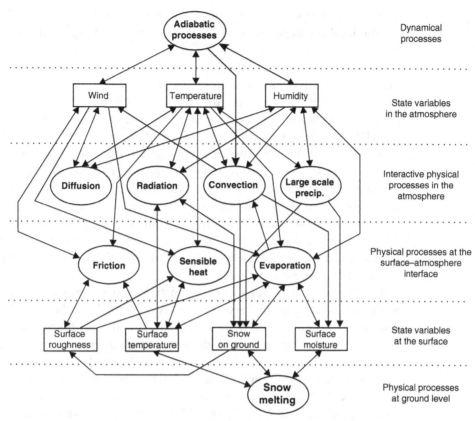

Figure 9.1   Physical interactions in a model. (From an ECMWF drawing by Louis, J.-F.)

The main physical processes to be parameterized when using the primitive equations are:

- radiation and its interaction with the various atmospheric constituents, especially clouds;
- exchanges of heat, humidity, and momentum with the Earth's surface and their vertical redistribution under the effects of dry and moist turbulence;
- large-scale precipitation;
- convection and its associated precipitation;
- the effect of sub-grid orography on the resolved part of the flow.

The various processes listed above interact in a fairly complicated fashion (as Figure 9.1 shows) involving many feedback loops. The effect of clouds is an example of negative (self-regulating) feedback: greater cloud cover entails a reduction in sunshine and evaporation, which in turn tends to reduce the amount of cloud cover. The effect of snow cover provides an example of a positive (amplifying) feedback loop: greater snow cover increases the reflection of solar radiation, thereby reducing the input of heat at ground level and contributing to further cooling of the atmosphere, which in turn favours the formation of snow. It is important therefore to correctly simulate the various feedback loops, above all avoiding the creation of unrealistic positive feedback loops.

Given the huge scope of the subject matter, we have opted to give relatively simple examples of parameterization of the main physical processes so as to bring out clearly the interaction with the dynamical part of the model. These examples correspond approximately to parameterizations used in versions of the ARPEGE and ALADIN operational models employed by Météo-France in the late 1990s and early 2000s, several of them (the use of energy-conserving multiphasic barycentric system and the processing of sub-grid orographic effects) still running in later versions. Owing to the very particular position of the parameterization of the deep convection in the modelling landscape, we have preferred to present a quite modern view, but only of a parameterization concept, with limited hints for its practical application. The set of schemes presented deals with the various interactions arising from sub-grid scale physical processes for a *column of atmosphere*, meaning there is no direct interaction between two adjacent columns.

The choice of the examples of parameterization schemes and their presentation was an opportunity to emphasize the variety of methods for the development of physical parameterizations: algebraic simplification and algorithmic splitting to process the radiative transfer equation (radiation); analytical approximate representation of the outcome of a statistical process (turbulence); use of the algebraic consequences of simple but accurate observed physical effects (large-scale precipitation); use of the framework of conceptual schemes for very complex phenomena to suggest a numerical handling (convection); heuristic algorithmic translation from larger scales to sub-grid scales of some observed local balances and remote retroactions (effects of sub-grid orography). As parameterization is a constantly evolving field of study, care has been taken to indicate, especially for the schemes still in use, the improvements or alternative solutions that may be found together with the relevant references to the scientific literature.

Because of the large number of variables and parameters involved in physical parameterizations, it has been impossible to use notations that are entirely unambiguous; we have therefore adopted the symbols commonly used in the literature on these parameterizations, avoiding where possible those that might cause confusion with the symbols used to describe the dynamical part of the models.

9.2 Equations for a multi-phase moist atmosphere

Equations (2.1), (2.2), (2.3), and (2.5) describing the behaviour of the atmosphere and the various forms of primitive equations that were seen subsequently were for a dry atmosphere (the constants C_p and R then being relative to dry air). An additional equation (2.4) made it possible to describe the evolution of water vapour (although the term M remained to be clarified). To deal with the atmosphere more realistically, it is necessary to take into account all of the constituents of the parcel under consideration: dry air, water vapour, liquid water, and ice. Such multi-phase treatment entails modifying the equations used so far (and possibly introducing new ones) in order to deal with the evolution of the parcel of atmospheric air formed by a mixture of dry air and of water in its different phases.

A fairly general formulation of the equations, which has the advantage of being applicable without detailed knowledge of the microphysical processes describing the transformations of the different phases of water in the atmosphere, was proposed by Catry et al. (2007). This formulation is based on a barycentric approach to the multi-phase moist air parcel (its velocity is the sum, weighted by mass, of the velocities of each of its constituents). This way of dealing with the moist air parcel is used both for determining a vertical coordinate (of the normalized pressure or hybrid type examined in Chapter 7) and for establishing the laws of conservation of the various phases; in addition, it is based on a number of simplifying assumptions:

- the atmosphere is assumed to be in thermodynamic equilibrium at each instant;
- the condensed phases of water, described by mean values of specific concentration, are assumed to have zero volume;
- the dry air and water vapour obey the Boyle-Mariotte and Dalton laws;
- the specific heat capacities are assumed to be independent of temperature;
- the various atmospheric constituents are all at the same temperature;
- no dry air flow offsets the mass loss/gain due to precipitation/evaporation at the surface.

Armed with these assumptions and a very general scheme describing the interactions among the various water phases, we obtain the evolution equations for the various components of moist air and the thermodynamic equation, whether for primitive equation models or nonhydrostatic models. We confine ourselves here to giving the equations for the primitive equation models and leave it to the reader to consult the paper cited for the details of calculation.

9.2.1 Schematic framework of interaction among constituents

With the formulation proposed by Catry et al. (2007), the transformations of the various phases of water in the atmosphere may be schematized as follows: we consider a parcel of moist air of unit mass and q_d, q_v, q_l, q_r, q_i, and q_s are the respective specific concentrations of dry air, water vapour, liquid water in suspension, precipitating liquid water (rain), ice in suspension, and precipitating ice (snow). These concentrations therefore verify the relation:

$$q_d + q_v + q_l + q_r + q_i + q_s = 1.$$

Given the assumptions made, the gas constant R relative to the moist air mixture is written:

$$R = q_d R_d + q_v R_v, \tag{9.1}$$

R_d and R_v being the gas constants relative to dry air and to water vapour.

The precipitating fluxes of rain and snow are written:

$$P_l = -\rho_r w_r \text{ and } P_i = -\rho_s w_s,$$

ρ_r and ρ_s being the respective densities of rain and snow, and w_r and w_s the corresponding vertical velocities relative to the parcel's centre of mass. These fluxes, like the pseudo-fluxes to be defined, are counted positively downwards.

The transformations between water vapour and its various condensed states are transfers due to microphysical processes that may be interpreted, via a mass-weighted vertical integration, as *pseudo-fluxes*: P_l' represents the transfer between water vapour and liquid water in suspension due to the processes of condensation and evaporation; P_l'' represents the transfer between liquid water in suspension and rain due to auto-conversion processes; P_l''' represents the transfer between rainwater and water vapour due to the evaporation of liquid precipitation; P_i' represents the transfer between water vapour and ice in suspension due to the processes of freezing and sublimation; P_i'' represents the transfer between ice in suspension and snow due to the processes of auto-conversion; P_i''' represents the transfer between snow and water vapour due to the sublimation of snow precipitation. The interactions among the various phases of water via pseudo-fluxes are outlined in Figure 9.2. All other changes from one water species to another can be reconstructed from the eight chosen ones, without loss of generality or of accuracy for the thermodynamic computations.

Within the framework of the barycentric approach, in order to eliminate any acceleration of the parcel, the downgoing flux of liquid P_l and ice P_i precipitation relative to the parcel's centre of mass must be offset by a rising flux of non-precipitating phases within the parcel. This principle is used for writing the evolution equations of the various components of moist air.

This schematic framework is used to describe theoretically, by means of pseudo-fluxes, the interactions between the various constituents of the moist air parcel that correspond to any given microphysics scheme; however, the system is greatly simplified in the case where we adopt simplified precipitation schemes without any condensates in suspension.

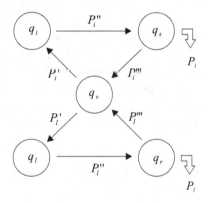

Figure 9.2   Schematic representation of the microphysical scheme for phase-changes of water in the atmosphere. (After Catry et al., 2007)

9.2.2 The equations in conservative form

Since there is no mass flux in the barycentric approach, the continuity equation remains unchanged and is written using the general vertical coordinate introduced in Chapter 7:

$$\frac{\partial}{\partial s}\left(\frac{\partial p}{\partial t}\right) = -\nabla_s \cdot \left(\frac{\partial p}{\partial s}\mathbf{V}\right) - \frac{\partial}{\partial s}\left(\dot{s}\frac{\partial p}{\partial s}\right), \tag{9.2}$$

the boundary conditions being:

at the top where $s = 0$: $\dot{s}\dfrac{\partial p}{\partial s} = 0$,

at the bottom where $s = 1$: $\dot{s}\dfrac{\partial p}{\partial s} = g(P_L + P_I - E)$.

P_L, P_I, and E representing respectively the flux of liquid precipitation, the flux of solid precipitation, and the flux of surface evaporation (E being counted positively upwards). It follows that the surface pressure tendency equation obtained by integrating the continuity equation is written:

$$\frac{\partial p_s}{\partial t} = -\int_0^1 \nabla\cdot\left(\frac{\partial p}{\partial s}\mathbf{V}\right)ds - g(P_L + P_I - E), \tag{9.3}$$

and the vertical velocity equation:

$$\dot{s}\frac{\partial p}{\partial s} = -\int_0^s \nabla\cdot\left(\frac{\partial p}{\partial s}\mathbf{V}\right)ds + \frac{\partial p}{\partial p_s}\int_0^s \nabla\cdot\left(\frac{\partial p}{\partial s}\mathbf{V}\right)ds + g\frac{\partial p}{\partial p_s}(P_L + P_I - E). \tag{9.4}$$

The conservation equations of the various components of the parcel are obtained by writing that the precipitation fluxes must be offset by the sum of the compensating fluxes for each of the non-precipitating components. This leads to the following set of equations:

$$\left.\begin{aligned}
\frac{dq_d}{dt} &= g\frac{\partial}{\partial p}\left[\frac{q_d(P_l + P_i)}{1 - q_r - q_s} - J_{q_d}\right], \\[4pt]
\frac{dq_v}{dt} &= g\frac{\partial}{\partial p}\left[P_l''' + P_i''' - P_l' - P_i' + \frac{q_v(P_l + P_i)}{1 - q_r - q_s} - J_{q_v}\right], \\[4pt]
\frac{dq_l}{dt} &= g\frac{\partial}{\partial p}\left[P_l' - P_l'' + \frac{q_l(P_l + P_i)}{1 - q_r - q_s} - J_{q_l}\right], \\[4pt]
\frac{dq_r}{dt} &= g\frac{\partial}{\partial p}\left[P_l'' - P_l''' - P_l\right], \\[4pt]
\frac{dq_i}{dt} &= g\frac{\partial}{\partial p}\left[P_i' - P_i'' + \frac{q_i(P_l + P_i)}{1 - q_r - q_s} - J_{q_i}\right], \\[4pt]
\frac{dq_s}{dt} &= g\frac{\partial}{\partial p}\left[P_i'' - P_i''' - P_i\right].
\end{aligned}\right\} \tag{9.5}$$

In these equations, the quantities J_{qd}, J_{qv}, J_{ql}, and J_{qi} represent respectively the turbulent fluxes for dry air, water vapour, cloud liquid, and solid water; these fluxes verify the relation:

$$J_{q_d} + J_{q_v} + J_{q_l} + J_{q_i} = 0.$$

The thermodynamic equation (conservation of energy) relative to the multiphase mixture is written, account being taken of the differential heat transport by precipitation:

$$\frac{\partial}{\partial t}\left(C_p T\right) = -g\frac{\partial}{\partial p}\left[(C_l - C_{p_d})P_l T + (C_i - C_{p_d})P_i T - (\hat{C} - C_{p_d})(P_l + P_i)T\right]$$

$$-g\frac{\partial}{\partial p}\left[J_s + J_{rad} - L_l(T_0)(P_l' - P_l''') - L_i(T_0)(P_i' - P_i''')\right]. \tag{9.6}$$

In this equation, the specific heat of the mixture at constant pressure is given by:

$$C_p = C_{pd}q_d + C_{pv}q_v + C_l(q_l + q_r) + C_i(q_i + q_s), \tag{9.7}$$

where the quantities C_{pd}, C_{pv}, C_l and C_i, are the specific heats at constant pressure for dry air, water vapour, liquid water, and ice, respectively, and the quantity $\hat{C}$, the specific heat of all the non-precipitating phases, which is written:

$$\hat{C} = \frac{C_{p_d}q_d + C_{p_v}q_v + C_l q_l + C_i q_i}{1 - q_r - q_s}.$$

The quantities $L_l(T_0)$ and $L_i(T_0)$ represent the latent heat of vaporization of liquid water and the latent heat of sublimation of ice at the temperature $T_0 = 0$ K; the quantities J_s and J_{rad} represent the turbulent heat flux and the radiation flux. The fully conservative form of (9.6) is obtained by substituting the impact of the differences between $L_{l/i}(T)$ and $L_{l/i}(T_0)$ through terms in $T\partial(P_{l/i})/\partial p$ to be recombined with those describing the above-mentioned heat transport by precipitation.

The parameterization problem therefore comes down to determining the various fluxes at the interlayer surfaces of the model, whose vertical divergence provides the physical tendencies for the evolution equations: the radiation flux J_{rad}, the turbulent heat flux J_s, the precipitation fluxes P_l and P_i, and the pseudo-fluxes associated with phase changes for the thermodynamic equation; the turbulent fluxes J_{qv}, J_{ql}, and J_{qi}, the precipitation fluxes P_l and P_i, and the pseudo-fluxes associated with microphysics for the evolution equations of the various water phases; and the turbulent fluxes of momentum $\mathbf{J_V}$, for the evolution equation of the horizontal wind.

9.3 Radiation

9.3.1 General points

Allowing for the effects of electromagnetic radiation in a prediction model consists in calculating the energy fluxes at the interlayer surfaces of the model. It is then possible to calculate from these the corresponding heat sources and sinks for those layers. The fluxes at the surface (lower boundary of the model's lowest layer) are of capital importance, as they are mainly responsible for the evolution of temperature in the vicinity of the ground and for its more or less marked diurnal variation. The sources of electromagnetic radiation to be considered are the Sun, the atmosphere, and the Earth.

The Sun radiates roughly like a black body whose surface temperature is thought to be 6000 K. It emits radiation ranging from 0.2 μm to 4 μm with the following energy distribution:

- 8% in the ultraviolet ($\lambda < 0.4$ μm),
- 41% in the visible (0.4 μm $< \lambda < 0.7$ μm),
- 51% in the infrared ($\lambda > 0.7$ μm).

Solar radiation is partly absorbed in the atmosphere by atmospheric gases: oxygen, nitrogen, and above all ozone, which is present in the upper atmosphere, absorb almost all of the ultraviolet, whereas water vapour and carbon dioxide in the troposphere absorb the infrared. Solar radiation also undergoes scattering; in the absence of clouds, this scattering is caused by atmospheric gases (Rayleigh scattering) and aerosols. Aerosols also absorb a small part of the solar radiation.

Clouds act as a diffusive medium because of the presence of water droplets and ice crystals that modify the atmosphere's diffuse transmission and reflection; clouds also absorb a fraction of solar radiation. All considered, the overall solar radiation received at ground level when the sky is overcast is lower than what is received when the sky is clear.

The Earth absorbs some of the incident radiation and reflects the other part. This reflection depends on the surface albedo A_T, the value of which itself depends on the nature of the surface (sea, bare ground, vegetation, snow) and the wavelength. The Earth radiates too and, given its temperature, close to 288 K, emission is in the domain of wavelengths ranging from 5 μm to 100 μm (infrared). Terrestrial radiation is a function of surface temperature T_S and surface emissivity e_T (which itself depends on the temperature and nature of the terrain) and is written $R_T = e_T \sigma T_S^4$, σ being the Stefan-Boltzmann constant. It should be noted that R_T is not equal to the surface flux of thermal radiation (see below equation (9.16)). Terrestrial radiation is almost entirely absorbed by the atmosphere because of the presence of carbon dioxide, water vapour, and ozone.

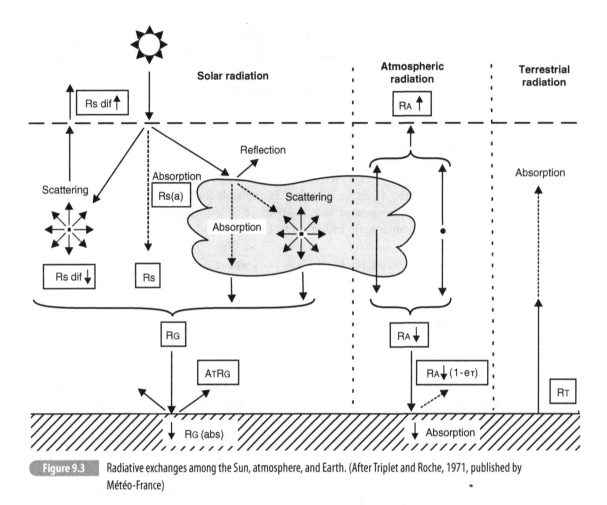

Figure 9.3 Radiative exchanges among the Sun, atmosphere, and Earth. (After Triplet and Roche, 1971, published by Météo-France)

By Kirchhoff's law, the layers of atmosphere containing clouds, aerosols, water vapour, carbon dioxide, and ozone emit an equal relative amount of radiation to that which they absorb; that radiation is in the infrared because of the temperature of the layers. It is necessary, therefore, to take account of all the radiative exchanges: those with space, which dominate, but also exchanges among layers of the atmosphere, and exchanges between the layers and the Earth's surface.

Figure 9.3 shows all of the processes and the complexity of the interactions that must be taken into account when determining radiation fluxes in the atmosphere and at the Earth's surface.

9.3.2 Allowance for the effects of radiation in the atmosphere

The basic equation used for calculating radiation fluxes is the radiative transfer equation, which describes the evolution of the spectral luminance of radiation scattered and absorbed in the atmosphere. It is a differential relation that must be integrated over the angular domain (solid angle) under consideration, over all the wavelengths, in the vertical, and over the optical depth of the interacting body (characterizing its density and its efficiency in absorbing and scattering) and by taking account of all the constituents of the atmosphere (atmospheric gases, aerosols, clouds).

Theoretically it can be fully calculated if we know the nature, distribution, and temperature of the various atmospheric constituents with precision. Now, while the distribution of water vapour is known, we must settle for the average or climatological distributions for the other constituents. However, even assuming these conditions hold, the time required for the complete computation is incompatible with the constraints of an operational forecast model. Parameterizing radiation consists therefore in introducing simplifying assumptions to alleviate the computations, while allowing us to compute sufficiently realistic fluxes at the scale of one time step of the forecast model.

The principles of parameterization of radiation in atmospheric models have been very thoroughly reviewed by Stephens (1984). Among the various schemes used in operational forecast models are those proposed by Geleyn and Hollingsworth (1979), Fouquart and Bonnel (1980), Morcrette (1991), Ritter and Geleyn (1992), and Mlawer et al. (1997).

For the sake of simplicity, we have opted here to present an updated version of a particularly economical scheme derived from the work of Geleyn and Hollingsworth (1979); this scheme was used in the operational models at Météo-France in the course of the 1990s and allows us to calculate energy fluxes at the interfaces of each of the model's layers for both solar radiation and thermal radiation.

9.3.3 Two-flux approximation and integration over a layer

Angular integration for the diffuse flux is performed by using a mean zenith angle, the value of which is determined empirically; the inverse of its cosine is called the *diffusivity factor*. Assuming isotropic scattering in the atmosphere, we use Eddington's approximation (also known as *two-flux approximation* or *two-stream approximation*), which is used to describe the scattered radiation field using only the downward flux $F_\downarrow$ and the upward flux $F_\uparrow$. For solar radiation, we must add to these two fluxes the flux S of parallel solar radiation. For infrared radiation (better known as *thermal* radiation) we must add the terms corresponding to thermal radiation emission $\pi B(T)$, where $B(T)$ is the Planck function. The approximation of the two fluxes giving the solar radiation flux and the thermal radiation fluxes for a layer of atmosphere of temperature T and optical depth δ comes down to a system of differential equations with respect to δ, that is written:

$$\left.\begin{aligned}
\frac{\partial F_\uparrow}{\partial \delta} &= \alpha_1 F_\uparrow - \alpha_2 F_\downarrow - \alpha_3 \frac{S}{\mu_0} - (\alpha_1 - \alpha_2)\,\pi B(T), \\[2mm]
\frac{\partial F_\downarrow}{\partial \delta} &= \alpha_2 F_\uparrow - \alpha_1 F_\downarrow + \alpha_4 \frac{S}{\mu_0} + (\alpha_1 - \alpha_2)\,\pi B(T), \\[2mm]
\frac{\partial S}{\partial \delta} &= -\alpha_5 S.
\end{aligned}\right\} \tag{9.8}$$

Solar radiation S at the top of the atmosphere is given by $S_0 = I_0\mu_0$, where I_0 is the solar constant corrected for seasonal variations and μ_0 is the cosine of the zenith angle. This formulation is valid for solar fluxes and thermal fluxes insofar as the coefficients α_i are suitably chosen to keep only the relevant terms corresponding to each type of radiation, as below:

$\alpha_1 = 2[1 - \varpi(1 - \beta)]$,

$\alpha_2 = 2\beta\varpi$,

$\alpha_3 = \beta_0\varpi$ for solar radiation, $\alpha_3 = 0$ for thermal radiation,

$\alpha_4 = (1 - \beta_0)\varpi$ for solar radiation, $\alpha_4 = 0$ for thermal radiation,

$\alpha_5 = 1/\mu_0$ for solar radiation, $\alpha_5 = 0$ for thermal radiation,

$\pi B(T) = 0$ for solar radiation.

In the expression of these coefficients, ϖ represents the single scattering albedo, β the fraction of backscattered radiation, and β_0 the fraction of parallel solar rays that are scattered upwards (function of μ_0).

In fact, since a good deal of the scattered radiation maintains a direction very close to the impeding one, the parameters leading to the α_i values are algebraically rescaled to consider this fraction as *non-scattered*. This more exact approximation is called the δ-*two-stream approximation*.

Integrating system (9.8) with respect to solar radiation for a layer situated between levels $k - 1$ (top) and k (base), of optical depth $\Delta\delta_k$, yields an expression for outgoing and incoming fluxes that is written:

$$
\begin{pmatrix} S_{\tilde{k}} \\ F_{\downarrow\tilde{k}} \\ F_{\uparrow\widetilde{k-1}} \end{pmatrix} = \begin{pmatrix} a_1 & 0 & 0 \\ a_2 & a_4 & a_5 \\ a_3 & a_5 & a_4 \end{pmatrix} \begin{pmatrix} S_{\widetilde{k-1}} \\ F_{\downarrow\widetilde{k-1}} \\ F_{\uparrow\tilde{k}} \end{pmatrix},
\tag{9.9}
$$

which is supplemented by the following boundary conditions:

$$\text{at the top: } F_{\downarrow 0} = 0 \text{ et } S_0 = I_0\mu_0, \tag{9.10}$$

$$\text{at the surface: } F_{\uparrow\tilde{N}} = A_T(\mu_0)S_{\tilde{N}}{!}^* \tilde{A}_T \overset{\uparrow}{F}_{\tilde{N}}, \tag{9.11}$$

where $A_T(\mu_0)$ denotes the albedo for parallel solar radiation (dependent on μ_0) and $\tilde{A}_T$ the albedo for diffuse solar radiation.

The coefficients in relations (9.9) take the values:

$$
\begin{aligned}
a_1 &= e^{-\alpha_s \Delta\delta_k}, \\
a_2 &= -a_4\gamma_2 - a_5\gamma_1 a_1 + \gamma_2 a_1, \\
a_3 &= -a_5\gamma_2 - a_4\gamma_1 a_1 + \gamma_1, \\
a_4 &= E(1-A^2)(1-E^2A^2)^{-1}, \\
a_5 &= A(1-E^2)(1-E^2A^2)^{-1},
\end{aligned}
\qquad\qquad (9.12)
$$

$$
\text{with: } E = e^{-\varepsilon\Delta\delta_k}, \quad A = \alpha_2/(\alpha_1 + \varepsilon),
$$

$$
\varepsilon = (\alpha_1^2 - \alpha_2^2)^{1/2}
$$

$$
\text{and: } \gamma_1 = \frac{\alpha_3 - \mu_0(\alpha_1\alpha_3 + \alpha_2\alpha_4)}{1 - \varepsilon^2\mu_0^2}, \quad
\gamma_2 = \frac{-\alpha_4 - \mu_0(\alpha_1\alpha_4 + \alpha_2\alpha_3)}{1 - \varepsilon^2\mu_0^2}.
$$

Their meanings are:

a_1: transmission factor for parallel solar radiation,

a_2: forward scattering factor of solar radiation,

a_3: backward scattering factor of solar radiation,

a_4: transmission factor of diffuse radiation,

a_5: reflection factor of diffuse radiation.

For thermal fluxes, in order to make the formulation of Kirchhoff's law more explicit, it is convenient to replace the fluxes F by the *modified fluxes* $F^* = \pi B(T) - F$, and the thermal version of system (9.8) is then written:

$$
\left.
\begin{aligned}
\frac{\partial F^*_\uparrow}{\partial\delta} &= \alpha_1 F^*_\uparrow - \alpha_2 F^*_\downarrow + \pi\frac{\partial B(T)}{\partial\delta}, \\
\frac{\partial F^*_\downarrow}{\partial\delta} &= \alpha_2 F^*_\downarrow - \alpha_1 F^*_\uparrow + \pi\frac{\partial B(T)}{\partial\delta}.
\end{aligned}
\right\}
\qquad (9.13)
$$

By considering isothermal layers and so taking $B(T)$ as constant inside each of them, with jumps at the transitions acting as sources for all F^* values, the modified fluxes at the layer interfaces may be written, taking the convention that the Planck functions correspond to the emitting layers and not to the receiving ones:

$$
F^*_{\uparrow\,\tilde k} = \pi B_{k+1} - F_{\uparrow\,\tilde k},
$$

$$
F^*_{\downarrow\,\tilde k} = \pi B_k - F_{\downarrow\,\tilde k}.
$$

Integrating system (9.13) relative to thermal radiation for a layer situated between levels $k-1$ (top of the layer) and k (base of the layer), of optical depth $\Delta\delta_k$, yields an expression for outgoing fluxes as a function of incoming fluxes that is written:

$$
\begin{pmatrix} F^*_{\downarrow\,\tilde k} \\ F^*_{\uparrow\,\widetilde{k-1}} \end{pmatrix}
=
\begin{pmatrix} a_4 & a_5 \\ a_5 & a_4 \end{pmatrix}
\begin{pmatrix} F^*_{\downarrow\,\widetilde{k-1}} \\ F^*_{\uparrow\,\tilde k} \end{pmatrix}
+
\begin{pmatrix} a_4 & a_5 \\ a_5 & a_4 \end{pmatrix}
\begin{pmatrix} \pi B_k - \pi B_{k-1} \\ \pi B_k - \pi B_{k+1} \end{pmatrix}.
\qquad (9.14)
$$

Coefficients a_4 and a_5 of system (9.14) describing thermal fluxes are analogous in meaning to the coefficients in system (9.9) for diffuse solar fluxes. This system is supplemented by the boundary conditions below:

$$\text{at the top: } F_{\downarrow\tilde{0}} = 0, \quad \text{that is: } F_{\downarrow\tilde{0}}^* = \pi B_{\tilde{0}}, \tag{9.15}$$

$$\text{at the surface: } F_{\uparrow\tilde{N}} = (1 - e_T)F_{\downarrow\tilde{N}} + e_T\pi B_S,$$

$$\text{that is: } F_{\uparrow\tilde{N}}^* = (1 - e_T)F_{\downarrow\tilde{N}}^* + (1 - e_T)(\pi B_S - \pi B_N), \tag{9.16}$$

where e_T is the thermal emissivity of the ground and B_S the Planck function corresponding to the surface temperature T_S.

Before introducing the complex effects of clouds, equation systems (9.9) and (9.14) must be written allowing for the effects of the three atmospheric gases (water vapour, carbon dioxide, and ozone) and of aerosols in calculating the optical depths, and therefore in calculating coefficients a_i. Because the matrices associated with the systems formed by equations (9.9) and (9.14) with the boundary conditions (9.10), (9.11), (9.15), and (9.16) are very sparse, the systems are solved simply by a targeted version of the Gaussian elimination; mathematically, this procedure corresponds exactly to the *adding method* (Van de Hulst, 1980) since it amounts to progressively determining the coefficients a_i for a combination of two adjacent layers by using the coefficients a_i found for each of the two layers.

9.3.4 Calculating optical depths and spectral integration

In order to process absorption as a function of wavelength, we construct simple absorption band models from which we define equivalent bandwidths. The computation of optical depths differs depending on whether we are considering gases or grey bodies. For gases, the variations of spectral absorption coefficients with respect to wavelength are such that, for spectral integration, we must use a nonlinear formulation with respect to the quantities of gas. However, for grey bodies (aerosols and clouds), the formulation is locally linear with respect to the quantity of absorbing and scattering body, although this may render the proportionality coefficients dependent on the instantaneous situation.

9.3.4.1 The case of gases

Optical depth $\Delta\delta$, reflecting the aggregate effect of optical depths that correspond to absorption by the various gases in question, is calculated from the scaled equivalent widths $w_{\Delta v}^g$ (each relative to the absorption band Δv and gas g) by the following formula:

$$\Delta\delta = \sum_{g=1}^{G} w_{\Delta v}^g \frac{1 + \sum_{n=1}^{N_g} (\zeta_{\Delta v}^g)_n (w_{\Delta v}^g)^n}{1 + \sum_{n=1}^{N_g} (\eta_{\Delta v}^g)_n (w_{\Delta v}^g)^n}. \tag{9.17}$$

This expression represents the sum of optical depths of the various gases in question, which are calculated individually by using the *Padé approximants* based on scaled equivalent widths $w_{\Delta v}^g$. G is the number of gases and N_g the number of Padé terms for a gas g. The coefficients $(\zeta_{\Delta v}^g)_n$ and $(\eta_{\Delta v}^g)_n$ are weighting coefficients, termed Padé coefficients, for band Δv and gas g. Water vapour, carbon dioxide, and ozone are the three main atmospheric gases to be handled, with a variable number of relevant spectral intervals for each gas.

The scaled equivalent widths $w_{\Delta v}^g$ for an absorption band Δv and gas g are computed by a formulation arising from spectral integration using the Malkmus (1967) band model, also allowing for a continuum term, and are given by:

$$w_{\Delta v}^g = \frac{a}{2b} \frac{u_r^g}{u_e^g} \left(\sqrt{1 + 4b \frac{(u_e^g)^2}{u_r^g}} - 1 \right) + c u_r^g. \tag{9.18}$$

In this formula, u_e^g is the effective quantity of gas g traversed. The incremental quantity of gas that a ray perpendicular to the layers passes through is given by $\Delta u^g = q_g \Delta p_k$, where q_g denotes the specific content of the gas in question and Δp_k the pressure thickness of layer k. For diffuse fluxes, it is suitable to take $\Delta u_e^g = 2\Delta u^g T^\phi$ (for thermal radiation) and $\Delta u_e^g = 2\Delta u^g T^\gamma$ (for solar radiation) to allow for the isotropic assumption underlying the two-flux approximation. However, for parallel downgoing solar radiation, $2\Delta u^g T^\gamma$ must be replaced by $\Delta u_*^g = \Delta u^g T^\gamma / \mu_0$ to allow for the inclination of the light rays in the atmosphere.

The *pressure-reduced* quantity u_r^g is obtained from u_e^g (or from u_*^g in the case of downward solar radiation) and takes into account variations of absorption intensity and of absorption bandwidths with respect to temperature T (as already for the *unreduced* values) and pressure p; it also integrates the diffusivity factor β, taking into account the effect of multiple scattering and of absorption throughout long optical paths, and follows from:

$$\Delta u_r^g = \Delta u_e^g p T^\alpha \beta / 2 \text{ for diffuse solar radiation,}$$

$$\Delta u_r^g = \Delta u_*^g p T^\alpha \text{ for parallel solar radiation,}$$

$$\Delta u_r^g = \Delta u_e^g p T^\varepsilon \beta / 2 \text{ for upward and downward thermal radiation,}$$

the exponents ϕ, γ, α, and ε varying with the type of gas and the chosen spectral interval.

The quantities a, b, and c occurring in formula (9.18) depend on the spectral interval and on the gas in question; they integrate the information on the relative widths and intensities of the characteristic spectral lines and on the corresponding reference temperature and reference pressure for the above scaling operations.

For water vapour in the thermal domain, the continuum term is more complex than for the other gases, which leads to Δu_r^g being expressed, for the sole term preceded by the factor c, as:

$$\Delta u_r'^g = \frac{c'}{c} \Delta u_r^g T^{-4} (1 + d_1 q T^{\varepsilon'} e^{d_2/T}); \tag{9.19}$$

in which formula, q denotes the specific humidity of the layer, d_1 and d_2, are specific constants, and we usually have $\varepsilon' \gg \varepsilon$.

9.3.4.2 The case of grey bodies

Optical depths for grey bodies in the atmosphere (liquid water and ice in clouds, aerosols) are obtained from linear formulas of the type:

$$\Delta\delta_{abs} = \sum_{c=1}^{C} K_{abs}^{c} q_c \ \text{ and } \ \Delta\delta_{sca} = \sum_{c=1}^{C} K_{sca}^{c} q_c,$$

where K_{abs}^{c}/K_{sca}^{c} are the absorption/scattering coefficients of the grey body c and q_c the specific content of the body c in the layer of atmosphere under consideration.

For clouds, the transmission and reflection factors for wide bands of the electromagnetic spectrum (solar and thermal parts) may be calculated either by having a bulk approach for the *saturation effect* (spectrally selective masking effect of intermediate layers) or by parameterizing it. From the values obtained for the absorption/scattering coefficients and for the diffusivity factor in both solar and thermal spectra, a weighted mean can be used to obtain intermediate values, which are used to calculate transmission and reflection factors so taking account of the saturation effect in an average or in a situation-dependent way. The correct choice of the spectral weighting function for the coefficients is the key to a correct accounting of the relevant average of the saturation effect, in both bulk and parameterized cases.

9.3.5 Integration over optical path and flux calculation

Relations (9.9) for solar radiation and (9.14) for thermal radiation, to which are added the boundary conditions (9.10) and (9.11) for solar radiation and (9.15) and (9.16) for thermal radiation, can be used to calculate the corresponding fluxes insofar as we are able to calculate the matrix coefficients by taking into account the different optical depths corresponding to each of the components. The problem is quite tricky in the presence of scattering. It may be solved, however, approximately but fairly economically by applying the *ideal optical path* method. The principle of this method consists in calculating exactly the optical depths of gas absorption for each layer in a simplified geometry, and then using that data to solve the linear systems of equations giving the fluxes which account for all effects, gaseous and grey, absorption and scattering.

9.3.5.1 The case of solar fluxes

For solar radiation, we first calculate optical depths from the top to the ground; these are used to obtain the downward parallel flux. The optical depths calculated from the ground up to the level in question, added to the optical depth calculated previously from the top to the ground, are used as input for computing the diffuse solar radiation. These depths are obtained without accounting for optical path lengthening due to multiple scattering, but are, of course, used with scattering taken into account. We therefore solve in one go the system comprising relations (9.9) and the boundary conditions (9.10) and (9.11).

9.3.5.2 The case of thermal fluxes

For thermal radiation, we must take into account the effects of multiple sources, which makes the problem even more complex. The optical depths traversed from a source to a given point may be varied thanks to scattering, and the saturation effect is then doubly non-uniform. The application of the ideal optical path method in the thermal case is based on a formulation using net exchange rates (NERs for short) and thus allowing us to express the radiation budget for a given layer of atmosphere as a sum of exchange terms with space, the surface, and all the other layers.

In the absence of scattering or surface reflection, the net fluxes are written, after integration over the entire spectrum:

$$
\left.
\begin{aligned}
F_{\tilde{k}} &= -\sigma T_S^4\, \tau(\tilde{k}, \tilde{N}) - \sum_{i=k+1}^{i=N} \sigma T_i^4 \left[\tau(\tilde{k}, \widetilde{i-1}) - \tau(\tilde{k}, \tilde{i}) \right] + \sum_{j=1}^{j=k} \sigma T_j^4 \left[\tau(\tilde{j}, \tilde{k}) - \tau(\widetilde{j-1}, \tilde{k}) \right], \\
F_{\widetilde{k-1}} &= -\sigma T_S^4\, \tau(\widetilde{k-1}, \tilde{N}) - \sum_{i=k}^{i=N} \sigma T_i^4 \left[\tau(\widetilde{k-1}, \widetilde{i-1}) - \tau(\widetilde{k-1}, \tilde{i}) \right] + \sum_{j=1}^{j=k-1} \sigma T_j^4 \left[\tau(\tilde{j}, \widetilde{k-1}) - \tau(\widetilde{j-1}, \widetilde{k-1}) \right],
\end{aligned}
\right\}
\tag{9.20}
$$

where $\tau(\tilde{i}, \tilde{j})$ represents the transmissivity of the thickness of atmosphere between the interlayer surfaces $\tilde{i}$ and $\tilde{j}$ (depicted in Figure 8.1), T_k, the temperature at level k, and T_S, the surface temperature.

The net budget R_{th} can then be written for the layer k as:

$$
\left.
\begin{aligned}
R_{th} = F_{\tilde{k}} - F_{\widetilde{k-1}} = {}& \sigma T_k^4 \left[\tau(\tilde{0}, \tilde{k}) - \tau(\tilde{0}, \widetilde{k-1}) \right] \quad \text{CTS} \\[2mm]
& + \left(\sigma T_S^4 - \sigma T_k^4 \right) \left[\tau(\tilde{k}, \tilde{N}) - \tau(\widetilde{k-1}, \tilde{N}) \right] \quad \text{EWS} \\[2mm]
& + \sum_{i=k+1}^{i=N} \left(\sigma T_i^4 - \sigma T_k^4 \right) \left[\tau(\tilde{k}, \widetilde{i-1}) - \tau(\widetilde{k-1}, \widetilde{i-1}) - \tau(\tilde{k}, \tilde{i}) + \tau(\widetilde{k-1}, \tilde{i}) \right] \\[2mm]
& + \sum_{j=1}^{j=k-1} \left(\sigma T_k^4 - \sigma T_j^4 \right) \left[\tau(\tilde{j}, \tilde{k}) - \tau(\tilde{j}, \widetilde{k-1}) - \tau(\widetilde{j-1}, \tilde{k}) + \tau(\widetilde{j-1}, \widetilde{k-1}) \right].
\end{aligned}
\right\}
\begin{aligned}
& (9.21) \\[6mm]
& \text{EBL}
\end{aligned}
$$

The net budget expression brings out the sum of three terms: one term for *cooling to space* (CTS), one for *exchange with surface* (EWS), and terms for *exchanges between layers* (EBL).

The first two terms can be calculated easily by solving the system formed by the relations (9.14) and the boundary conditions (9.15) and (9.16) and by adopting specific temperature profiles for the source terms in σT^4 (obtained after spectral integration of the monochromatic terms πB). The term for cooling to space (CTS) is obtained via a computation taking the *CTS profile* (σT_s^4 at the surface and everywhere in the atmosphere, which removes all the terms other than the CTS); the corresponding idealized optical depths $\Delta\delta_{gas}^{CTS}$ are then calculated from the top of the atmosphere down to the level in question. The exchange with surface (EWS) term is obtained via a computation taking the *EWS profile* (σT_S^4 at the surface and 0 everywhere else in the atmosphere, which eliminates all the exchanges other than EWS); the corresponding idealized optical depths $\Delta\delta_{gas}^{EWS}$ are then calculated from the surface up to the level in question. By adopting the *real profile* (σT_S^4 at the surface and σT_k^4 at level k), we obtain the fluxes resulting from all the exchanges, but the complete calculation implying solving $N(N+1)/2$ systems for all the optical depths is prohibitive. Given that most NER contributions result from cooling to space, from the exchange with the surface and from exchanges between adjacent layers (Eymet et al., 2004), it suffices to evaluate the input of the remaining exchange terms between layers as well as can be done by trying to bracket it between two extremes (Geleyn et al., 2005).

A minimum value for optical pathways can be obtained by assigning for each layer the smallest optical depth $\Delta\delta_{gas}^{min}$ along which it can be viewed from anywhere in the vertical; because of the saturation effect, that can only be the depth viewed either from the top or from the surface, which gives: $\Delta\delta_{gas}^{min} = \min(\Delta\delta_{gas}^{CTS}, \Delta\delta_{gas}^{EWS})$. Likewise, a maximum value can also be obtained by taking account of the fact that exchanges between adjacent layers dominate the EBL part: we assign for each layer the optical depth $\Delta\delta_{gas}^{min}$ along which it is seen from either of its adjacent layers.

E_1 and E_2 are the solutions to the system formed by relations (9.14) and the boundary conditions (9.15) and (9.16) by adopting respectively the CTS profile and EWS profile and the corresponding optical depths. The solutions obtained with the optical depths $\Delta\delta_{gas}^{min}$ yield solutions E_1^{min}, E_2^{min}, and E_3^{min} (with the real profile) respectively, from which a minimum value of the exchange-between-layers term can be calculated: $(EBL)_{min} = E_3^{min} - E_1^{min} - E_2^{min}$. By making the same calculations with the optical depths $\Delta\delta_{gas}^{max}$, we get the solutions E_1^{max}, E_2^{max}, and E_3^{max} (with the real profile), allowing us to calculate a maximum value of the exchange-between-layers term: $(EBL)_{max} = E_3^{max} - E_1^{max} - E_2^{max}$. The exchange-between-layers term in between $(EBL)_{min}$ and $(EBL)_{max}$ is evaluated by calculating $EBL = \alpha(EBL)_{min} + (1-\alpha)(EBL)_{max}$, parameter α being statistically adjusted from experimental data and being of course equal to zero in the case of adjacent layers.

9.3.6 Processing of clouds

The absorption and scattering properties of the atmosphere that are reflected by the values of the coefficients a_i depend on the nature of the components of the atmospheric column in question. The presence or absence of clouds leads to different fluxes and, in the general case, the cloudiness of a layer of atmosphere at level k is fractional (it is characterized by N_k ranging from 0 for clear sky to 1 for totally overcast sky), which assumes of course that the cloud extends over the entire layer and there are no lateral effects. However, there are several ways to take account of the effect of clouds, depending on whether it is considered that the cloud scene of a layer is independent of

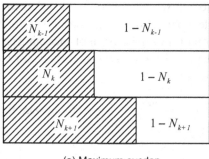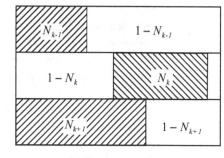

(a) Maximum overlap   (b) Random overlap

Figure 9.4   Two types of cloud cover. (a) maximum overlap, (b) random overlap.

that of adjacent layers (hypothesis of random overlap) or on the contrary dependent on that of the adjacent layers (hypothesis of maximum overlap), as shown in Figures 9.4a and 9.4b.

9.3.6.1 The random overlap hypothesis

In this instance, the coefficients taken into account for a partially cloudy layer $(a_i)^n$ are calculated from a weighted sum of coefficients $(a_i)^c$ obtained for a totally clear layer and coefficients $(a_i)^o$ for an overcast layer as below:

$$(a_i)^n = (1 - N_k)(a_i)^c + N_k(a_i)^o.$$

9.3.6.2 The maximum-random overlap hypothesis

Under the *maximum-random overlap* assumption, adjacent clouds are in a maximum overlap situation while cloudy parts separated by some volume of clear sky overlap randomly. We must then simultaneously process two systems for partial fluxes that are written, in the case of solar fluxes:

$$\begin{pmatrix} (S_{\tilde{k}})^c \\ (F_{\downarrow\tilde{k}})^c \\ (F_{\uparrow\widetilde{k-1}})^c \end{pmatrix} = \begin{pmatrix} (a_1)^c & 0 & 0 \\ (a_2)^c & (a_4)^c & (a_5)^c \\ (a_3)^c & (a_5)^c & (a_4)^c \end{pmatrix} \begin{pmatrix} \lambda_2 (S_{\widetilde{k-1}})^c + v_2 (S_{\widetilde{k-1}})^o \\ \lambda_2 (F_{\downarrow\widetilde{k-1}})^c + v_2 (F_{\downarrow\widetilde{k-1}})^o \\ \lambda_1 (F_{\uparrow\tilde{k}})^c + v_1 (F_{\uparrow\tilde{k}})^o \end{pmatrix}. \qquad (9.22)$$

$$\begin{pmatrix} (S_{\tilde{k}})^o \\ (F_{\downarrow\tilde{k}})^o \\ (F_{\uparrow\widetilde{k-1}})^o \end{pmatrix} = \begin{pmatrix} (a_1)^o & 0 & 0 \\ (a_2)^o & (a_4)^o & (a_5)^o \\ (a_3)^o & (a_5)^o & (a_4)^o \end{pmatrix} \begin{pmatrix} \mu_2 (S_{\widetilde{k-1}})^c + \xi_2 (S_{\widetilde{k-1}})^o \\ \mu_2 (F_{\downarrow\widetilde{k-1}})^c + \xi_2 (F_{\downarrow\widetilde{k-1}})^o \\ \mu_1 (F_{\uparrow\tilde{k}})^c + \xi_1 (F_{\uparrow\tilde{k}})^o \end{pmatrix}. \qquad (9.23)$$

System (9.22) gives the outgoing clear fluxes as a function of a weighted sum of incoming clear and cloudy fluxes in the clear part of the layer, whereas system (9.23) gives the outgoing cloudy fluxes as a function of a weighted sum of incoming clear and cloudy fluxes in the overcast part of the layer. The weighting coefficients for incoming fluxes in the clear part of the layer are given by:

$$\lambda_1 = \frac{1 - \max(N_k, N_{k+1})}{1 - N_{k+1}}, \quad v_1 = 1 - \frac{\min(N_k, N_{k+1})}{N_{k+1}} \quad \text{for upward fluxes,}$$

$$\lambda_2 = \frac{1 - \max(N_k, N_{k-1})}{1 - N_{k-1}}, \quad v_2 = 1 - \frac{\min(N_k, N_{k-1})}{N_{k-1}} \quad \text{for downward fluxes,}$$

and those for the incoming fluxes in the overcast part of the layer by:

$$\mu_1 = 1 - \lambda_1, \; \xi_1 = 1 - v_1 \text{ for upward fluxes,}$$

$$\mu_2 = 1 - \lambda_2, \; \xi_2 = 1 - v_2 \text{ for downward fluxes.}$$

Of course, the complete fluxes are written by summing the fluxes for the clear sky and for the cloudy sky.

9.3.6.3 Calculation of optical depths allowing for cloudiness

A more rational way of working than having a bulk approach to the saturation effect (see Paragraph 9.3.4.2) is to process it together with the cloud geometry when calculating the transmission factors (Mašek, 2006). Based on experimental data and many calculations made for monochromatic radiation, we compute, for the various optical depths of cloud constituents, average *broad band* absorption and scattering coefficients $k_{sat}^{abs/sca}$ (in the solar domain and thermal domain) that take the saturation effect into account. By comparing these values with the averaged values obtained without the saturation effect $k_0^{abs/sca}$, we obtain *saturation factors*:

$$c_{sat}^{abs/sca} = k_{sat}^{abs/sca} / k_0^{abs/sca}.$$

Adjusting for the various optical depths allows us to express the saturation factors corresponding to an optical depth $\Delta\delta$, for the solar and thermal domains, from the formula:

$$c_{sat}^{abs/sca}(\Delta\delta) = \frac{1}{\left(1 + \Delta\delta / \Delta\delta_{crit}\right)^\mu},$$

and to determine the values of the parameters $\Delta\delta_{crit}$ and μ (value between 0 and 1) depending on the spectral interval and on the type of process (i.e. absorption or scattering).

These saturation factors are calculated by using an effective optical depth $\Delta\delta_k^{eff}$ (relative to layer k) from non-saturated optical depths $\Delta\delta_k^0$ and from the cloud cover of other layers taken into account as below:

$$\Delta\delta_k^{eff} = \Delta\delta_k^0 + \sum_{i\neq k} f(N_k, N_i)\Delta\delta_i^0$$

The function $f(N_k, N_i)$, which takes a value between 0 and 1, depends on the hypothesis chosen for the distribution of clouds; for random cloud overlap, we choose:

$$f(N_k, N_i) = (N_i)^p, \text{ where } p \text{ is a positive integer,}$$

and for maximum-random cloud overlap:

$$f(N_k, N_i) = [\min(1, N_i/N_k)]^p, \text{ where } p \text{ is a positive integer.}$$

9.3.6.4 Calculating cloudiness

The cloudiness of a layer results from the presence of water in liquid droplets or ice crystals that appear with condensation phenomena related to ascending currents at the grid scale (large-scale precipitation) or at sub-grid scale (convection). Where the parameterization of condensation and precipitation phenomena fails to give the quantities of liquid water and ice in suspension directly, an estimate may be made diagnostically from the model variables. These quantities of liquid water and ice are also required for calculating the optical depths of clouds.

For large-scale precipitation, a surplus of water vapour is calculated as below:

$$\Delta q_{ex}^{ls} = q - Hu_{cr}\, q_{sat}(T, p),$$

where q is the specific humidity, q_{sat} the saturation specific humidity, and Hu_{cr} is a critical humidity depending on altitude according to a profile of the type depicted in Figure 9.5.

In the case of convective precipitation, the surplus of water vapour is given by:

$$\Delta q_{ex}^{conv} = g\frac{\partial P}{\partial p}\Delta t,$$

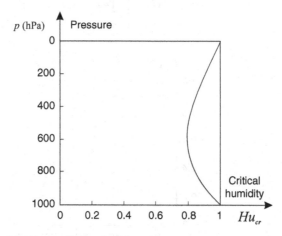

Figure 9.5 Example of a critical humidity Hu_{cr} profile for calculating surplus water vapour.

where P is the convective precipitation flux.

For both large-scale and convective precipitation, the content of condensed water in suspension is given by:

$$q_c^{ls/conv} = q_{max}\left(1 - e^{-\alpha_s \frac{\Delta q_{ex}^{ls/conv}}{q_{max}}}\right);$$

(9.24)

in this formula, q_{max}, which represents the maximum content of condensed water in suspension in the layer, is a constant, while α_s is a variable scaling factor also introduced for avoiding full proportionality of q_c with respect to q_{sat}.

The total content of condensed water in suspension is the sum of the contents of condensed water resulting from grid scale and sub-grid scale precipitation:

$$q_c^{tot} = q_c^{ls} + q_c^{conv}.$$

Cloudiness N is then obtained by means of the semi-empirical formula proposed by Xu and Randall (1996):

$$N = \left(\frac{q}{q_{sat}}\right)^r\left[1 - e^{-\frac{\alpha q_c^{tot}}{(q_{sat}-q)^\delta}}\right],$$

(9.25)

α, δ, and r being adjusted constants.

Liquid and solid phases may be separated by using a discrimination function $h(T)$, shown in Figure 9.6, and by writing:

$$q_l = q_c[1-h(T)] \text{ and } q_i = q_c h(T),$$

the discrimination function being defined as:

$$h(T) = \begin{cases} 1 - e^{-\frac{1}{2}\left(\frac{T-T_{00}}{\Delta T_0}\right)^2} & \text{if } T < T_{00}, \\ 0 & \text{otherwise.} \end{cases}$$

(9.26)

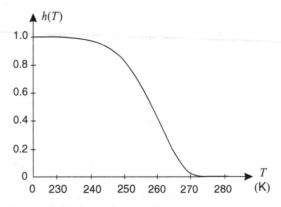

Figure 9.6   Discrimination function of liquid and solid phases of water with respect to temperature.

T_{00} is the triple-point temperature of water and ΔT_0 allows for the co-existence of the liquid and solid phases in the vicinity of the triple point; the value of this constant corresponds in principle to the difference between the triple-point temperature and the temperature of the maximum difference between the two saturation curves.

9.4 Boundary layer and vertical diffusion

9.4.1 General points

The *atmospheric boundary layer* (ABL) is the part of the atmosphere located in the vicinity of the ground, from 0 to about 1500 m, where turbulent motions contribute to exchange momentum (friction), sensible heat (conduction), and water vapour (evaporation) between the surface and the atmosphere. The ABL may be subdivided into two parts: the *surface boundary layer* (SBL), the vertical extent of which is of the order of 50 to 100 m and in which the effects of the Coriolis force may be ignored, and the *planetary boundary layer* (PBL), in which those effects must be taken into account.

The exchanges of momentum, dry static energy $s = C_p T + gz$, and moisture that result from turbulence are determined by turbulent fluxes, defined as:

$$\mathbf{J_V} = \rho \overline{w'V'}, \quad J_s = \rho \overline{w's'}, \quad J_q = \rho \overline{w'q'}.$$

These fluxes (counted here positively upwards) involve the mean values of the products of fluctuations of vertical velocity w by the corresponding fluctuations for the components of the wind u, v, dry static energy, and specific humidity q; dry static energy, noted s (not to be confused with the general vertical coordinate introduced in Chapter 7), is used here as a conservative thermodynamic variable and plays a similar role to potential temperature.

When we use the Lorenz variable arrangement (introduced in Subsection 8.3.8), it is practical to calculate the fluxes on the interlayer surfaces (including the ground surface) since their effect on the prognostic variables is obtained by calculating the flux divergences on the levels of the model.

Parameterization of the boundary layer therefore consists in determining the values of these turbulent fluxes at ground level and throughout the atmosphere as a function of the values of the prognostic variables supplied by the dynamical part of the model; near the surface, these fluxes also depend on the values of the ground variables supplied by a surface model (to be specified below) and of the characteristic parameters of the ground.

9.4.2 Parameterization of turbulent surface fluxes

Here, to simplify matters, we work within the reference system where the second component of the wind is zero, so that only the first component of the wind u is used for the calculations involving momentum. By using the new variables u^*, s^*, and q^*, which are commonly used for characterizing a turbulent boundary layer (De Moor, 2006), turbulent fluxes of momentum, dry static energy, and humidity are written in the form:

$$\overline{\rho w' u'} = -\rho u^{*2}, \quad \overline{\rho w' s'} = -\rho u^* s^*, \quad \overline{\rho w' q'} = -\rho u^* q^*. \qquad (9.27)$$

As turbulence in the vicinity of the surface is greater when the wind modulus is higher and the gradients higher, we express the fluxes by means of transfer coefficient formulas:

$$\rho u^{*2} = \rho C_D |u_N| u_N, \quad \rho u^* s^* = \rho C_H |u_N|(s_N - s_S), \quad \rho u^* q^* = \rho C_E |u_N|(q_N - q_S). \qquad (9.28)$$

The quantities C_D, C_E, and C_H (which are all dimensionless) are the transfer coefficients relative to momentum flux, sensible and latent heat fluxes respectively; the quantities u_N, s_N, and q_N, are the large-scale variables at the lowest level of the model, while $u_S = 0$, s_S and q_S are surface values, the evolution equations of which are supplied by a surface/atmosphere interface model. The problem is therefore to determine the values of these coefficients, which depend on the nature of the surface and on the flow characteristics, as well as on the height above the ground of the lowest model level.

We now assume that the lowest level of the model is located within the SBL, so that the results yielded by the Monin-Obukhov theory (1954) can be applied. This theory, based on dimensional analysis, applies if the boundary layer is horizontally homogeneous, quasi-stationary, and if the effects of the pressure force and the Coriolis force may be ignored, which is the case in the SBL; it postulates that the fluxes are constant and the vertical gradients are related to the fluxes via *universal functions* as below:

$$\left. \begin{array}{l} \dfrac{\partial u}{\partial z} = \dfrac{u^*}{\kappa(z + z_0)} \Phi_1\left(\dfrac{z + z_0}{L_M}\right), \\[3mm] \dfrac{\partial s}{\partial z} = \dfrac{s^*}{\kappa(z + z_{0H})} \Phi_2\left(\dfrac{z + z_{0H}}{L_M}\right), \quad \dfrac{\partial q}{\partial z} = \dfrac{q^*}{\kappa(z + z_{0H})} \Phi_2\left(\dfrac{z + z_{0H}}{L_M}\right) \end{array} \right\}, \qquad (9.29)$$

where $L_M = \bar{s} u^{*2}/g\kappa s^*$ is the Monin-Obukhov length, a quantity that characterizes the vertical stability of the atmosphere, $\bar{s}$ is a mean value of s in the SBL, z_0 and z_{0H} are the roughness lengths (parameter characterizing surface heterogeneity), z is altitude, and κ is von Karman's constant, the value of which is close to 0.4.

These universal functions are determined empirically by adjusting experimental data using analytical functions. The functions obtained for the dry static energy flux (equivalent to the sensible heat flux) and for the latent heat flux are identical. Among the most commonly used functions are those proposed by Businger et al. (1971) and that are shown in Figure 9.7. However, the calculation of fluxes requires the inversion of functions Φ_1 and Φ_2 and integration in the vertical, which are calculations that cannot be made analytically in the general case.

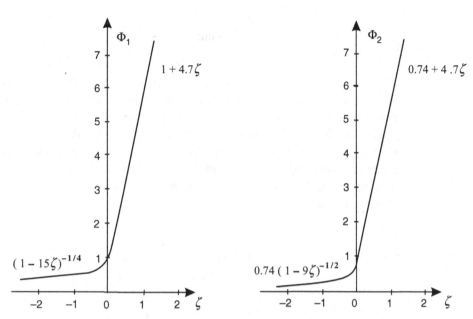

Figure 9.7 Shape of the Φ functions as a function of $\zeta = z/L_M$. (a) Φ_1 function for the momentum flux; (b) Φ_2 function for sensible heat and latent heat fluxes (after Businger J. A. et al. (1971). *J. Atmos. Sci.*, 28, 184 © Amer. Met. Soc.).

If we take as our target the case of neutrality for vertical stability (corresponding to $S_N = S_S$ and $s^* = 0$, therefore to an infinite Monin-Obukhov length), it can be seen that the value of the function Φ_1 is close to 1, and the vertical integration of equations (9.29) yields *logarithmic profiles*. We therefore obtain for wind and static energy:

$$u = \frac{u^*}{\kappa} \ln(1 + z/z_0) \quad \text{and} \quad s - s_s = \frac{s^*}{\kappa} \ln(1 + z/z_{0H}). \tag{9.30}$$

The parameters z_0 and z_{0H} characterize surface heterogeneity. Over the sea, their values depend on wave height and so on the surface wind (Charnock, 1955); over land, their values depend on the terrain (bare ground, wooded land, urbanization, mountain).

By identification with the formula given in (9.28), we obtain the value of C_{ND} and an asymptotic limit for C_{NH}, which are transfer coefficients for the neutral case:

$$C_{ND} = \left[\frac{\kappa}{\ln(1 + z/z_0)} \right]^2 \quad \text{and} \quad C_{NH} = \frac{\kappa^2}{\ln(1 + z/z_0)\ln(1 + z/z_{0H})}. \tag{9.31}$$

An approximate analytical formulation of the solution for stable and unstable cases was proposed by Louis (1979): it provides an expression of transfer coefficients taking account of vertical stability of the atmosphere (via the Richardson number *Ri*), of the nature of the terrain (via z_0 and z_{0H} values), and of the altitude of the lowest model level z, by writing:

$$C_D = C_{ND} f_M \left[Ri, (1 + z/z_0) \right] \quad \text{and} \quad C_H = C_{NH} f_H \left[Ri, (1 + z/z_{0H}) \right], \tag{9.32}$$

$$\text{with: } Ri = \frac{g}{\bar{s}} \frac{\partial s / \partial z}{\|\partial u / \partial z\|^2}.$$

This last, dimensionless quantity Ri is the Richardson number characterizing the vertical stability of the atmosphere. It is negative for an unstable atmosphere, zero for a neutral atmosphere, and positive for a stable atmosphere.

The functions f_M (for the momentum flux) and f_H (for the sensible and latent heat fluxes) are chosen so as to comply with the behaviour of the Φ functions both around the neutral case and for the asymptotic cases of high stability and high instability. We thus obtain for the transfer coefficients the expressions:
in the stable case:

$$C_D = \left[\frac{1}{1 + 2bRi / \sqrt{1 + \dfrac{d}{k} Ri}} \right] C_{ND}, \tag{9.33}$$

$$C_H = \left[\frac{1}{1 + 3bRi\sqrt{1 + dkRi}} \right] C_{NH}, \quad C_E = C_H, \tag{9.34}$$

and in the unstable case:

$$C_D = \left[1 - \frac{2bRi}{1 + 3bc_M C_{ND} \sqrt{(1 + z / z_0)|Ri|}} \right] C_{ND}, \tag{9.35}$$

$$C_H = \left[1 - \frac{3bRi}{1 + 3bc_H C_{NH} \sqrt{(1 + z / z_{0H})|Ri|}} \right] C_{NH}, \quad C_E = C_H. \tag{9.36}$$

The parameters b, c_M, c_H, d, and k determine the transfer coefficient values respectively in the vicinity of neutrality, in the unstable asymptotic regime, and in the stable asymptotic regime.

The variation of transfer coefficients with respect to the Richardson number for a given value of $1 + z/z_0$ is illustrated in Figure 9.8.

However, to obtain correct profiles of the coefficients for high positive values of Ri corresponding to weak mixing along the guidelines explained by Zilitinkevich et al. (2008), it is necessary to replace Ri in the stable regime by a corrected Richardson number Ri' given by the formula:

$$Ri' = \frac{Ri}{(1 + \alpha Ri / Ri_{ref})^{1/\alpha}}, \tag{9.37}$$

where Ri_{ref} is a tuned reference value of the Richardson number and α a coefficient the value of which is adjusted differently for the momentum flux ($\alpha = 1$) and the sensible and latent heat fluxes ($\alpha = 3$) .

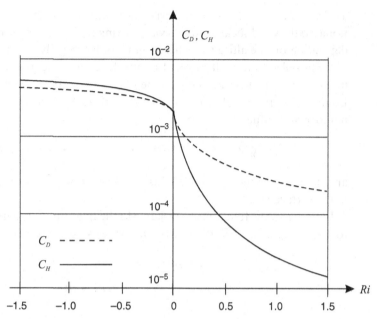

Figure 9.8 Coefficients C_D and C_H as a function of the Richardson number Ri, for a value of $1 + z/z_0$ equal to 5500, with $b = 5$, $c_M = c_H = 5$ and $d = k = 5$.

9.4.3 Planetary boundary layer fluxes

In the PBL, fluxes may be expressed by a formulation involving *exchange coefficients*, as follows:

$$\overline{\rho w'u'} = -\rho K_M \frac{\partial u}{\partial z}, \quad \overline{\rho w's'} = -\rho K_H \frac{\partial s}{\partial z}, \quad \overline{\rho w'q'} = -\rho K_E \frac{\partial q}{\partial z}. \tag{9.38}$$

Determining fluxes therefore comes down to determining the exchange coefficients K.

At the top of the surface boundary layer, in neutral regime (corresponding to $\phi_1 = 1$), the Monin-Obukhov formulation allows us to write:

$$\overline{\rho w'u'} = -\rho u^{*2} = -\rho \kappa^2 (z + z_0)^2 (\partial u / \partial z)^2,$$

$$\overline{\rho w's'} = -\rho u^* s^* = -\rho \kappa^2 (z + z_0)(z + z_{0H})(\partial u / \partial z)(\partial s / \partial z).$$

Comparison with the exchange coefficient formulation yields:

$$K_M = \kappa^2 (z + z_0)^2 |\partial u / \partial z| \quad \text{and} \quad K_H = \kappa^2 (z + z_0)(z + z_{0H}) |\partial u / \partial z|,$$

The exchange coefficients K_M and K_H are themselves determined by using Prandtl's theory (1932), which generalizes the foregoing formulation by writing:

$$K_M = \ell_M^2 \|\partial \mathbf{V} / \partial z\| \quad \text{and} \quad K_H = \ell_M \ell_H \|\partial \mathbf{V} / \partial z\|. \tag{9.39}$$

These coefficients, which characterize the intensity of turbulent exchanges, are proportional to the wind shear modulus via a mixing length $\ell(z)$, known as the *Prandtl length*, depending on the altitude z (this length can be roughly interpreted as the mean distance travelled by a small parcel of atmospheric air under the effect of random vertical fluctuations). To take account of the effects of vertical stability of the atmosphere, Louis (1979) proposed weighting these coefficients by a function of the Richardson number by writing:

$$K_M = \ell_M^2 \|\partial \mathbf{V} / \partial z\| \cdot f_M(Ri), \; K_H = \ell_M \ell_H \|\partial \mathbf{V} / \partial z\| f_H(Ri), \; K_E = K_H, \quad (9.40)$$

and by choosing the function f so as to ensure continuity with the expressions of flux in the surface layer.

In this case, the functions f_M, relative to the quantity of momentum, and f_H, relative to sensible and latent heat, are written, for the stable case:

$$f_M(Ri)_{stable} = \cfrac{1}{1 + 2bRi / \sqrt{1 + \cfrac{d}{k} Ri}}, \quad (9.41)$$

$$f_H(Ri)_{stable} = \cfrac{1}{1 + 3bRi\sqrt{1 + dkRi}}, \quad (9.42)$$

and for the unstable case:

$$f_M(Ri)_{unstable} = 1 - \cfrac{2bRi}{1 + 3bc_M \sqrt{\cfrac{|Ri|}{27}\left(\cfrac{\ell_M}{z + z_0}\right)^2}}, \quad (9.43)$$

$$f_H(Ri)_{unstable} = 1 - \cfrac{3bRi}{1 + 3bc_H \sqrt{\cfrac{|Ri|}{27}\left(\cfrac{\ell_H}{z + z_{0H}}\right)^2}}, \quad (9.44)$$

the parameters b, c_M, c_H, d, and k being the same as those for defining the transfer coefficients in equations (9.33) to (9.36). Here again, in the stable case, it is necessary to replace R_i by R_i' according to (9.37).

System closure requires us to define a vertical profile for the mixing length $\ell(z)$. This may be expressed by the extension of a formulation proposed by Blackadar (1962), that introduces an asymptotic mixing length:

$$\ell(z) = \cfrac{\kappa(z + z_0)}{1 + \kappa(z + z_0)/\lambda}\left[\beta + \cfrac{1 - \beta}{1 + (z + z_0)^2 / Z^2}\right], \quad (9.45)$$

λ being an asymptotic mixing length (λ_M for the momentum and $\lambda_H = \lambda_M \sqrt{3 d/2}$ for sensible and latent heat), Z being a characteristic height for adjusting the decrease of λ with respect to altitude, and β a form factor for vertical diffusion, allowing the mixing length to be reduced with altitude. Formulated in this way, the mixing length $\ell(z)$ increases from the ground, where it has a value close to kz_0, then peaks before falling off in the free atmosphere up to an asymptotic value $\beta\lambda$.

It is also possible to add to the model a new prognostic variable characterizing turbulence, *turbulent kinetic energy* e_T with its own evolution equation, and to express the exchange coefficients K using this variable as below:

$$K = \alpha.\ell(z)\sqrt{e_T},$$

α being a proportionality coefficient.

Noteworthy among the schemes developed using this approach are those proposed by Bougeault and Lacarrère (1989) and by Cuxart et al. (2000).

9.4.4 Allowance for shallow convection

The scheme proposed for parameterization of turbulence in the ABL, based on an evaluation of vertical stability in a dry atmosphere, proves inadequate to reproduce the behaviour of *convective boundary layers* properly. The exchanges at the top of the ABL resulting from shallow convection are inhibited because of the insufficient exchange coefficient values K; we observe moistening of the boundary layer together with drying in the free atmosphere, leading to a collapse in the height of the boundary layer.

To get around this problem, Geleyn (1987) proposed a parameterization scheme for shallow convection consisting in calculating the exchange coefficients using a Richardson number that was modified to allow for the humidity profile as below:

$$Ri_{(modified)} = \frac{g}{C_p T} \frac{\partial s / \partial z + L \min\left(0, \, \partial(q - q_{sat}) / \partial z\right) \delta_h}{\left\| \partial u / \partial z \right\|^2}, \tag{9.46}$$

where q_{sat} represents saturation specific humidity and δ a binary indicator taking the value 1 if the moist static energy $h = C_p T + Lq + gz$ diminishes with altitude and 0 in the opposite case. The second term in the numerator of expression (9.46) is always negative and tends to reduce vertical stability, and therefore to increase vertical exchanges; the term δ_h prevents this correction from being activated where it is not necessary, for example in the stratosphere. Lastly, a specific anti-fibrillation treatment must be introduced in order to cope with the risk of time step to time step alternating choices for the 'min' function.

This way of regulating the activation of exchanges between the boundary layer and the atmosphere by correcting the Richardson number for cloudy air is not the only way to allow for the effects of shallow convection; more sophisticated parameterization schemes have been developed, generally based on a mass flux approach (Bechtold et al., 2001) and/or by modifying entrainment at the top of the boundary layer (Grenier and Bretherton, 2001).

9.4.5 The evolution of surface parameters

Determining turbulent fluxes requires knowing the surface values of the wind $\mathbf{V}_S$, static energy s_S (and so temperature T_S), and specific humidity q_S. While the surface wind is zero, the surface temperature and moisture even result from physical processes involving not only the dynamic model variables in the lower layers but also the parameters characterizing the surface (oceanic or continental surface, ground with or without plant cover).

For oceanic surfaces, insofar as we are interested only in forecasts for a few days ahead on the synoptic scale, it can be assumed that the surface temperature obtained by analysing sea temperature at the initial time varies only slightly; we then set $T_S = T_{sea}$, the sea surface temperature (commonly termed *SST*) being assumed to be locally constant throughout the forecast period, and we consider the atmosphere as saturated at the surface, leading us to take $q_S = q_{sat}(T_{sea}, p_s)$.

For land surfaces the problem is more complex since the surface temperature and moisture vary continuously because of atmospheric forcing of heat content and soil moisture. Detailed allowance for physical soil processes means considering the soil as a stack of layers, each being characterized by its temperature and its water content, and then processing the various interactions among those layers and the moisture transfers brought about by vegetation.

Among the various schemes describing the interactions among the atmosphere, surface, and soil, we should mention the ISBA (*Interaction Surface Biosphere Atmosphere*) scheme developed by Noilhan and Planton (1989) and revised by Noilhan and Mahfouf (1996), that takes into account rainwater intercepted by vegetation, and plant transpiration. We have opted here to present only a highly simplified parameterization scheme of interactions among the soil, the surface, and the atmosphere, confining ourselves to a description of the soil using two layers (a surface layer and a deep layer) and without introducing here the complexity linked to the solid water phase.

9.4.5.1 Evolution of surface temperature

For continental surfaces, we can calculate the surface equilibrium temperature T_S by writing that the sum of energy fluxes penetrating an infinitesimally thin layer of soil in contact with the atmosphere must be zero. The budget equation is written:

$$R_G \downarrow (1 - A_T) + e_T R_A \downarrow \, - G - e_T \sigma T_S^4 - H_S - H_L = 0. \tag{9.47}$$

In this equation, $R_G \downarrow (1-A_T)$ is the net solar radiation (parallel + scattered), $e_T R_A \downarrow$ the net atmospheric radiation, and G the heat flux in the ground (these fluxes being counted positively downwards). $e_T \sigma T_S^4$ is the flux emitted by the Earth's surface, $H_S = J_s = \overline{\rho w's'}$, the sensible heat flux, and $H_L = LJ_q = \rho L \overline{w'q'}$, the latent heat flux (these fluxes being counted positively upwards). The albedo for solar radiation A_T and thermal emissivity e_T are local surface characteristics and σ is the Stefan-Boltzmann constant.

By following the method proposed by Bhumralkar (1975) and Blackadar (1976), we can parameterize the heat flux within the soil by writing that it must be proportional to the soil temperature gradient:

$$G = -\lambda_S \partial T/\partial z,$$

λ_S being the thermal conductibility coefficient of the soil, and calculate the vertical gradient of G to determine the evolution of temperature in the soil:

$$C_S \partial T/\partial t = -\partial G/\partial z,$$

C_S being the specific heat of the soil.

These two relations combine to give the diffusion equation for the evolution of temperature in the soil:

$$\frac{\partial T}{\partial t} = \frac{\lambda_S}{C_S} \frac{\partial^2 T}{\partial z^2}. \tag{9.48}$$

Assuming that at the surface ($z = 0$) the temperature T_S exhibits a sinusoidal variation with a period $\tau_1 = 24$ hours and that at a certain depth it takes the value T_p, characteristic of the deep layer and varying more slowly, integrating equation (9.48) leads to the expression for the flux within the soil:

$$G = \sqrt{\frac{C_S \lambda_S \pi}{\tau_1}} \left[\frac{\tau_1}{2\pi} \frac{\partial T_S}{\partial t} + T_S - T_p \right].$$

By including this value in the budget equation (9.47), we finally get for the evolution of surface temperature:

$$\frac{\partial T_S}{\partial t} = C_{Soil} \left[R_G \downarrow (1 - A_T) + e_T R_A \downarrow - e_T \sigma T_S^4 - H_S - H_L \right] - \frac{2\pi}{\tau_1}(T_S - T_p), \tag{9.49}$$

with: $C_{Soil} = 2 \sqrt{\dfrac{\pi}{\lambda_S C_S \tau_1}}$.

The soil constant C_{Soil} varies with the nature of the surface and takes different values depending on whether it is bare ground, vegetation, or snow cover.

This formulation, which gives an evolution equation for surface temperature T_S, is called *force-restore*: *forcing* results from all fluxes except the flux within the soil, the effect of which is parameterized using a *restore* term towards a deep layer temperature T_p. It avoids complex processing of heat diffusion within the soil by using a multi-layer description.

The influence of surface temperature T_S on the deep layer temperature T_p (which is assumed to vary slowly) may be taken into account using an evolution equation of T_p, based on a single restore term toward the surface temperature T_S:

$$\frac{\partial T_p}{\partial t} = \frac{2\pi}{\tau_2}(T_S - T_p), \tag{9.50}$$

the time constant τ_2 being of the order of about 5 days.

9.4.5.2 Evolution of soil moisture

The surface moisture can be determined by using a force-restore formulation of the type proposed by Deardorff (1977). This formulation is analogous to that used previously for temperature and describes the evolution of two layers, each characterized by their relative water content (per unit volume) w: we thus distinguish a surface layer of thickness e_1, with water content w_S, and a deep layer of thickness e_2 and water content w_p.

The evolution equation for the relative water content of the surface layer is written:

$$\frac{\partial w_S}{\partial t} = \frac{C_1}{\rho_w e_1}(P_T - E) - \frac{C_2}{\tau_1}\left(w_S - w_p \frac{w_S^{max}}{w_p^{max}}\right). \tag{9.51}$$

The algebraic sum of moisture fluxes (total precipitation $P_T = P_L + P_I$ and evaporation $E = H_L / L = J_q = \rho \overline{w'q'}$) supplies the water amount of the surface layer and constitutes the forcing term, while the restore term to the scaled value of w_p characteristic of the deep layer involves the time constant τ_1 of 24 hours, given the importance of the diurnal cycle; ρ_w denotes the density of water; C_1 and C_2 are two constants that may be calibrated for different types of soil and vegetation cover.

The equation reflecting the water amount supply to the deep layer is written (unlike in the case of temperature, the deep layer encompasses the surface one):

$$\frac{\partial w_p}{\partial t} = \frac{1}{\rho_w e_2}(P_T - E). \tag{9.52}$$

The relative water contents w_S and w_p have upper limits that are the maximum values w_S^{max} and w_p^{max} beyond which the layers are considered to be waterlogged, leading to the surplus water running off.

For calculating the evaporation flux E, two cases are distinguished depending on whether the ground is bare or supports vegetation.

For bare ground, we write simply:

$$E = C_H |\mathbf{V}_S|\left[Hu(w_S / w_S^{max}) q_{sat}(T_S, p_s) - q_N\right], \tag{9.53}$$

while for vegetation, we adopt the Halstaed formulation (1954), which takes account of the effects of evapotranspiration by vegetation:

$$E = C_H |\mathbf{V}_S| Hu(w_S / w_S^{max})\left[q_{sat}(T_S, p_s) - q_N\right], \tag{9.54}$$

the surface relative humidity Hu being a function of the scaled water content of the surface layer:

$$Hu\left(\frac{w_S}{w_S^{max}}\right) = \frac{1 - \cos\left[\pi(w_S / w_S^{max})\right]}{2}. \tag{9.55}$$

By including the *Veg* parameter, as the proportion of the grid covered with vegetation, the two formulations (9.53) and (9.54) combine to give the evaporation flux:

$$E = Veg \cdot Hu\left(\frac{w_S}{w_S^{max}}\right)\left[q_{sat}(T_S, p_s) - q_N\right] + (1 - Veg)\left[Hu\left(\frac{w_S}{w_S^{max}}\right)q_{sat}(T_S, p_s) - q_N\right]. \quad (9.56)$$

9.4.6 Allowance for fluxes, vertical diffusion

As already said, the choice of the position of the variables in the vertical (in compliance with the Lorenz arrangement) implies determining the various fluxes on the interlayer surfaces, which allows us to calculate the *physical tendencies* of the various parameterizations on the levels of the model for each of the variables.

The effect of turbulent fluxes, expressed using the formulation in exchange coefficients K, is obtained, for a given layer k of thickness Δp_k, by calculating the vertical divergence of those fluxes; for a generic conservative variable X, the discretization in central differences in the vertical is written:

$$\left(\frac{\partial X}{\partial t}\right)_k = \frac{g}{\Delta p_k}\left[\left(\overline{\rho w' X'}\right)_k - \left(\overline{\rho w' X'}\right)_{k-1}\right]. \quad (9.57)$$

Given the expression of fluxes using the exchange coefficient K:

$$\overline{\rho w' X'} = -\rho K \frac{\partial X}{\partial z} = -g\rho K \frac{\partial X}{\partial \Phi},$$

we obtain the diffusion equation:

$$\left(\frac{\partial X}{\partial t}\right)_k = -\frac{g^2}{\Delta p_k}\left[(\rho K)_{\overline{k}}\left(\frac{X_{k+1} - X_k}{\Phi_{k+1} - \Phi_k}\right) - (\rho K)_{\widetilde{k-1}}\left(\frac{X_k - X_{k-1}}{\Phi_k - \Phi_{k-1}}\right)\right]. \quad (9.58)$$

In the event the time integration scheme for the dynamical part is of the explicit type, the time step used may be relatively short, but we must be sure even so that it provides a stable scheme for the diffusion equations of the (9.58) type. Review of the various time integration schemes for the classical diffusion equation, where K is a constant (for example Richtmyer and Morton, 1967), shows that the centred explicit scheme is to be ruled out, as it is unconditionally unstable. The forward off-centred explicit scheme (for which $\partial X/\partial t$ is replaced by $[X(t + \Delta t) - X(t - \Delta t)]/2\Delta t$, the right-hand side being expressed at $t-\Delta t$), is conditionally stable and cannot be used if the time step exceeds a critical value.

When the time integration scheme is of the semi-implicit (or semi-Lagrangian semi-implicit) type, allowing a longer time step to be used for the dynamical part, it is then necessary to solve the diffusion equations and the evolution equations of the surface variables implicitly. In this case, the right-hand sides must be expressed at time $t + \Delta t$, which therefore complicates the way of solving these diffusion equations. Examination of equation (9.58) written for k ranging from 2 to $N-1$ shows that it provides an implicit relation between the variables X_k calculated at three levels in succession. For $k=1$, the turbulent fluxes at the top of the atmosphere are zero and for $k = N$, the surface fluxes H_S and H_L, defined from relations (9.28) contain nonlinear terms. The expression C_p given in (9.7), restricted to a mixture of dry air and water vapour, leads us to write the surface static energy S_S as:

$$s_S = C_p T_S = C_{p_d}\left(1 + \frac{C_{p_v} - C_{p_d}}{C_{p_d}} q_S\right) T_S,$$

while the surface specific humidity q_S is written:

$$q_S = Hu q_{sat}(T_S, p_s).$$

The right-hand side of the surface temperature T_S evolution equation (9.49) contains nonlinear terms: the surface turbulent fluxes and the Earth's radiation flux. To solve the diffusion equations and the evolution equation for surface temperature implicitly, we have to replace the nonlinear terms by linear expressions obtained by a 1st-order expansion with respect to T_S. This gives:

$$q_S(t + \Delta t) = q_S(t - \Delta t) + \left(\frac{\partial q_S}{\partial T_S}\right)_{t-\Delta t}\left[T_S(t + \Delta t) - T_S(t - \Delta t)\right],$$

$$s_S(t + \Delta t) = s_S(t - \Delta t) + \left(\frac{\partial s_S}{\partial T_S}\right)_{t-\Delta t}\left[T_S(t + \Delta t) - T_S(t - \Delta t)\right],$$

$$e_T \sigma\left[T_S(t + \Delta t)\right]^4 = e_T \sigma\left[T_S(t - \Delta t)\right]^4 + 4 e_T \sigma\left[T_S(t - \Delta t)\right]^3\left[T_S(t + \Delta t) - T_S(t - \Delta t)\right].$$

Substituting the expressions $q_S(t + \Delta t)$, $s_S(t + \Delta t)$, and $e_T \sigma[T_S(t + \Delta t)]^4$ into the diffusion equations of the (9.58) type and into the surface temperature evolution equation (9.49) leads, for each vertical column, to a linear system involving the variables q_k, T_k, T_S, and u_k and v_k, taken at time $t + \Delta t$. Notice also that by ordering the specific humidity and temperature variables as below,

$$(q_1, q_2, \dots, q_k, \dots, q_N, T_S, T_N, \dots, T_k, \dots, T_2, T_1),$$

the implicit equations giving values at time $t+\Delta t$ for all variables amount to tridiagonal systems, which can be efficiently solved by the Gaussian elimination method.

When solving the implicit system, fluxes are expressed at time $t + \Delta t$, whereas the diffusion coefficients depending on the flux are still expressed at time $t - \Delta t$. In the vicinity of neutrality, the transition from low stability with strong gradients to high stability with weak gradients may lead to the fibrillation phenomenon, which is manifested by the development of a wave with period $4\Delta t$ (when using the leapfrog scheme) and an oscillating vertical structure. We can avoid this phenomenon by expressing the gradients as a fixed weighted average of values at times $t - \Delta t$ and $t + \Delta t$ but this process leads to a numerical scheme of lower order accuracy. To maintain as far as can be done the high-order accuracy of the numerical scheme, Girard and Delage (1990) proposed introducing this correction only when it is absolutely essential in view of the flux characteristics; lastly, the weighting coefficient may be determined optimally (Bénard et al., 2000).

9.5 Precipitation resolved at the grid scale

9.5.1 General points

Precipitation is the rain or snow, resulting from changes of phase of water vapour, that falls under the effect of gravity. These phase changes occur with modifications of the thermodynamic state of a layer of atmosphere (characterized by its pressure p, temperature T, and specific humidity q). Large-scale precipitation results in part from the effects of atmospheric dynamics at the grid scale of the model; it does not take into account precipitation resulting from sub-grid scale vertical motion. When the grid spacing is of a few tens of kilometres, this distinction corresponds to the synoptic and sub-synoptic scales on the one hand and to the convective scale on the other. However, with the advent of nonhydrostatic models with grid spacing of a few kilometres, the boundary is tending to shift towards the smaller scales.

The mechanisms by which precipitation forms are a matter for microphysics and are relatively complex to implement. Schemes describing interactions among the various phases of water, processed as prognostic variables transported by the dynamic model, have been proposed by Sundquist (1978) and more recently by Lopez (2002).

The highly simplified scheme presented here uses just a single prognostic variable characterizing the moisture of an atmospheric layer. It is based on the fact that we observe little supersaturation in the atmosphere: almost all condensed water precipitates immediately and only a small fraction remains present in the form of droplets or crystals in suspension making up the cloud. We also have the empirical Marshall-Palmer law (Marshall and Palmer, 1948), that gives the distribution of raindrops and snowflakes as a function of their diameter: this can be used to calculate the evaporation of precipitation or the melting of snow when crossing an atmospheric layer. Precipitation corresponding to the processes resolved at the grid scale may then be calculated within the framework proposed by Kessler (1969). If we refer to the schematic framework presented in Subsection 9.2.2, the system of equations (9.5) is then greatly simplified, since it comes down to a single equation describing the variation of specific humidity, q_v, in the moist air parcel.

9.5.2 Calculating precipitation in an atmospheric layer

The principle for calculating precipitation resolved at the grid scale is based on the elimination of supersaturation that occurs given the thermodynamic state of the moist air parcel under consideration. On the thermodynamic plots in Figure 9.9, simultaneously representing the temperature T and specific humidity q of an atmospheric layer

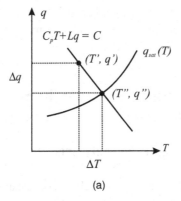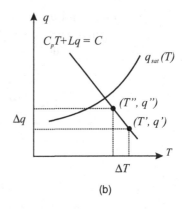

$$(a) \qquad\qquad\qquad (b)$$

Figure 9.9 Condensation and evaporation of large-scale precipitation.

of average pressure p, we have plotted the curve corresponding to the saturation specific humidity $q_{sat}(T, p)$. The state of an atmospheric layer when starting precipitation processing is characterized on these plots by the position of the point (T', q') relative to the curve $q_{sat}(T, p)$.

If the layer is saturated (as in Figure 9.9a), the surplus water vapour condenses as rain or snow and precipitates immediately into the underlying layer; the final state of the layer is then represented by the point $(T'' = T_w, q'' = q_{sat}(T'', p) = q_w)$, wet bulb temperature, and corresponding specific humidity, located on the curve $q_{sat}(T, p)$. Indeed, the layer where the condensation occurs also heats up under the effect of the release of latent heat. The elimination of supersaturation comes down to bringing the point (T', p') back to the curve $q_{sat}(T, p)$; as the evaporation and condensation processes occur with constant energy, the state point moves along the equation curve $C_pT + Lq = C$, C being a constant.

If the layer is not saturated (as in Figure 9.9b), the incoming precipitation evaporates in part or in full and therefore cools the layer. The fraction of the precipitation that is not evaporated may then melt or freeze, depending on its nature and the temperature of the layer, before being transmitted into the underlying layer.

When the layer's temperature T is below the triple-point temperature T_{00}, the separation between liquid and solid phases occurs when the precipitation is generated by using the discrimination function $h(T)$ already introduced in Paragraph 9.3.6.4 and shown in Figure 9.6.

This partition between liquid and solid phases is provisional, though, and is to be used to calculate the effects of evaporation and of the melting or freezing of precipitation; it also allows us to take into account the gradual melting of snow when $T > T_{00}$.

9.5.3 Evaporation of precipitation during fall

The empirical Marshall-Palmer law, giving the number N of raindrops of diameter D per unit volume, is written:

$$N(D) = N_0 e^{-\lambda D}, \tag{9.59}$$

where λ characterizes the intensity of precipitation and N_0 is a constant.

It is assumed too that the fall speed of raindrops of diameter D reaches its limit value W immediately, as given by the formula:

$$W = aD^\alpha, \text{ where } \alpha \text{ is a function of } p \text{ and of } T.$$

The precipitation flux for all the drops, denoted P here, is therefore written:

$$P = \int_0^\infty aD^\alpha N_0 e^{-\lambda D} \frac{\pi D^3}{6V_l} dD, \tag{9.60}$$

with V_l being the specific volume of water and α an adjustable coefficient.

The evaporation of a raindrop of diameter D over time is given by:

$$\frac{\partial M}{\partial t} = -bD^\beta \rho(q_w - q), \tag{9.61}$$

where ρ is the density of moist air, b a function of p and of T, and β an adjustable coefficient.

By writing that the variation of precipitation flux between two layers separated by dz is equal to the loss by evaporation, for all of the drops we get the relation:

$$\frac{dP}{dz} = \int_0^\infty bD^\beta \rho(q_w - q)N_0 e^{-\lambda D} dD. \tag{9.62}$$

After calculating the integrals that appear in relations (9.60) and (9.62), eliminating the parameter λ between these two relations and using the pressure vertical coordinate, we finally obtain the relation:

$$\frac{dP}{dp} = -\frac{bN_0\Gamma(\beta+1)}{g}\left[\frac{6V_l P}{\Gamma(\alpha+4)aN_0\pi}\right]^{\frac{\beta+1}{\alpha+4}}(q_w - q).$$

In the previous formula, Γ represents the gamma function; the values of the coefficients α and β can be obtained from experimental data. After adjusting these coefficients and calculating values of a and b by using the standard atmosphere profiles for pressure p and temperature T in the troposphere, the relation takes the form:

$$\frac{d\sqrt{P}}{dp} = -\frac{1}{p^2}C_{evap}(q_w - q), \text{ with } C_{evap} = 4.8 \ 10^6 \text{ SI.} \tag{9.63}$$

This relation indicates that evaporation in the layer of atmosphere is greater when the layer considered is less saturated, which is quite logical; we also remark that, contrary to first intuition, but in line with precipitation data processing, the relation is not linear in P.

9.5.4 The melting of snow as it falls

Here we deal with the melting of snow as it precipitates through the layers of the atmosphere. The rate of change in the mass of melting ice M_i is given by:

$$\frac{\partial M_i}{\partial t} = -b' D^\beta \rho \, (T - T_{00}), \text{ with } b' = b\frac{\gamma}{\mathcal{D}}\frac{C_p}{L_f} = bB. \tag{9.64}$$

The parameters γ and $\mathcal{D}$ represent the thermal diffusion coefficient and the molecular diffusion coefficient; L_f is the latent heat of fusion: B is therefore a constant.

As the snow melts, the proportion of snow in the precipitating mixture is characterized by the ratio r_i; the expressions for the mass of snow M_i and for the snow flux P_i are given by:

$$P_i = r_i P \text{ and } M_i = r_i M.$$

The rate of change in the snow flux for a layer of thickness dz is written:

$$\frac{dP_i}{dz} = \frac{d(r_i P)}{dz} = r_i \frac{dP}{dz} + P\frac{dr_i}{dz}.$$

By writing that the rate of change in the proportion of snow between two levels dz apart is equal to the relative quantity of snow disappearing by fusion, we get:

$$\frac{dP_i}{dz} = -r_i \frac{\partial M}{\partial t} - \left(\frac{\partial M_i}{\partial t}\right)_f.$$

By writing in the same way for the entire mixture $dP/dz = -\partial M/\partial t$, we obtain, for a raindrop of diameter D, the relation:

$$P(D)\frac{dr_i}{dz} = bBD^\beta \rho(T - T_{00}),$$

and for all of the drops:

$$\left[a\frac{\pi}{6V_i}\int_0^\infty D^\alpha D^3 e^{-\lambda D}\,dD\right]\frac{dr_i}{dz} = bB\rho(T - T_{00})\int_0^\infty D^\beta e^{-\lambda D}\,dD. \tag{9.65}$$

After calculating the integrals appearing in relations (9.60) and (9.65), eliminating the parameter λ between these two relations and using the pressure vertical coordinate, we finally get the relation:

$$\frac{dr_i}{dp} = -\frac{1}{p^2}C_{melt}\frac{(T - T_{00})}{\sqrt{P}}, \text{ with } C_{melt} = 2.4\ 10^4 \text{ SI}. \tag{9.66}$$

This relation shows that more snow melts when the temperature difference $T - T_{00}$ is higher, which is what we would expect.

9.5.5 The calculation process in the various layers

To calculate the evolution of precipitation flux while crossing a layer k, two cases are to be considered depending on whether the layer is supersaturated or not.

Where $q_k > q_{wk}$, condensation occurs; the precipitation flux $P_{\tilde{k}}$ reaching the base of layer k is obtained by adding to the precipitation flux $P_{\tilde{k-1}}$ arriving at the top of the layer the flux corresponding to the quantity of condensed water vapour:

$$P_{\tilde{k}} = P_{\tilde{k-1}} + \frac{1}{g} \frac{(q_k - q_{wk})}{\Delta t} (p_{\tilde{k}} - p_{\tilde{k-1}}). \tag{9.67}$$

To calculate the variation of the proportion of snow Δr_i in the precipitating flux, we must first evaluate the quantity of snow that can form, given the temperature, before calculating the effects of melting and freezing.

When $T < T_{00}$, precipitation occurs as snow and the proportion of snow $(r_i)_{\tilde{k}}$ in the precipitating flux $P_{\tilde{k}}$ increases; we therefore calculate a provisional proportion of snow $(r_i')_{\tilde{k}}$:

$$(r_i')_{\tilde{k}} = \frac{1}{P_{\tilde{k}}} \left[(r_i)_{\tilde{k-1}} P_{\tilde{k-1}} + P_{\tilde{k}} - P_{\tilde{k-1}} \right].$$

The fictitious proportion of snow $(r_f)_k$ relative to level k used to calculate the rates of melting or freezing and the evaporation of precipitation also increases; we therefore calculate a fictitious proportion of provisional snow $(r_f')_k$ by taking into account the solid/liquid differentiation supplied by the function $h(T)$:

$$(r_f')_k = \frac{1}{P_{\tilde{k}}} \left[(r_f)_{k-1} P_{\tilde{k-1}} + h(T_k) (P_{\tilde{k}} - P_{\tilde{k-1}}) \right].$$

When $T_k \geq T_{00}$, precipitation occurs as rain and the provisional proportion of snow $(r_i')_{\tilde{k}}$ in the flux diminishes, as does the fictitious proportion of provisional snow $(r_f')_k$ relative to level k:

$$(r_i')_{\tilde{k}} = \frac{1}{P_{\tilde{k}}} \left[(r_i)_{\tilde{k-1}} P_{\tilde{k-1}} \right] \quad \text{and} \quad (r_f')_k = \frac{1}{P_{\tilde{k}}} \left[(r_f)_{k-1} P_{\tilde{k-1}} \right].$$

The final proportion of snow $(r_i)_{\tilde{k}}$ in the precipitating flux and the final fictitious proportion of snow $(r_f)_k$ relative to level k are calculated by using formula (9.66), in which the melting coefficient C^*_{melt} is modified to take account of the fact that snow melts much faster than rain freezes; this coefficient is then written:

$$C^*_{melt} = C_{melt}(1 - r_f') + \sigma_{melt} C_{melt} r_f' \tag{9.68}$$

where σ_{melt} designates the ratio of the snow melting rate to the raindrop freezing rate.

We therefore obtain after discretization:

$$(r_i)_{\tilde{k}} = (r'_i)_{\tilde{k}} - C^*_{melt} \frac{T - T_{00}}{\frac{1}{2}\left(\sqrt{P_{\widetilde{k-1}}} + \sqrt{P_{\tilde{k}}}\right)} \left(\frac{1}{P_{\widetilde{k-1}}} - \frac{1}{P_{\tilde{k}}}\right)$$

and:

$$(r_f)_k = (r'_f)_k - C^*_{melt} \frac{T - T_{00}}{\frac{1}{2}\left(\sqrt{P_{\widetilde{k-1}}} + \sqrt{P_{\tilde{k}}}\right)} \left(\frac{1}{P_{\widetilde{k-1}}} - \frac{1}{P_{\tilde{k}}}\right).$$

The values of $(r_i)_{\tilde{k}}$ and $(r_f)_k$ are initialized for the first layer where precipitation occurs, given the temperature of this layer: if $T < T_{00}$, snow only is formed and we write: $(r_i)_{\tilde{k}} = 1$ and $(r_f)_k = h(T_k)$, whereas if $T \geq T_{00}$, water alone is formed and we write: $(r_i)\tilde{k} = 0$ and $(r_f)_k = 0$.

When $q \leq q_{sat}$, there is no condensation and all or part of the flux reaching the top of the layer $P_{\widetilde{k-1}}$ must be evaporated. $P_{\tilde{k}}$ is calculated from (9.63) in which the evaporation coefficient C^*_{evap} is modified to allow for the fact that snow evaporates relatively much faster than water, owing to its slower fall speed. This is written:

$$C^*_{evap} = C_{evap}(1 - r'_f) + \sigma_{evap} C_{evap} r'_f \tag{9.69}$$

where σ_{evap} is the ratio of the evaporation rate of snow to the evaporation rate of rain.

After discretization, the flux $P_{\tilde{k}}$ reaching the base of layer k is given by:

$$\sqrt{P_{\tilde{k}}} = \sqrt{P_{\widetilde{k-1}}} + C^*_{evap}(q_k - q_{w_k})\left(\frac{1}{P_{\widetilde{k-1}}} - \frac{1}{P_{\tilde{k}}}\right). \tag{9.70}$$

There is no reason to modify the provisional proportion of snow (r'_i) in the flux nor the provisional fictitious proportion of snow $(r'_f)_k$ relative to level k, and we write:

$$(r'_i)_{\tilde{k}} = (r_i)_{\widetilde{k-1}} \text{ and } (r'_f)_k = (r_f)_{k-1}.$$

To obtain the final proportions, we proceed as before to calculate the effect of the melting or freezing of the remaining precipitation.

The precipitation fluxes in solid and liquid form reaching the base of layer k within one time step are written:

$$P_l = (1 - r_i)P_{\tilde{k}} \text{ and } P_i = r_i P_{\tilde{k}}. \tag{9.71}$$

The modifications of enthalpy resulting from the condensation of water vapour and from the evaporation of precipitation in layer k are obtained by using the evolution equation (9.6), after calculating the relevant pseudo-fluxes: P'_l and P'_i, corresponding to the transformation of vapour into liquid water or ice, and P'''_l and P'''_i, corresponding to the evaporation of precipitation. It should be noted that formulating this equation in terms of *flux divergence* means discretization can be done in keeping with the Green-Ostrogradski theorem when integrating in the vertical.

9.5.6 The need for more detailed schemes

Such a scheme, based on the use of specific humidity as the only prognostic variable, cannot of course account for the complexity of the microphysics behind the formation and evolution of precipitation. The adjunction of prognostic variables for liquid water and ice means far more detailed schemes can be used (already cited in Subsection 9.5.1), and cloud properties, which interact with radiation, can be defined more precisely. Besides, the use of models with increasingly fine grid spacing means that new effects must be allowed for such as the horizontal component of the raindrop/snowflake displacement during fall and the screen effect of mountains, or the different rates of fall of the various condensates. This in turn requires prognostic equations for the specific amount of falling water species (raindrops and snowflakes at least). The handling of their budget is part of system (9.5), while their thermodynamic properties are the same as those of the corresponding suspended cloud species. The difficulty in handling these particular prognostic variables is twofold. One must first ensure local numerical stability for the numerous and complex parameterized exchange processes between them and all other components. Second, the sedimentation (due to the vertical displacement) must be accounted for, both in absolute terms in the corresponding equation of (9.5) and in its relative influence on the intensity of the above-mentioned exchanges. The latter delicate problem was generally treated via an *advective method* (approximations along trajectories during one time step). Recently, Geleyn et al. (2008) proposed a more exact and at least equally stable solution, through the replacement of the advective method by a *statistical-type* one.

9.6 Convection

9.6.1 General points

Convection in the atmosphere may be defined as '*thermally direct circulations which result from the action of gravity upon an unstable vertical distribution of mass*', as Emanuel (1994) put it. Given the characteristic scales of convective motion, ranging from a few metres for thermals to several hundred metres for updrafts and downdrafts of moist convection, such motion cannot be taken into account directly by the dynamics of present-day models; this is especially true for the primitive equation models (because of the assumptions leading to the hydrostatic relation, as set out in Subsection 2.2.3) but remains true for nonhydrostatic models with grid spacing of a few kilometres. So it is essential to parameterize the effects of convection that are reflected by exchanges of moisture, heat, and momentum with the environment and by precipitation (then termed *convective precipitation*). The difficulty in parameterizing the effects of convection lies in the fact that it is an interactive process, involving a wide range of scales: it can be considered that large-scale circulation helps to create zones amenable to the triggering of convection, just as it may be thought that it is the effect of convection that bring about changes in the large-scale circulation pattern.

Ever since primitive equation models have been used for weather forecasting, many parameterization methods have been proposed and they have been reviewed in great detail by Arakawa (2004). The *convective adjustment method* suggested by Manabe and Strickler (1964) consists in adjusting, where instability occurs, temperature and moisture profiles to neutral and saturated profiles; no other cloud model is required and the only constraint is the conservation of moist static energy in the vertical. In the approach proposed by Kuo (1965, 1974), the input of moisture to a buoyant air column from vertical diffusion at its base and from horizontal convergence along its depth is redistributed in the vertical in the form of both generation of precipitation and environment moistening by the modelled cloud (so-called Kuo-type closure). The redistribution takes into account the differences between temperature and moisture profiles resolved by the model and the saturated pseudo-adiabatic profiles of the cloud developing in this environment. Further to Arakawa and Schubert (1974), many schemes have used a *cloud mass flux* formulation, a quantity representing the intensity of updrafts at the convective scale; by this approach, the effects of large-scale convection can be evaluated after determining the temperature and moisture profiles of the cloud and after defining how to evaluate the mass flux (closure hypothesis). Schemes using a mass flux formulation include those proposed by Bougeault (1985), Tiedtke (1989), Kain and Fritsch (1990), and by Bechtold et al. (2001).

9.6.2 The problem of causality for convection

Convection being an interactive physical process covering a wide range of scales, any conceptual separation of scales is somewhat pointless; such separation is effective in practice, though, because of the finite spatial resolution of the model. It is also difficult to say what proportion of the mean profile observed is already the result of convection and, under the circumstances, it is pointless to try to ground a formulation in *forcing* and *response* terms. It is more helpful therefore to seek to solve the following problem: on a given scale in the atmosphere or in a column of the model, what is the modification of state variables resulting from convective updrafts on smaller scales? In this way we are led to look more especially at the source of energy of the convective processes and to define, as Piriou (2005) proposes, the *rate of buoyant convective condensation* (BCC):

$$BCC = \begin{cases} -\left(\dfrac{dq}{dt}\right)_{cond}, & \text{if } \left(\dfrac{dq}{dt}\right)_{cond} < 0 \text{ and if the parcel is unstable,} \\ 0, & \text{otherwise,} \end{cases}$$

where q denotes the specific humidity of the parcel.

This quantity represents the source of condensates due to moist convection; it is by definition a positive or zero quantity, whether the convection is the source of precipitation or not. All of the other processes involved in convection, such as convective precipitation, downdrafts, and the effect due to feedback of gravity waves generated by the release of latent heat, may be seen as the consequences of condensation due to convective instability within updrafts. The characteristic scale of condensation due to convective instability is therefore of the same order of magnitude as that of the updrafts, which is 10 km or so for the largest of them. The parameterization of convection will therefore now consist in defining the condensation rates due to convective instability at sub-grid scale given the forcing resolved at the grid scale. It is on this principle that Piriou et al. (2007) proposed a parameterization framework that we have chosen to outline here.

9.6.3 The basic equations

Following the approach taken by Yanai et al. (1973), the effects of convection at scales resolved by the model on the evolution of dry static energy s, specific humidity q, and horizontal wind $\mathbf{V}$ are reflected by the equations below:

$$\left.\begin{aligned} Q_1^c = Q_1 - Q_R &= LC_N - \frac{\partial \overline{\omega' s'}}{\partial p}, \\ Q_2^c = -\frac{Q_2}{L} &= -C_N - \frac{\partial \overline{\omega' q'}}{\partial p}, \\ \mathbf{Q}_3^c &= -\frac{\partial \overline{\omega' \mathbf{V}'}}{\partial p}. \end{aligned}\right\} \tag{9.72}$$

Q_1 is the total heat input per unit time while Q_R is the input of radiation only; Q_2 is the energy loss per unit time corresponding to the water vapour deficit. The quantities $Q_1^c = Q_1 - Q_R$, $Q_2^c = -Q_2/L$, and $\mathbf{Q}_3^c$ therefore represent respectively the tendencies of static energy, specific humidity, and momentum due to convection within the grid. On the right-hand side of equations (9.72), C_N represents the net rate of condensation per unit time (condensation minus evaporation).

Supposing that clouds cover a fraction σ of the total mesh surface, the Reynolds decomposition of the turbulent flux $\overline{\omega' \psi'}$ of variable ψ is used to express this as a function of ψ_c and ψ_e, which are values of ψ for the cloud and the environment respectively, and of vertical velocity within the cloud ω_c:

$$\overline{\omega' \psi'} = \sigma \omega_c (\psi_c - \psi_e).$$

The quantity $\omega^* = \sigma \omega_c$ is used to define the *convective mass flux* M_c (being counted positively for convective updrafts):

$$M_c = -\omega^* = g\rho w^*.$$

To give a central role to *BCC*, the net condensation rate is broken down into its various components:

$$C_N = C_{BCC} + C_{UCC} - E_C - E_p;$$

C_{BCC} is the rate of condensation due to unstable updrafts, C_{UCC} the condensation rate due to stable updrafts (such as the anvils of cumulonimbus clouds), E_C the evaporation rate of cloud condensates, and E_p the evaporation rate of precipitation. All of these quantities are defined as positive.

To be thorough, we should also add to the right-hand side of the equation for dry static energy an additional term $-W$ representing the heat loss resulting from the falling precipitation (melting of snow, exchange of sensible heat between raindrops and the surrounding air).

Equations (9.72) are then written:

$$\left.\begin{aligned}
Q_1^c &= L(C_{BCC} + C_{UCC} - E_C - E_P) - W - \frac{\partial\left[\omega^*(s_c - s_e)\right]}{\partial p}, \\[2mm]
Q_2^c &= -(C_{BCC} + C_{UCC} - E_C - E_P) - \frac{\partial\left[\omega^*(q_c - q_e)\right]}{\partial p}, \\[2mm]
\mathbf{Q}_3^c &= -\frac{\partial\left[\omega^*(\mathbf{V}_c - \mathbf{V}_e)\right]}{\partial p},
\end{aligned}\right\} \tag{9.73}$$

subscripts c and e referring to cloud and environment values, respectively.

The transport terms and the microphysics terms can then be coupled by writing that the terms C_{BCC}, C_{UCC} are E_C proportional to the vertical velocity in pressure coordinates ω^*; we therefore write:

$$C_{BCC} + C_{UCC} - E_C = -\omega^*(\widehat{C}_{BCC} + \widehat{C}_{UCC} - \widehat{E}_C).$$

The quantities $\widehat{C}_{BCC}$, $\widehat{C}_{UCC}$, and $\widehat{E}_C$ now represent the (convective and non-convective) condensation and evaporation rates of the cloud condensates per unit thickness in the vertical. There is no reason why the terms for the rates of evaporation E_p and heat loss W due to the falling precipitation should be proportional to ω^*.

We finally obtain the equations for the *microphysics and transport convection scheme* (MTCS) proposed by Piriou et al. (2007):

$$\left.\begin{aligned}
Q_1^c &= -\omega^* L(\widehat{C}_{BCC} + \widehat{C}_{UCC} - \widehat{E}_C) - LE_P - W - \frac{\partial\left[\omega^*(s_c - s_e)\right]}{\partial p}, \\[2mm]
Q_2^c &= -\omega^*(-\widehat{C}_{BCC} - \widehat{C}_{UCC} + \widehat{E}_C) + E_P - \frac{\partial\left[\omega^*(q_c - q_e)\right]}{\partial p}, \\[2mm]
\mathbf{Q}_3^c &= -\frac{\partial\left[\omega^*(\mathbf{V}_c - \mathbf{V}_e)\right]}{\partial p}.
\end{aligned}\right\} \tag{9.74}$$

Having established the system of equations (9.74), the following operations remain to be performed:

- define a simplified microphysics scheme for calculating the condensation terms $\hat{C}_{BCC}$, $\hat{C}_{UCC}$, and the cloud condensate evaporation term $\hat{E}_c$ as well as the profiles $S_c(p)$, $q_c(p)$, and $\mathbf{V}_c(p)$ within the cloud;
- define a scheme for calculating the vertical velocity profile within the cloud, $\omega_c(p)$;
- choose a closure relation to express $\omega^* = -M_c$ as a function of the vertical velocity within the cloud ω_c;
- couple the *purely convective* part of computations with a complete microphysical handling of the outcome of *BCC* (cloud-geometry, auto-conversion, sedimentation, collection, evaporation, melting ... in the most sophisticated case) in order to link (9.74) with physically realistic values for the terms E_p and W;
- add a treatment of convective downdrafts similar in spirit to that of MTCS for the convective updrafts and consistent with the microphysical handling just outlined.

9.6.4 The characteristic profiles of the cloud and the influence of microphysics

The cloud profiles can be calculated by making an air parcel rise from the lowest level of the model adiabatically up to the condensation level, then by following the moist adiabat taking into account the effects of entrainment of the surrounding moist air, which comes down to satisfying the equation:

$$\frac{\partial q_c}{\partial z} = \left(\frac{\partial q_c}{\partial z}\right)_{cond} + \varepsilon(q_e - q_c), \tag{9.75}$$

where q_c is the specific humidity of the cloud, q_e the specific humidity at the grid scale, and ε the entrainment rate, which remains the most heuristic variable in parameterization schemes choosing the MTCS approach.

The condensation rate $(\partial q_c / \partial z)_{cond}$, which is a negative quantity, is determined in such a way that the parcel evacuates the surplus water vapour, allowing it to remain saturated relative to water or ice (given the temperature). The quantity $-(\partial q_c / \partial p)_{cond}$ then represents the condensation rate $\hat{C}_{UCC}$ or $\hat{C}_{BCC}$, depending on whether the parcel is stable or unstable. The cloud condensates in liquid and solid form may be calculated in the course of vertical integration of equation (9.75) from the $h(T)$ partition (9.26), so as to also yield $\hat{E}_C$. In the simplest case, precipitation is immediately transferred to the lower layer where it may evaporate; the evaporation rate E_P of falling precipitation is found for instance from the formulation by Kessler (1969) already seen in Subsection 9.5.3. But even then, the algorithmic problem for an optimal coupling between the ascending computation of the cloud properties and the descending computation of changes in the precipitation fluxes is far from trivial. Gerard et al. (2009) solve this issue by externalizing the complete microphysical computations (at any degree of sophistication) and by applying them to the sum of all sources of condensation, not only to the BCC-type ones. This way of handling matters also simplifies the path towards an MTCS-like treatment of the convective downdrafts.

The upward vertical velocity is calculated by vertical integration of the work of the various forces acting on the parcel (buoyancy, friction, and possibly nonhydrostatic effects). Siebesma et al. (2003) and Bretherton et al. (2004) have shown that this variation may be written as a function of buoyancy B and of the entrainment coefficient ε as below:

$$d\left(\frac{w_c^2}{2}\right) = w_c dw_c = (aB - b\varepsilon w_c^2)dz, \tag{9.76}$$

a and b being two constants.

Buoyancy B is expressed with the help of the potential temperature of the cloud θ_c and of the environment θ_e in the form $B = g(\theta_c - \theta_e)/\theta_e$.

It should be noted that relation (9.76) is obtained from the vertical velocity evolution equation, which comes from a nonhydrostatic formulation of updraft dynamics.

As for the wind, a very simple solution is to consider that inside the cloud, the wind $\mathbf{V}_c$ maintains a constant value, obtained by calculating the average wind $\mathbf{V}_e$ from the free convection level (cloud base) to the neutral convection level (cloud top). Other more detailed parameterization schemes have been proposed for momentum transport resulting from convection, notably by Kershaw and Gregory (1997), with account being taken of both entrainment and the impact of (nonhydrostatic) pressure differences between the cloud and its environment.

9.6.5 The triggering of instability

When equation (9.76) is integrated in the vertical, we are in fact calculating the work done. Initially, if the layer is stable, this work is negative and contributes to Convective INhibition (*CIN*) representing the energy required to overcome the buoyancy force in stable layers. When the work is positive, it contributes to Convective Available Potential Energy (*CAPE*), which is the total energy that can be converted into kinetic energy; in case $CAPE > CIN$, the convection scheme may be triggered; it can then be shown that the maximum vertical velocity possibly reached within the cloud corresponds to kinetic energy $(w_c^{max})/2 = CAPE - CIN$.

9.6.6 The closure relation

Closure of the system of equations consists in linking $\omega^* = -M_c$ to the vertical velocity within the cloud ω_c via a parameter σ characterizing the fraction of the mesh surface where convection is active, that is $\omega^* = \sigma\omega_c$; if a constant value is chosen for σ, we thus introduce a CAPE type closure, but other types of closure may also be contemplated (Craig, 1996; Plant, 2010).

In fact the first operational application of the MTCS equations, described in Gerard et al. (2009), relies on a prognostic version of the Kuo-type closure described in Subsection 9.6.1. As already hinted at, it also has a specific handling of the links between updraft, microphysics, and downdraft, and it uses the Kershaw and Gregory (1997) approach for momentum convective transport.

9.6.7 Prospects

The relatively straightforward framework just described provides, via the condensation rate due to convective instability, means of evaluating the feedback of convective processes on environmental forcing resolved at the grid scale. It allows us to define which clouds develop in a given environment and the feedback effect of those clouds on their environment. By isolating the terms of microphysics and transport, this concept can be used in various forecast models, ranging from large scale to mesoscale. When the horizontal resolution of the model is increased, we can contemplate using a very detailed scheme for microphysics and dealing with transport terms only marginally insofar as the complementing process is properly handled on the scale resolved by the model and the interaction with the *resolved precipitation* scheme is correctly accounted for. This is why schemes based on the MTCS concept presented here, and that are in principle able to process dry convection, nonprecipitating cloud convection, and precipitating convection, should make it possible in the coming years to unify the handling of the various convective processes in models.

9.7 Effect of sub-grid orography

Orographic gravity waves are generated when a stable atmospheric flow passes over mountains; under some circumstances the waves may propagate vertically and then be dissipated or absorbed in the upper atmosphere. Simplistic allowance for the effect of *sub-grid mountains* on the large-scale flow consisted initially in locally reinforcing the roughness length and in using for the dynamics an *envelope orography* (average orography at grid scale, plus a fraction of sub-grid variance); this approach had the drawback of dissipating energy at the base of the atmosphere without ensuring vertical transport of the momentum before dissipation at altitude. More realistic parameterization of this effect now involves calculating the momentum flux resulting from uneven orography and its propagation and possible deposition in the vertical. To correctly handle the effect of orography on large-scale flow, we must also take into account the blocking effect of the sub-grid mountain under certain circumstances, and that is reflected by additional deceleration and by deviation of part of the resolved flow. To parameterize all these effects, Catry et al. (2008) proposed a coherent set of solutions the main characteristics of which are set out below. Assuming that the resolved mountainous effects are correctly described by the dynamical part of the model, we shall below, for the sake of simplicity, drop the term 'sub-grid' in front of 'mountains'.

9.7.1 The wave momentum flux induced by orography

Following arguments similar to those developed by Boer et al. (1984), the wave momentum flux $\boldsymbol{\tau}_S$ induced by surface topography may be expressed, for sufficiently strong effects, as:

$$\boldsymbol{\tau}_S = -\rho_S N_S \mathbf{V}_S K_g h, \tag{9.77}$$

ρ_S being the density close to the ground,
N_S, the Brunt-Väisälä frequency close to the ground: $[(g/s)(\partial S/\partial z)]^{1/2}$,
$\mathbf{V}_S$ the wind close to the ground,
K_g, an adjustment coefficient,
h, the standard deviation of orography not resolved by the model's grid.

As horizontal wind cancels out at the surface, it is suitable to adopt average values for N_S and $\mathbf{V}_S$, calculated over a characteristic thickness of relief on the sub-grid scale H (*effective height of the obstacle*) proportional to h. If we have information about the orientation of the orography, the wind $\mathbf{V}_S$ may be modified to be replaced by a fictitious value $\mathbf{V}_S^f$ taking into account the angle of attack of the flux relative to the obstacle.

The linear theory of mountain waves (Eliassen and Palm, 1961) specifies that the momentum flux $\boldsymbol{\tau}$ remains constant in the vertical so long as the wave amplitude δ (vertical displacement of isentropic surfaces) remains low relative to the wavelength L, and it may take the form:

$$\boldsymbol{\tau} = -\rho N \mathbf{U} \frac{\delta^2}{L}, \tag{9.78}$$

N being the Brunt-Väisälä frequency and $\mathbf{U}$ the projection of the wind $\mathbf{V}$ on the effective surface wind given by:

$$\mathbf{U} = \frac{\mathbf{V} \cdot \mathbf{V}_S}{\|\mathbf{V}_S\|^2} \mathbf{V}_S. \tag{9.79}$$

The wave propagates vertically with an amplitude that increases because of the diminishing density of the air and sometimes because of the turning of the wind direction leading to a lower (or even zero) scalar product in the numerator of expression (9.79). Eventually it reaches a *critical level* where its amplitude is of the same order of magnitude as its wavelength and it breaks and loses momentum; thus damped it continues to propagate in the vertical in such a way that δ remains proportional to $\|\mathbf{U}\|/N$ (the saturation condition of Lindzen, 1981). Assuming the wave, of amplitude δ_S and wavelength proportional to U_S/N_S, is saturated at ground level, a provisional value of the momentum flux in the upper air is written as a function of its value close to the ground as below:

$$\boldsymbol{\tau} = -\rho \frac{U^2}{LN} \frac{\delta_S^2 N_S^2}{U_S^2} \mathbf{U} = \frac{\rho}{\rho_S} \left(\frac{U}{U_S}\right)^3 \frac{N_S}{N} \boldsymbol{\tau}_S = \Gamma \boldsymbol{\tau}_S, \tag{9.80}$$

U_S and N_S being the values of $U = \|\mathbf{U}\|$ and of N close to the surface.

The momentum flux on each interlayer surface k is calculated from the value of this flux close to the ground, $\tau_{\tilde{N}} = \tau_S$, as follows:

$$\tau_{\tilde{k}} = \min(\Gamma_{\tilde{k}}\, \tau_S, \tau_{\widetilde{k+1}}) \text{ for } k = N-1,\ldots,1 \text{ with: } \Gamma_{\tilde{k}} = \frac{\bar{\rho}_{\tilde{k}}}{\rho_S}\left(\frac{\bar{U}_{\tilde{k}}}{U_S}\right)^3 \frac{N_S}{N_{\tilde{k}}},$$

$\bar{\rho}_{\tilde{k}}$, $\bar{U}_{\tilde{k}}$, and N_k representing the values of ρ, U, and N on the interlayer surface k.

9.7.2 Effects of resonance and trapping of the wave

The type of non-increasing profile thus obtained for Γ is shown in Figure 9.10a. Modification may however be made to the profile of Γ to take account of the phenomena of resonance and of trapping of the gravity wave.

The composition of a wave that is partially reflected at a critical level with the incident wave may contribute to strengthening or depleting the momentum flux. Given the position of the critical level, we can determine a phase shift θ (depending on the vertical integral of the quantity N/U) and calculate a resonance function $f(\theta)$ corresponding to the ratio of amplification or damping of the momentum flux. Given the value of this function and the principle of non-growth of the flux with altitude, the Γ profiles are modified as in Figure 9.10b (dotted curves).

We can also find levels where the value of the squared Brunt-Väisälä frequency N^2 becomes negative, preventing the wave from propagating above this level: the wave is trapped. It should be tested therefore whether such a level of trapping is to be found.

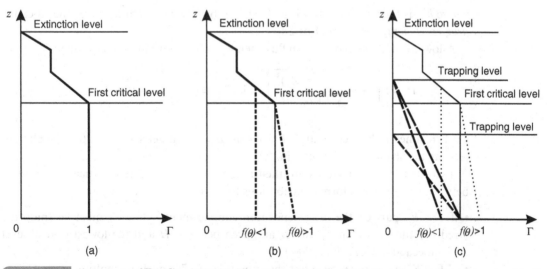

Figure 9.10 Various types of Γ profiles obtained in different cases: (a) standard case with neither resonance nor trapping; (b) with resonance; (c) with trapping, with or without resonance. (From a drawing by Catry, 2006, personal communication)

If it is lower than the critical level, there cannot be any resonance; if it is above the critical level, the possibility of damping resonance must be contemplated and the Γ profiles modified consequently so the wave is extinguished at the trapping level; Figure 9.10c shows the new Γ profiles obtained in both cases (the curves with long dashes are for the case where the trapping level is above the critical level and the short dashed curve for the opposite case). The Brunt-Väisälä frequency is obtained from the static stability of the atmosphere; in order to allow for moist atmospheric conditions, the static stability of the atmosphere must be modified in a similar manner to that described in Subsection 9.4.4 with regard to shallow convection.

9.7.3 Consequences of partial blocking of flow

This parameterization of the effects of gravity waves does not cover all of the physical effects of mountains on a stratified flow. Given the stability of the atmosphere, only part of the stream flows over the mountain, creating gravity waves. In addition, the blocking of the lower part of the flow forcing it to split around the mountain induces two forces: a drag force known as *form drag* in the opposite direction of the flow and a *lift force* perpendicular to the flow and responsible for its deviation.

The flow is considered to be blocked if the inverse Froude number, given by $F = N_S H / U_S$ (where H denotes the effective height of the obstacle), exceeds a critical value F_c (of the order of 0.5). This leads to defining a blocking height $Z_b = H \max (0, 1 - F_c/F)$ below which the flow is indeed blocked and subjected to form drag and lift (Lott and Miller, 1997). The momentum flux corresponding to form drag is written:

$$\boldsymbol{\tau}_b = -C_d \int_0^{Z_b} \rho l(z) \frac{\mathbf{V} \|\mathbf{V}\|}{2} \, dz, \qquad (9.81)$$

where $l(z)$ is the width of the obstacle viewed by flow at height z and C_d the so-called *drag coefficient*, of the order of 2 to 4.

Allowance for this momentum flux then leads to modifying the Γ profile:

$$\Gamma' = \Gamma \left[1 + a_d \sqrt{\frac{(1 - z/Z_b)^3}{1 + \alpha z/H}} \right], \quad \text{with} \quad a_d = C_d \frac{\|\mathbf{V}_S^f\|}{HN_S} \max(0, \, 1 - F_c/F). \qquad (9.82)$$

z is here a height between 0 and H, whereas α is a parameter between 0 and 1 characterizing the shape of the mountain.

Two further modifications are then necessary to get a better vertical consistency between the wave and form contributions to the total drag deposition:

- if $F > F_c$, part of the drag deposition enhancement below Z_b must be smoothly redistributed between Z_b and H and even between H and the lowest critical level (Scinocca and McFarlane, 2000);
- if $F \leq F_c$, a_d becomes zero but the expression (9.77) must be multiplied by F/F_c in order to retrieve the correct formulation for the case of pure linear weak gravity waves (cf. (9.78) and the explanation introducing (9.80)).

The effect of lift does not in principle involve any work since the force applied is perpendicular to the flow; taking it into account amounts to modifying the Coriolis parameter f by adding between the surface and the height H an extra term f' expressed as (Lott, 1999):

$$f' = f \left[L_t \frac{\alpha^2 \dfrac{1-z/H}{1+\alpha z/H}}{(H/h)\left[(1+\alpha)\ln(1+\alpha)-\alpha\right]} \right],$$

(9.83)

L_t being a numerical coefficient of the order of 1 and H/h being of the order of 3.

In fact, practice has shown that the lift effect correctly fulfils its main role (replacing the envelope orography for the simulation of the volume effect of stagnant air in the sub-grid valleys) only if its direction is made orthogonal to that of the geostrophic wind, and so more representative of the unperturbed flow than the actual low-level wind.

The tendency of the horizontal wind resulting from the effect of orography at subgrid scale is obtained by calculating the vertical divergence of the momentum flux. When large time steps are used (with semi-implicit semi-Lagrangian models), it is then essential to process the terms quasi-implicitly.

9.8 Horizontal diffusion

The introduction of horizontal diffusion into numerical prediction models means we can reduce small-scale *noise* that is manifested by the occurrence of small waves whether locally or throughout the working domain. This behaviour results from a number of imperfections deriving from the model itself or from the numerical processing.

Whether we use the spectral method or the finite difference method, the model's resolution is necessarily limited and the nonlinear interactions generate small waves that, because they cannot be represented properly, accumulate in the upper part of the spectrum. This behaviour may be construed as being due to the absence of any actual physical parameterization of turbulent transfer in the horizontal. The decrease in resolution in the vertical in the stratospheric part of the model and the *reflective* boundary condition often adopted at the upper limit of the working domain are also factors that contribute to the creation of small-scale noise over the entire domain. Moreover, the presence of high-gradient orography and the triggering of physical processes when a threshold is exceeded contribute to the feeding of local small-scale noise that may amplify in the course of integration of the model and even engender numerical instability.

To prevent the development of small-scale noise in models, it is necessary to introduce an additional term into the equations to deal with *horizontal diffusion* (sometimes also called lateral mixing). Generally, so as to be sufficiently selective and to reduce the smallest waves in the main, we add to the evolution equations iterated Laplacian terms ($\nabla^{2n_{iter}}$, Laplacian operator virtually applied n_{iter} times) multiplied by a suitable horizontal diffusion coefficient K_{DH}. In this way for a generic variable X, we write:

$$\frac{\partial X}{\partial t} = \cdots + K_{DH}(-1)^{n_{iter}-1} \nabla^{2n_{iter}} X. \tag{9.84}$$

This diffusion term is introduced into the evolution equations for the horizontal wind, dry static energy, and specific humidity; it has the effect of smoothing the unevenness that may occur in the horizontal fields representing these quantities. This term must not, however, be directly applied to variables exhibiting very high gradients on the levels of the model in mountainous areas. It should be mentioned in this respect that this parameterization of horizontal diffusion is the only one, among all of those presented, not to have an effect confined to the atmospheric column alone, since it has an effect on adjacent columns and consequently throughout the working domain.

Discretization must be done meticulously to ensure the numerical stability of calculations. For explicit processing, the centred scheme cannot be employed since it is unconditionally unstable; the forward off-centred scheme (the diffusion terms being calculated at time $t - \Delta t$) is conditionally stable, but the value of the diffusion coefficient must not exceed a threshold value so as to meet the stability criterion for the diffusion equation (see Richtmyer and Morton, 1967), which may limit its efficiency for small-wave filtering.

So as to be able to work with larger time steps, as is the case when the dynamical part of the model is processed in a semi-implicit semi-Lagrangian way, diffusion must be handled quasi-implicitly. In the case of a grid point model, as the Laplacian term leads to the equivalent of an implicit linear system for $X(t + \Delta t)$, the diffusion equation is processed iteratively. Once the linear system has been solved, it is then easy to calculate the tendencies due to horizontal diffusion and to add them to the tendencies due to the dynamical and physical parts of the model.

In the case of a spectral model, the calculation is much easier since the implicit processing of a diffusion equation in spectral space comes down to a simple division of spectral coefficients $X_n^m(t + \Delta t)$ by a factor depending on the horizontal diffusion coefficient K_{DH}, the global wavenumber n, and the time step Δt. However, contrary to finite difference processing, spectral processing of horizontal diffusion terms is global and does not allow the intensity of diffusion to vary locally by taking into account local properties of the flow.

For global processing of diffusion while keeping the possibility of increasing diffusion locally a mixed method has been proposed recently (Váňa et al., 2008) in the context of semi-Lagrangian semi-implicit processing applied to a spectral model. Global diffusion is performed by using the spectral method, while local diffusion is performed by modifying the characteristics of the interpolation operator at the origin points of the parcel trajectories so as to reinforce its smoothing character. Local diffusion may then be made dependent on flux dynamic characteristics as proposed by Sadourny and

Maynard (1997). This unconditionally stable scheme can easily be adapted to the case of grid point models.

9.9 Validation of physical parameterizations

As stated at the beginning of this chapter, physical parameterizations are strongly interactive and introduce many parameters that must be adjusted. Adjustment is particularly difficult because of the presence of many feedback loops, some of which may easily lead locally to instabilities in the model. As the physical parameterizations proposed deal with a column of atmosphere, a first control is to check that the whole operates correctly in the context of a one-dimensional model (1D model). There are, however, more elaborate means that can be implemented to test and validate the physical parameterizations.

Elementary validation of a set of physical parameterizations (or even a change in some particular parameterization or other) consists in checking that the prediction is not deteriorated. This check may be made by calculating the various classical statistical parameters (bias, standard deviation of error) in a number of test situations.

If we work with a global prediction model, we can use *zonal mean diagnostics* (averages over the latitude circles at all levels); in actual fact, zonal averages vary only slightly over periods of a few days. If we obtain high values for zonal average tendencies, that means the model has systematic bias, which can be localized from the zonal average diagnostics. We may also be able to perform very long integrations of the model to check that the different variables remain relatively well adjusted to the climatological values.

Mention should also be made of the paramount importance of measurement campaigns organized mostly through the cooperation of many countries and devised to study in the field a number of physical processes, the effects of which physical parameterizations attempt to simulate. During these measurement campaigns and then during the exploitation of the data, comparison between the simulated and observed data may yield useful information about the validity of solutions, especially if the observed data provide insight into the key features of the parameterizations.

Since the 2000s, new tools consisting in very high resolution models such as *cloud resolving models* (in short CRM) or *large eddy simulation* (in short LES) models, able to deal explicitly with most sub-grid scale motions and with water phase changes, have proved to be helpful in improving parameterization of the physical processes. Once they have been calibrated within a particular environment by using experimental data from measurement campaigns, they are able to provide a number of typical varied simulations that enable us to assess physical quantities at the resolved scale and to better tune the parameterization schemes under a large variety of configurations.

Local control of standard meteorological parameters provided by the prediction model (such as temperature and humidity at 2 m or wind at 10 m, obtained by interpolating from values at the lowest level of the model and from surface values) using data observed at weather stations often allows us to spot malfunctions in the model in a certain context. It is important to point out that the values predicted by the model and the observed values are compared fairly systematically in the context of *data assimilation* (sequence of short-term analyses and forecasts). Insofar as we have a tool for determining the parameters that are directly observable from the model's variables (what is termed an *observation operator*), any unexpected deviation between forecast and observation can be detected and analysed for correction.

Operational forecasting

10.1 Introduction

An atmospheric numerical prediction model is the main component in a much larger whole termed a *numerical prediction suite*, which comprises all of the processes implemented for making operational weather forecasts. Among these processes we distinguish the acquisition of meteorological data, the objective analysis (and more generally the observation data assimilation), the forecasting process using one or more atmospheric models, the determination of weather parameters at local scale, and lastly the display of all the available information.

In order to grasp the behaviour of a prediction model, we need to deploy a set of verification methods that allow us to compare predictions with the real situation. There are a good number of quality indices that provide statistical information on the model's behaviour. It is essential to verify models so as to monitor their imperfections and try to rectify them.

The quality of weather forecasts obtained in a purely deterministic way is mainly related to the accuracy with which the initial state is defined, which has vindicated the major effort made in recent years to make the best of all available observations. However, uncertainty about data remains, which is why the 1990s saw the development of *ensemble forecasting*: the aim is to make a set of forecasts using several probable initial states or even several models. This technique has been made possible by increasingly powerful computer systems and allows us to estimate the *probability density function* of the forecast parameters.

10.2 Meteorological observations

10.2.1 The Global Earth Observation System

The measurements of meteorological parameters are made by the Global Observing System (GOS) covering the entire Earth. This system is composed of various means of measurement of meteorological parameters as shown in Figure 10.1. These sets of measurements can nonetheless be grouped into two subsystems: the *in situ observation network* and the *remote sensing observation network*.

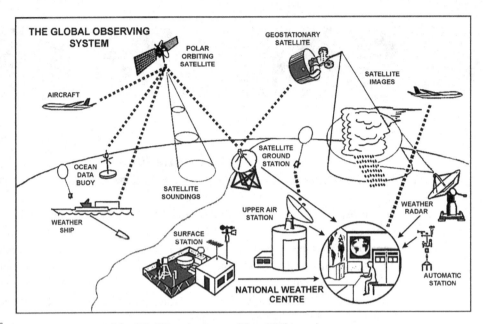

THE GLOBAL OBSERVING SYSTEM

POLAR ORBITING SATELLITE

GEOSTATIONARY SATELLITE

SATELLITE IMAGES

AIRCRAFT

SATELLITE GROUND STATION

OCEAN DATA BUOY

SATELLITE SOUNDINGS

WEATHER RADAR

WEATHER SHIP

UPPER AIR STATION

SURFACE STATION

AUTOMATIC STATION

NATIONAL WEATHER CENTRE

Figure 10.1 The various components of the Global Observing System. (After a WMO image)

10.2.2 *In situ* observations

The *in situ* observation subsystem comprises a network of fixed stations that measure the main meteorological parameters (pressure, temperature, horizontal wind, humidity) at given reference times. This network is far from being evenly distributed since the stations are mostly located on the mainland and on islands insofar as such places remain accessible. A large number of surface weather stations (some 10 000) make measurements every hour, whereas a small number of upper air weather stations (about 500) make vertical soundings of the atmosphere every 12 or 24 hours.

This observation network is supplemented by a set of measurements made at fixed times by ships at sea. This is referred to as the *synoptic network* as the measurements are made at what are called *synoptic* times. In order to work synchronously, all the world's meteorologists use *coordinated universal time* or *Greenwich time* (UTC); thus the synoptic reference times are on the hour for surface measurements and at 00 UTC and 12 UTC for upper air measurements. The measurements made by this network are therefore termed *synoptic observations*.

The development of automated telecommunications now means we can receive measurements, practically in real time, of surface parameters collected by sets of drifting buoys and a sample of pressure, temperature, and wind measurements made onboard airliners (throughout the flight including during take-off and landing); these observations, which have multiplied in recent years, obviously only provide data along the courses of the buoys or aircraft flight paths and are not made at synoptic times: they are therefore called *asynoptic observations*.

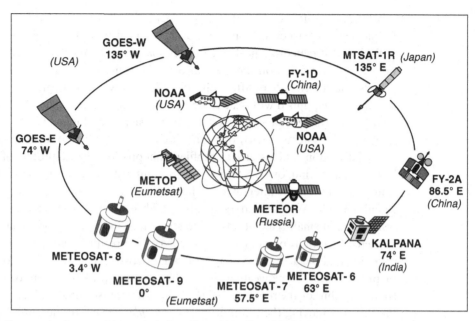

Figure 10.2   The operational system of meteorological satellites in 2007. (From EUMETSAT)

10.2.3 Remote sensing observations

Meteorological satellites, which first appeared in the 1960s, not only provide images of the Earth and of cloud cover but also enable us to measure electromagnetic radiation at the top of the atmosphere over an ever enhanced range of spectral channels. These are spatial remote sensing systems.

Sun-synchronous polar orbiting satellites travel in a low orbit (about 800 km altitude) and sweep the Earth's surface in the course of its daily rotation; in this way the satellite sees a given zone twice a day: once on its ascending pass and 12 hours later on its descending pass (Figure 10.2). In this way, with a set of two satellites orbiting in quadrature, any zone of the Earth is swept four times per day by the system. The satellite measures energy radiances in various channels of the electromagnetic spectrum from which the temperature of the various layers of the atmosphere can be evaluated by a process known as *inversion* (Smith, 1968). Polar orbiting satellites therefore yield *asynoptic* measurements that are made at the moment the satellite overflies a given zone.

Satellites in *geostationary* equatorial orbit at an altitude of 36 000 km remain virtually motionless relative to the Earth and cover a very broad zone permanently. A ring of five satellites is enough to cover the Earth's surface if we omit the polar regions (Figure 10.2). Some of the satellites are equipped with radiometers for measuring radiances. Moreover, shape recognition tools can calculate the motion of clouds or of water vapour structures from successive snapshots (in the visible, infrared, and water vapour channels) and infer the horizontal wind from them. Unlike polar orbiting satellites, geostationary satellites provide measurements for the same zone of the Earth at high frequencies (at least every hour) and so therefore at synoptic times.

While most meteorological satellites use *passive* systems analysing the Earth's own or atmospheric radiation there are also satellites fitted with *active* systems that emit signals and analyse them after reflection. In this way oceanographic polar orbiting satellites are equipped with scatterometers that can determine the speed of the surface wind by analysing a backscattered microwave signal modulated by the ocean waves.

Alongside systems carried onboard satellites, we must also mention ground-based remote detection systems such as profilers that provide vertical profiles of temperature, humidity, and wind, or GPS (global positioning system) receivers that can evaluate the water vapour content integrated in the vertical from the delay in propagation of the signals emitted. Weather radars are also highly effective remote sensing tools for analysing certain small-scale atmospheric structures: analysis of the signal that is backscattered by raindrops can be used to determine precipitation within a limited radius of action and the digitalized images of radar networks can be used to construct a field of precipitation. Some weather radars are equipped with a system for measuring the frequency shift of the return signal (Doppler effect) associated with the speed at which the target is moving (here raindrops). A *Doppler radar* can measure the radial wind and the wind field can therefore be reconstructed by analysing the signals received by several Doppler radars. Doppler radars therefore provide a remote measurement system for wind at small scale, with a resolution of the order of a few kilometres, and are an extremely promising tool for initializing mesoscale forecasting models.

10.3 Objective analysis and data assimilation

10.3.1 Introduction

The initial state of a model is ascertained by using all of the available data, which are distributed unevenly in space and time. Meteorologists who interpolate weather data by hand to draw charts of meteorological parameters (e.g. sea level pressure) make a *subjective analysis*; determining an initial state on a regular grid by calculation also makes it possible to map the various parameters by using appropriate programmes and justifies the name of *objective analysis* that is given to this operation.

In the context of operational forecasting, this operation, if performed regularly, can update the forecast by taking into account freshly arrived data. We thus obtain a series of snapshots of the state of the atmosphere at regular intervals (typically 3, 6, or 12 hours) by introducing the observation data. This process of retrieval of the atmosphere in its three spatial dimensions and its time dimension then takes the name of *intermittent data assimilation*. When applied regularly at each time integration step of the model, the process is known as *continuous data assimilation*. It is easy to see that the continuous system is far more suitable than the intermittent system for taking into account asynoptic data.

Data assimilation is a relatively complex process, since the variables must be given values that are as close as possible to the observed data while continuing to verify the model's equations. Given the decisive impact of the accuracy of the initial state on the skill of predictions, the effective implementation of this process has called for considerable scientific investment and requires very substantial computing resources. While previously it took a comparatively short time to determine the initial state compared with the time taken to make the prediction with the model, the situation has now been completely reversed.

The presentation below does not aim to describe all of the methods used for objective analysis, which have been featured in numerous publications (Lorenc, 1981; Daley, 1991; Talagrand, 1997; Courtier, 1997); it is limited to setting out succinctly the principles of the various methods (which may be seen as specific instances of a very general optimization problem) by using the very didactic approach and the notations employed by Bouttier and Courtier (1999).

10.3.2 The successive correction method

The methods of interpolation by adjusting a series of functions (Gilchrist and Cressman, 1954) soon proved deficient both for retrieving meteorological fields in data sparse areas and for imparting to them a coherent structure with regard to the major balances observed within the atmosphere.

The successive correction method initially proposed by Bergthorsson and Döös (1955) and then improved upon by Cressman (1959) consists in modifying an earlier forecast as background field (or first guess) at the grid points using a weighted sum of differences between the background field and the observations at the observation points.

By calling $x_b(i)$ the background field value of a parameter from a prediction of the model at grid point i, $x_b(j)$ its value at observational point j and $y(j)$ the value of the parameter observed at point j, the value analysed $x_a(i)$ at grid point i by using p_i observations is given by:

$$x_a(i) = x_b(i) + \frac{\sum_{j=1}^{p_i} w_{i,j}\left[y(j) - x_b(j)\right]}{\sum_{j=1}^{p_i} w_{i,j}}, \quad \text{with: } w_{i,j} = \max\left[0, \frac{R^2 - d_{i,j}^2}{R^2 + d_{i,j}^2}\right], \qquad (10.1)$$

where $d_{i,j}$ measures the distance between the grid point i and the observation point j and R is a radius of influence beyond which the observations are not taken into account (Figure 10.3). The weighting function $w_{i,j}$ is 1 if points i and j are at the same location and decreases towards 0 as the distance $d_{i,j}$ approaches R.

This procedure may be repeated; the field analysed then replaces the background field for a new iteration during which the weighting function is usually modified (by reducing the value of the radius R).

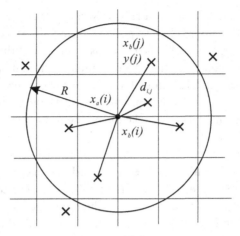

Figure 10.3   The location of the various parameters and variables used with the successive correction method.

This method, variously refined, has been used for operational forecasting but has been somewhat limited by the arbitrary character of the choice of weighting functions and the absence of any guarantee as to the aspect of the fields obtained and their coherence with respect to the major balances such as the hydrostatic relation or geostrophic balance.

10.3.3 The statistical approach and error modelling

In order to take advantage of the various information with which the analysis is performed (background, observations, physical balances), all of which is tainted by error to some degree, we need to represent the quality of the data mathematically using error statistics that can be obtained by repeating analysis experiments many times. We should first, then, define the notations used and the statistical parameters that are to be calculated.

$\mathbf{X}_t$ is the state vector, of dimension n, describing the actual atmosphere (subscript t for truth), given the discretization of the model; $\mathbf{X}_b$ and $\mathbf{X}_a$ are the state vectors, both of dimension n, describing the background (subscript b) and analysis (subscript a). The vector $\mathbf{Y}$ of dimension p describes the set of observations available for the time of the analysis and $\mathcal{H}$ is the observation operator – meaning that the vector $\mathcal{H}(\mathbf{X}_t)$, of dimension p, represents the *pseudo-observation* induced by the state vector $\mathbf{X}_t$.

The error vectors for the background $\varepsilon_b = \mathbf{X}_b - \mathbf{X}_t$, observations $\varepsilon_o = \mathbf{Y} - \mathcal{H}(\mathbf{X}_t)$, and analysis $\varepsilon_a = \mathbf{X}_a - \mathbf{X}_t$ have as their respective mean values (or *bias*) $\overline{\varepsilon_b}$, $\overline{\varepsilon_o}$, and $\overline{\varepsilon_a}$; their *covariance matrices* are respectively written:

$$\mathbf{B} = \overline{\left(\varepsilon_b - \overline{\varepsilon_b}\right)\left(\varepsilon_b - \overline{\varepsilon_b}\right)^{\mathrm{T}}}, \ \mathbf{R} = \overline{\left(\varepsilon_o - \overline{\varepsilon_o}\right)\left(\varepsilon_o - \overline{\varepsilon_o}\right)^{\mathrm{T}}} \ \text{and} \ \mathbf{A} = \overline{\left(\varepsilon_a - \overline{\varepsilon_a}\right)\left(\varepsilon_a - \overline{\varepsilon_a}\right)^{\mathrm{T}}},$$

where the superscript ^{T} after a vector or matrix indicates the transposition operation.

The error statistics (bias and covariance) depend on the atmospheric dynamics and the observation networks as well as the knowledge we may have *a priori* about errors. They may be estimated from repeated data assimilation experiments by a model, but some approximation is inevitable when error estimates cannot be directly observed.

10.3.4 Statistical interpolation by the least squares approach

The *least squares estimation*, also called *Best Linear Unbiased Estimator* (BLUE), enables us to solve an *optimal control problem* in which we try to estimate the analysed state having the best possible quality by taking into account the accuracy of all the available information.

Assuming:

- that in the vicinity of the background, the observation operator $\mathcal{H}$ can be linearized, that is, for a state $\mathbf{X}$ close to the background $\mathbf{X}_b$ we can write:

$$\mathcal{H}(\mathbf{X}) - \mathcal{H}(\mathbf{X}_b) \approx \mathbf{H}(\mathbf{X} - \mathbf{X}_b), \text{ where } \mathbf{H} \text{ is a linear operator,}$$

- that there are no trivial errors (matrices $\mathbf{B}$ and $\mathbf{R}$ are positive definite),
- that the background errors and observational errors are unbiased and uncorrelated with each other, that is:

$$\overline{(\mathbf{X}_b - \mathbf{X}_t)} = 0, \ \overline{(\mathbf{Y} - \mathcal{H}(\mathbf{X}_t))} = 0 \ \text{ and } \ \overline{(\mathbf{X}_b - \mathbf{X}_t)(\mathbf{Y} - \mathcal{H}(\mathbf{X}_t))^{\mathrm{T}}} = 0,$$

then the best unbiased linear estimation that minimizes error variance is written:

$$\mathbf{X}_a = \mathbf{X}_b + \mathbf{K}\,[\mathbf{Y} - \mathcal{H}(\mathbf{X}_b)], \tag{10.2}$$

$\mathbf{K}$ being a weight matrix that is written:

$$\mathbf{K} = \mathbf{B}\mathbf{H}^{\mathrm{T}}(\mathbf{H}\mathbf{B}\mathbf{H}^{\mathrm{T}} + \mathbf{R})^{-1}. \tag{10.3}$$

With this weight matrix, the covariance matrix of the analysis error is written:

$$\mathbf{A} = (\mathbf{I} - \mathbf{K}\mathbf{H})\mathbf{B}, \tag{10.4}$$

where $\mathbf{I}$ represents the unit matrix.

The least squares estimate may also be obtained by solving the following variational problem of optimization: find the value of $\mathbf{X}$ that minimizes the *cost function*:

$$J(\mathbf{X}) = (\mathbf{X} - \mathbf{X}_b)^{\mathrm{T}}\mathbf{B}^{-1}(\mathbf{X} - \mathbf{X}_b) + [\mathbf{Y} - \mathcal{H}(\mathbf{X})]^{\mathrm{T}}\mathbf{R}^{-1}[\mathbf{Y} - \mathcal{H}(\mathbf{X})], \tag{10.5}$$

written in condensed form:

$$J(\mathbf{X}) = J_b(\mathbf{X}) + J_o(\mathbf{X}), \tag{10.6}$$

where J_b is the term for deviation from the background and J_o the term for deviation from observations.

10.3.5 Optimal interpolation

The *optimal interpolation* method is derived directly from the least squares approach and has been widely used and is thoroughly documented (Gandin, 1963; Lorenc, 1981; Daley, 1991). It consists in using relations (10.2) and (10.3) (rewritten below) and in taking advantage of the simplified calculation of the weight matrix $\mathbf{K}$ obtained when selecting a reduced number of observations, p_i, to analyse a value at a grid point.

$$\mathbf{X}_a = \mathbf{X}_b + \mathbf{K}\,[\mathbf{Y} - \mathcal{H}(\mathbf{X}_b)], \tag{10.7}$$

with:

$$\mathbf{K} = \mathbf{B}\mathbf{H}^{\mathrm{T}}(\mathbf{H}\mathbf{B}\mathbf{H}^{\mathrm{T}} + \mathbf{R})^{-1}. \tag{10.8}$$

Each of the elements $x_a(i)$ of $\mathbf{X}_a$ is determined as below:

- select a relatively small number of observations, p_i, using a geographical proximity criterion;
- calculate deviations between the observations and the values induced by the background, $[\mathbf{Y} - \mathcal{H}(\mathbf{X}_b)]_j$ at the observation points: this is a column vector of dimension p_i;
- calculate p_i covariances between the background errors in the model space and the background errors in the observation space (extracted from row i of the matrix $\mathbf{B}\mathbf{H}^{\mathrm{T}}$), calculate the background error covariance matrix in the observation space (submatrix of $\mathbf{H}\mathbf{B}\mathbf{H}^{\mathrm{T}}$, of dimension $p_i \times p_i$), and the observation error covariance matrix (submatrix of $\mathbf{R}$ of dimension $p_i \times p_i$);
- calculate the inverse of the positive definite matrix $(\mathbf{R} + \mathbf{H}\mathbf{B}\mathbf{H}^{\mathrm{T}})$ restricted to the selected observations (matrix of dimension $p_i \times p_i$);
- multiply the elements of row i of matrix $\mathbf{B}\mathbf{H}^{\mathrm{T}}$ by the p_i columns of the inverted matrix to obtain row i of the weight matrix $\mathbf{K}$ (that is, the p_i weights corresponding to the p_i observations used for analysing $x_a(i)$).

Notice the relationship between formula (10.7) given for optimal interpolation and formula (10.1) proposed for successive corrections. This analogy was pointed out by Royer (1976) and Bratseth (1986), who had proposed using statistical information to calculate the weighting function in the successive correction method.

Optimal interpolation can be used to calculate the covariance matrix of the analysis error:

$$\mathbf{A} = (\mathbf{X}_a - \mathbf{X}_t)(\mathbf{X}_a - \mathbf{X}_t)^{\mathrm{T}} = (\mathbf{I} - \mathbf{K}\mathbf{H})\mathbf{B}(\mathbf{I} - \mathbf{K}\mathbf{H})^{\mathrm{T}} + \mathbf{K}\mathbf{R}\mathbf{K}^{\mathrm{T}}, \tag{10.9}$$

which, where $\mathbf{K}$ has been calculated exactly, comes down to:

$$\mathbf{A} = (\mathbf{I} - \mathbf{K}\mathbf{H})\mathbf{B}. \tag{10.10}$$

The objective analysis of meteorological fields for initializing numerical prediction models made wide use of optimal interpolation. However, optimal interpolation can be used to analyse a variable of the model by using various types of observed parameters (known as *multivariate* analysis) only if these parameters are linearly related to the variable to be analysed. Moreover, a selection of available data is essential to obtain a matrix $(\mathbf{R} + \mathbf{HBH}^{\mathrm{T}})$ for inversion of a reasonable size and that is well conditioned (allowance for two observations that are too close together may entail ill-conditioning of the matrix). For these reasons this method now tends to be superseded by variational methods.

10.3.6 The 3D variational method

In the second half of the 1980s, a very general variational formulation of least square estimation (which came down to minimizing a functional) was proposed, the solution of which may be obtained by using optimal control methods (Lewis and Derber, 1985; Le Dimet and Talagrand, 1986; Talagrand and Courtier, 1987).

In the 3D variational approach (3DVar for short), we seek to minimize, at a given reference time, an objective criterion defined as a function of the problem. Minimization relates to a quadratic function quantifying the deviations from the available information which are the observations and the background field weighted by their respective error standard deviations. If the errors in question have a Gaussian distribution, then the objective function we try to minimize is a sum of quadratic terms analogous to the expression already mentioned in (10.5), with the extra factor 1/2:

$$J(\mathbf{X}) = \frac{1}{2}(\mathbf{X} - \mathbf{X}_b)^{\mathrm{T}}\mathbf{B}^{-1}(\mathbf{X} - \mathbf{X}_b) + \frac{1}{2}[\mathbf{Y} - \mathcal{H}(\mathbf{X})]^{\mathrm{T}}\mathbf{R}^{-1}[\mathbf{Y} - \mathcal{H}(\mathbf{X})]. \quad (10.11)$$

To these first two terms, representing the deviation from the background $J_b(\mathbf{X})$ and the deviation from the observations $J_o(\mathbf{X})$, can be added an extra term $J_c(\mathbf{X})$ characterizing a specific constraint (to impose a certain balance between the dynamic variables for example).

The solution to the problem of minimization from which the analysed value $\mathbf{X}_a$ can be obtained is reached iteratively by means of a steepest descent algorithm using the calculation of the cost function gradient:

$$\nabla_{\mathbf{X}} J(\mathbf{X}) = \mathbf{B}^{-1}(\mathbf{X} - \mathbf{X}_b) - \mathbf{H}^{\mathrm{T}}\mathbf{R}^{-1}[\mathbf{Y} - \mathcal{H}(\mathbf{X})]. \quad (10.12)$$

A few iterations (shown geometrically in Figure 10.4) are enough for the norm $\|\nabla_{\mathbf{X}} J(\mathbf{X})\|$ to reach a small enough value, corresponding to an acceptable minimum value. In practice, it is the background that is used as a starting point for minimization. The variational approach is particularly valuable because calculating the term for deviation from observations $J_o(\mathbf{X})$ requires knowledge of the observation operator $\mathcal{H}$ alone. In the event the observations are radiances, say, the operator $\mathcal{H}$ is the radiative transfer model that computes radiances from the model's variables. The constraint of the model lies in the fact that the observation operator must be continuous and differentiable so $\mathcal{H}$ can be calculated.

If the minimum of the cost function $J(\mathbf{X})$ is reached for $\mathbf{X} = \mathbf{X}_a$, then the gradient $\nabla_{\mathbf{X}} J(\mathbf{X}_a)$ is zero and expression (10.12) leads to the relation:

$$\mathbf{B}^{-1}(\mathbf{X}_a - \mathbf{X}_b) = \mathbf{H}^{\mathrm{T}}\mathbf{R}^{-1}[\mathbf{Y} - \mathcal{H}(\mathbf{X}_a)], \quad (10.13)$$

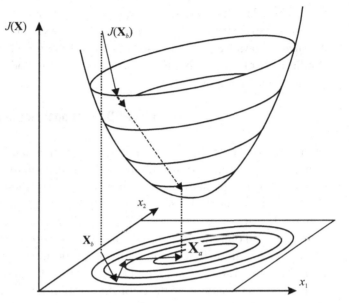

Figure 10.4 Geometric representation of minimization of a cost function depending on two variables. (After Bouttier and Courtier, 1999)

which enables us to calculate the covariance matrix of the analysis error:

$$\mathbf{A} = (\mathbf{B}^{-1} + \mathbf{H}^{\mathrm{T}}\mathbf{R}^{-1}\mathbf{H})^{-1}. \tag{10.14}$$

Notice that the covariance matrix of the analysis error (when minimization has reached its goal) is the inverse of the matrix representing the *Hessian matrix* (the term used to designate the gradient of the gradient) of the cost function $J(\mathbf{X})$;

$$\nabla_{\mathbf{x}}[\nabla_{\mathbf{x}}(J(\mathbf{X}))] = \mathbf{B}^{-1} + \mathbf{H}^{\mathrm{T}}\mathbf{R}^{-1}\mathbf{H}. \tag{10.15}$$

10.3.7 The incremental variant

The costs of such a process may be cut by introducing a *reduced state variable*, $\mathbf{X}'$, obtained from $\mathbf{X}$ by eliminating the small-scale components beyond a certain threshold; the reduced state variable $\mathbf{X}'$ is a lower resolution restriction of the state variable $\mathcal{H}(\mathbf{X}')$, which can then be linearized in the vicinity of the background at reduced resolution, $\mathbf{X}'_b$, by writing:

$$\mathrm{H}(\mathbf{X}') \approx \mathrm{H}(\mathbf{X}'_b) + \mathbf{H}_r(X' - X'_b), \tag{10.16}$$

with $\mathbf{H}_r$ representing the observation operator linearized around $\mathbf{X} = \mathbf{X}'_b$.

To obtain a coherent representation between the complete problem and the reduced resolution problem when $\mathbf{X} = \mathbf{X}'$, the deviations from observations at reduced resolution must be written:

$$\mathbf{Y} - \mathcal{H}(\mathbf{X}') \approx \mathbf{Y} - \mathcal{H}(\mathbf{X}_b') - \mathbf{H}_r(\delta\mathbf{X}') = \mathbf{d}' - \mathbf{H}_r(\delta\mathbf{X}'), \qquad (10.17)$$

with:

$$\mathbf{d}' = \mathbf{Y} - \mathcal{H}(\mathbf{X}_b') \text{ and } \delta\mathbf{X}' = \mathbf{X}' - \mathbf{X}_b'.$$

The functional to be minimized is then written:

$$J_r(\delta\mathbf{X}') = \frac{1}{2}(\delta\mathbf{X}')^{\mathrm{T}} \mathbf{B}^{-1}(\delta\mathbf{X}') + \frac{1}{2}[\mathbf{d}' - \mathbf{H}_r(\delta\mathbf{X}')]^{\mathrm{T}} \mathbf{R}^{-1}[\mathbf{d}' - \mathbf{H}_r(\delta\mathbf{X}')], \quad (10.18)$$

and the minimization is performed by calculating the gradient:

$$\nabla_{\delta\mathbf{X}'} J_r(\delta\mathbf{X}') = (\mathbf{B}^{-1} + \mathbf{H}_r^{\mathrm{T}} \mathbf{R}^{-1}\mathbf{H}_r)\delta\mathbf{X}' - \mathbf{H}_r^{\mathrm{T}} \mathbf{R}^{-1}\mathbf{d}'. \qquad (10.19)$$

After reduced resolution minimization, the deviation $\delta\mathbf{X}'$ related to the largest scales is added to the background $\mathbf{X}_b$ to give the new value for the state variable $\mathbf{X}$:

$$\mathbf{X} = \mathbf{X}_b + \delta\mathbf{X}'.$$

This way of working, which is termed the *incremental approach* (Courtier et al., 1994), allows us to reduce the cost of the procedure since the analysis increment $\delta\mathbf{X}'$ is evaluated in a space of reduced dimension compared with that of the model's state variable.

Despite the difficulties in suitably modelling the background error covariance matrix $\mathbf{B}$, the 3D variational method with its incremental variant is highly effective: it can take into account observed data that differ from the model's state variables, insofar as a continuous and differentiable observation operator $\mathbf{H}$ can be defined.

10.3.8 4D variational assimilation

The 3D variational method consists in performing a minimization at a given time. This principle may be extended to four dimensions by including time. We then speak of four-dimensional variational assimilation (4DVar for short). By this approach, the objective function to be minimized measures the distance between the model's trajectory and the available information (background and observations) within an assimilation interval (or time-window), typically of 6 hours, centred on the chosen reference time.

With this method the cost function to be taken into account is the sum of a term J_b characterizing the deviation between the state vector $\mathbf{X}(t_0)$ and the background $\mathbf{X}_b(t_0)$ at the initial time and a term J_o characterizing the sum of differences between the pseudo-observations $\mathbf{H}_k\mathbf{X}(t_k)$ deduced from the state vector and the observations $\mathbf{Y}(t_k)$, at all times of the assimilation window. Thus all the observations $\mathbf{Y}(t_k)$ available within the assimilation window can be taken into account insofar as we introduce a series of observation operators $\mathbf{H}_k$; of course, to be able to deduce the pseudo-observation value from state variable $\mathbf{X}(t_0)$, we need to integrate the model until the time of observation.

The cost function to be taken into account for all of the instants of the assimilation window is written:

$$J[\mathbf{X}(t_0)] = \frac{1}{2}[\mathbf{X}(t_0) - \mathbf{X}_b(t_0)]^T \mathbf{B}^{-1}[\mathbf{X}(t_0) - \mathbf{X}_b(t_0)]$$

$$+ \frac{1}{2}\sum_{k=0}^{k=N} [\mathbf{Y}(t_k) - \mathcal{H}_k \mathbf{X}(t_k)]^T \mathbf{R}_k^{-1}[\mathbf{Y}(t_k) - \mathcal{H}_k \mathbf{X}(t_k)]. \tag{10.20}$$

To get a quadratic cost function, it is assumed that the nonlinear model $\mathcal{M}$ that makes the transition from state vector $\mathbf{X}(t_k)$ to state vector $\mathbf{X}(t_{k+1})$ can be linearized. The corresponding linearized model $\mathbf{M}$ is known as the *tangent linear model*. It is also shown that the operator that makes the transition from gradient $\nabla_{x_{k+1}}J$, to gradient $\nabla_{x_k}J$, is the *adjoint model*, $\mathbf{M}^T$, which turns out to be the transpose of $\mathbf{M}$. Thus an iteration of the minimization algorithm requiring the calculation of the gradient implies an N time step forward integration of the tangent linear model followed by an N time step backward integration of the adjoint model. It is easy to understand that implementing the 4DVar method is equivalent to performing a large number of integrations over the period of assimilation.

The cost of the 4DVar model may be cut by using an incremental method, allowing us to work with a lower resolution state vector. In addition, the need to have a continuous and differentiable observation operator $\mathcal{H}_k$ leads to the use of a prediction model with a set of simplified physical parameterizations from which the discontinuous processes have been eliminated (Rabier et al., 2000) so as to calculate the tangent linear model $\mathbf{M}$ and its adjoint $\mathbf{M}^T$.

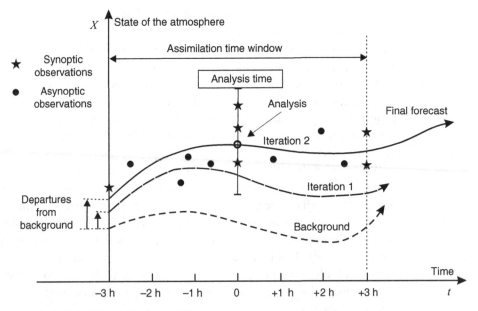

Figure 10.5   Schematic principle of 4D variational assimilation.

Figure 10.5 schematizes the operating principle of the 4DVar method: from the background, made up by a recent trajectory of the forecast model, and from observations distributed within the window, we can calculate the cost function $J(\mathbf{X}_0)$. This cost function characterizes the difference between the state vector of the model and the background at the initial time on the one hand, and the difference between the background and the observations on the other. By calculating the gradient $\nabla_{x_0} J$, we can determine the modifications to be made to X_0 to reduce the cost function, and the new values of $\mathbf{X}_0$ then allow us to calculate a new trajectory. The process is thus iterated until the minimum of the cost function is attained.

10.3.9 Kalman filtering

Linear Kalman filtering may be thought of as an application of statistical interpolation by least squares in the context of sequential data assimilation over a fixed period. However, the terminology is slightly modified:

- the background $\mathbf{B}$ and analysis $\mathbf{A}$ covariance error matrices are now termed $\mathbf{P}_f$ (for forecast) and $\mathbf{P}_a$, respectively;
- the model for passing from state vector $\mathbf{X}(t_k)$ to $\mathbf{X}(t_{k+1})$ is noted $\mathcal{M}_k$;
- the difference between the forecast made from the analysis at time t_k and the truth at time t_{k+1}, $\mathcal{M}_k[\mathbf{X}(t_k)] - \mathbf{X}_t(t_{k+1})$, is called the *model error* at time t_k, and the corresponding covariance matrix, which is assumed to be known, is noted $\mathbf{Q}(t_k)$.

It is further assumed that:

- the model error $\mathcal{M}_k[\mathbf{X}(t_k)] - \mathbf{X}_t(t_{k+1})$ is unbiased;
- the analysis error $\mathbf{X}_a(t_k) - \mathbf{X}_t(t_k)$ and model error $\mathcal{M}_k[\mathbf{X}(t_k)] - \mathbf{X}_t(t_{k+1})$ are uncorrelated;
- the variation in the model forecast can be expressed in the vicinity of a state given as a linear function of the initial state thanks to the tangent linear model $\mathbf{M}_k$, that is: $\mathcal{M}_k[\mathbf{X}(t_k)] - \mathcal{M}_k[\mathbf{X}_a(t_k)] \approx \mathbf{M}_k[\mathbf{X}(t_k) - \mathbf{X}_a(t_k)]$.

Under these circumstances, the optimal way (in the least squares sense) of assimilating the observations sequentially is by using *Kalman filtering* (KF for short), consisting in a recurrence over the times corresponding to observations t_k, defined by the following sequence of relations:

- forecast state: $\mathbf{X}_f(t_{k+1}) = \mathcal{M}_k \mathbf{X}_a(t_k)$, \hfill (10.21)

- forecast covariance: $\mathbf{P}_f(t_{k+1}) = \mathbf{M}_k \mathbf{P}_a(t_k) \mathbf{M}_k^{\mathrm{T}} + \mathbf{Q}(t_k)$, \hfill (10.22)

- Kalman gain: $\mathbf{K}(t_k) = \mathbf{P}_f(t_k) \mathbf{H}(t_k)^{\mathrm{T}} [\mathbf{H}(t_k) \mathbf{P}_f(t_k) \mathbf{H}(t_k)^{\mathrm{T}} + \mathbf{R}(t_k)]^{-1}$, \hfill (10.23)

- analysed state: $\mathbf{X}_a(t_k) = \mathbf{X}_f(t_k) + \mathbf{K}(t_k)\{\mathbf{Y}(t_k) - \mathcal{H}[\mathbf{X}_f(t_k)]\}$, \hfill (10.24)

- analysed covariance: $\mathbf{P}_a(t_k) = [\mathbf{I} - \mathbf{K}(t_k)\mathbf{H}(t_k)] \mathbf{P}_f(t_k)$. \hfill (10.25)

Relation (10.21) merely represents the evolution of the model's state vector from the analysed state to the forecast state and relation (10.22) is constructed by calculating the model error $\mathbf{X}_f(t_{k+1}) - \mathbf{X}_t(t_{k+1})$ from relation (10.21) (see Bouttier and Courtier, 1999 for details of the calculation). As for relations (10.23), (10.24), and (10.25), they are simply the transpositions of relations (10.3), (10.2), and (10.4) established in the context of optimal interpolation by least squares and are deduced by identifying $\mathbf{P}_f(t_k)$ with the background error covariance matrix $\mathbf{B}$ and by assuming that the matrix $\mathbf{K}$ is calculated so as to ensure optimality (minimization of error variance).

It is also shown with a supposedly perfect model (such that the model error covariance matrix $\mathbf{Q}(t_k)$ is zero), for the same assimilation window and the same set of observations, that the final state provided by Kalman filtering exactly matches the final state of the trajectory obtained upon completing the 4DVar assimilation process.

The principle of Kalman filtering may be generalized to the case where operators $\mathcal{H}$ and $\mathcal{M}_k$ are nonlinear, although the optimality of analysis and the equivalence with 4DVar then cease to be valid. To assimilate observed data in a weather forecasting model, we can define for each time t_k the tangent linear observation operator $\mathbf{H}(t_k)$ in the vicinity of the background $\mathbf{X}_b(t_k)$ and the tangent model $\mathbf{M}_k$ in the vicinity of the analysed state $\mathbf{X}_a(t_k)$, allowing us to write:

$$\mathcal{M}_k[\mathbf{X}(t_k)] - \mathbf{M}_k[\mathbf{X}_a(t_k)] \simeq \mathbf{M}_k[\mathbf{X}(t_k) - \mathbf{X}_a(t_k)].$$

In this event, the algorithm described by equations (10.21) to (10.25) is called the *extended Kalman filter*.

Kalman filtering is cumbersome to implement since, aside from the process of analysis at each time step, it implies calculating a forecast for the model state vector $\mathbf{X}_f$ and for the forecast error covariance matrix $\mathbf{P}_f$. Besides, storage of the various matrices requires substantial memory space. Thus this algorithm may be regarded as a reference for the development of data assimilation algorithms that attempt to approximate this theoretical model.

Calculation of the analysis $\mathbf{P}_a$ and background $\mathbf{P}_f$ error covariance matrices from equations (10.22) and (10.25) may, however, be replaced by an approximate calculation performed from a set of analyses and a set of backgrounds. By making several analyses from several disturbed backgrounds, we obtain a set of initial states from which to calculate the analysis error covariance matrix $\mathbf{P}_a$; by making a very short-term forecast for each of these analyses we similarly obtain a set of forecasts from which to calculate the background error covariance matrix $\mathbf{P}_f$. This procedure is then termed *ensemble Kalman filtering* (Evensen, 1994; Houtekamer and Mitchell, 1998), a simplified version of which, called *ensemble transform Kalman filtering*, was proposed by Bishop et al. (2001).

10.4 Initialization of data on starting the model

10.4.1 Early attempts at initialization

Initializing a primitive equation model consists in making corrections on the field supplied by the objective analysis that allow the model to start up free from high-frequency oscillations. When the fields supplied by the objective analysis are used directly, we observe oscillations during the first hours of integration of the model that are unrelated to what we actually observe in the real world. These oscillations are due to the propagation of inertia-gravity waves which, after a certain time, disperse and attenuate because of the dissipative processes in the model, as shown by the solid curve in Figure 10.6.

These inertia-gravity waves (analogous to waves that can be seen when a stone is thrown into a pond) result from the difficulty in precisely determining the wind (especially its divergent part, an order of magnitude lower than its rotational part) using measuring instruments. This is the main reason why Richardson's experiment resulted in failure because he had initialized his model directly with observed temperature and wind data. As Lynch (1994) showed, extrapolation at a 6-hour range, by using the time derivative calculated at the initial time, could indeed lead to an erroneous prediction of surface pressure, as Figure 10.6 shows.

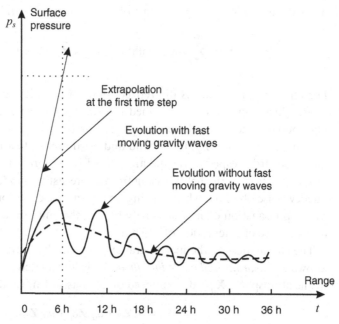

Figure 10.6 Evolution of surface pressure with and without fast inertia-gravity waves.

A number of methods based on diagnostic relations were proposed for initializing primitive equation models. The *nonlinear balance* equation, by which the wind field can be determined from the geopotential field, is obtained by assuming both that the wind field has zero divergence and that the divergence tendency is zero (Charney, 1955). This method, though, does not completely eliminate inertia-gravity waves. Hinkelmann (1959) then Phillips (1960) subsequently showed that the nonlinear balance equation was insufficient to eliminate inertia-gravity waves and suggested adding a divergent component of the wind obtained after solving the *quasi-geostrophic omega equation*.

10.4.2 The principle of using normal modes

It was only towards the late 1970s that a satisfactory solution could be proposed to the problem of initialization of primitive equation models thanks to the introduction of *normal mode analysis*. *Normal modes*, which were studied in the case of a spectral model on the sphere by Dickinson and Williamson (1972), are defined as eigenfunctions $\hat{\mathbf{Z}}_i (i = 1,...,N_p)$ of the operator $\mathcal{L}$ constituted by the *linearized model* (linearization being achieved by taking a vertical profile of standard temperature and a zero zonal wind as reference values). The corresponding eigenvalues are given by $i\sigma_i$, σ_i representing the eigenfrequencies of oscillation of the normal mode $\hat{\mathbf{Z}}_i$.

It is shown that by defining a scalar product $<\hat{\mathbf{Z}}_i, \hat{\mathbf{Z}}_j>$, where $<\hat{\mathbf{Z}}_i, \hat{\mathbf{Z}}_i> = \|\hat{\mathbf{Z}}_i\|^2$ (square of the norm associated with scalar product) represents the total energy of mode $\hat{\mathbf{Z}}_i$, the eigenmodes are orthogonal; they can be used, via normalization, to define an orthonormal basis on which a given state of the atmosphere can be decomposed as below:

$$\mathbf{X} = \sum_{i=1}^{i=N_P} a_i \hat{\mathbf{Z}}_i, \text{ with: } a_i = <\mathbf{X}, \hat{\mathbf{Z}}_i> \text{ for i = 1, ..., N}_P. \tag{10.26}$$

For an atmosphere that is discretized along the vertical using N levels, the normal modes divide into N subsets termed *vertical modes*. For each vertical mode, the normal modes divide into three subsets: the first is for slow-moving modes known as *Rossby modes*, noted $\hat{\mathbf{Z}}_R$; the second and third are for fast moving modes (eastwards and westwards, respectively) and are termed *inertia-gravity modes*, noted $\hat{\mathbf{Z}}_{G_1}$ and $\hat{\mathbf{Z}}_{G_2}$. Decomposition of the fields arising from a forecasting model shows that the high-frequency noise observed when starting up a primitive equations model is indeed caused by the propagation of inertia-gravity modes, the amplitude of which tends to decline in the course of integration of the model.

The elimination of the inertia-gravity waves in the initial state (Williamson, 1976), known as *linear normal mode initialization*, is formulated as below. The analysed state of the atmosphere $\mathbf{X}_a$ is decomposed on the basis of normal modes:

$$\mathbf{X}_a = a_R \hat{\mathbf{Z}}_R + a_{G_1} \hat{\mathbf{Z}}_{G_1} + a_{G_2} \hat{\mathbf{Z}}_{G_2}, \tag{10.27}$$

and the initialized state is obtained by keeping only the slow mode component:

$$\mathbf{X}_i = a_R \hat{\mathbf{Z}}_R. \tag{10.28}$$

This process does not really solve the problem because the model is nonlinear and from the first time steps nonlinear interaction soon recreates inertia-gravity modes that propagate.

The solution for obtaining evolutions without unrealistic propagation of inertia-gravity waves was proposed independently by Baer and Tribbia (1977) and by Machenhauer (1977). This technique, known as *nonlinear normal mode initialization*, can be used to define the initial state as a linear combination of Rossby modes and of inertia-gravity modes such that the inertia-gravity modes remain stationary.

In the most general case, the nonlinear evolution equation of a state of the atmosphere can be written as:

$$\frac{\partial \mathbf{X}}{\partial t} = \mathcal{L}(\mathbf{X}) + \mathcal{N}(\mathbf{X}), \tag{10.29}$$

where N represents the nonlinear part of the model.

The normal mode decomposition coefficients are obtained by calculating the scalar products:

$$a_n = <\mathbf{X}_a, \hat{\mathbf{Z}}_{jn}> \quad \text{for } n = R, G_1 \text{ and } G_2. \tag{10.30}$$

By projecting the evolution equation (10.29) into the normal mode space, we obtain an evolution equation for each of the components:

$$\frac{\partial a_n}{\partial t} = i\sigma_n a_n + <\mathcal{N}(\mathbf{X}_a), \hat{\mathbf{Z}}_n>. \tag{10.31}$$

We make the gravity waves stationary by imposing a zero derivative for the coefficients ($\partial a_n / \partial t = 0$), by choosing new values of a_{G_1} and a_{G_2} that satisfy the relations:

$$a_{G_1} = -\frac{<\mathcal{N}(\mathbf{X}), \hat{\mathbf{Z}}_{G_1}>}{i\sigma_{G_1}} \quad \text{and} \quad a_{G_2} = -\frac{<\mathcal{N}(\mathbf{X}), \hat{\mathbf{Z}}_{G_2}>}{i\sigma_{G_2}}. \tag{10.32}$$

Since the terms $<\mathcal{N}(\mathbf{X}), \hat{\mathbf{Z}}_{G_1}>$ and $<\mathcal{N}(\mathbf{X}), \hat{\mathbf{Z}}_{G_2}>$ are themselves functions of a_{G_1} and a_{G_2}, the solutions a'_{G_2} and a'_{G_2} are obtained by the iterative algorithm:

$$a_{G_1}'^{p+1} = -\frac{1}{i\sigma_{G_1}}\left(\frac{\partial a_{G_1}}{\partial t}\right)^{(p)} + a_{G_1}^p \quad \text{and} \quad a_{G_2}'^{p+1} = -\frac{1}{i\sigma_{G_2}}\left(\frac{\partial a_{G_2}}{\partial t}\right)^{(p)} + a_{G_2}^p. \tag{10.33}$$

At the end of the process, the initialized state is therefore written:

$$\mathbf{X}_a = a_R \hat{\mathbf{Z}}_R + a'_{G_1} \hat{\mathbf{Z}}_{G_1} + a'_{G_2} \hat{\mathbf{Z}}_{G_2}. \tag{10.34}$$

A given state of the atmosphere may be represented by a point in normal mode space that is geometrically schematized two-dimensionally (a Rossby mode R and a gravity mode G in Figure 10.7).

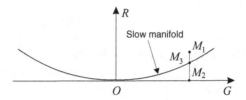

Figure 10.7 Simplified scheme of the slow manifold in a two-dimensional space and location of various initial states.

As Leith (1980) showed, the geometric locus of model states that evolve without gravity wave propagation is a hypersurface S, termed the *slow manifold*. The analysed state corresponds to point M_1, the result of *linear* initialization to point M_2 and the result of the nonlinear initialization process to point M_3 located on the slow manifold.

Once the problem was solved for spectral models on the sphere, the principle of the method was extended to models operating over limited areas. After modifying the model's linearization hypotheses (by fixing a mean value of the Coriolis parameter), Brière (1982) managed to adapt the normal mode nonlinear initialization method by using a decomposition into periodic functions. Based on intuitive considerations Bourke and McGregor (1983) defined iterative initialization methods in physical space that were generalized and justified by Juvanon du Vachat (1986, 1988) and Temperton (1988). Finally a new method for eliminating high-frequency oscillations based on the principle of nonlinear normal mode initialization and using a Laplace transform was proposed by Lynch (1985).

Normal mode analysis has allowed us to understand in detail the origin of the high-frequency oscillations in primitive equations models, but it remains relatively complex to implement and the iterative process for determining the initialized state does not always converge (Rasch, 1985). A new method, picking up on the dynamic initialization ideas, applied in particular by Edelmann (1972), was proposed by Lynch and Huang (1992). This method, which is very simple to implement, uses a *digital filter* for eliminating high-frequency oscillations (corresponding to fast inertia-gravity wave propagation) when integrating the model. From the initial time t_0, the purely dynamic model is integrated over a period $\Delta T/2$ forwards, then $\Delta T/2$ running backwards in time; in the course of these integrations, we calculate at each point the weighted sum of state variables $\mathbf{X}(t_k)$ with the weights h_k ($k = -N_T, ..., 0, ..., N_T$), as below:

$$\mathbf{X}_F^* = \frac{1}{2}h_0\mathbf{X}(t_0) + \sum_{k=1}^{N_T} h_{-k}\mathbf{X}(t_k) \text{ and } \mathbf{X}_B^* = \frac{1}{2}h_0\mathbf{X}(t_0) + \sum_{k=-1}^{-N_T} h_{-k}\mathbf{X}(t_k); \quad (10.35)$$

the initialized state is then obtained by calculating:

$$\mathbf{X}_0^* = \mathbf{X}_F^* + \mathbf{X}_B^*. \quad (10.36)$$

Tests by Lynch and Huang (1992) showed that the choice of weighting coefficients given by the *cardinal sine* function, also known as the *Lanczos window* (Lanczos, 1956), amounts to applying a low-pass filter for proper initialization for an integration period

ΔT of 6 hours; with this type of filtering, all oscillations whose period is less than 6 hours are eliminated when the model is integrated over time.

Digital filtering is therefore a highly effective alternative to nonlinear normal mode initialization. Ballish et al. (1992) have shown, in a simplified model, that applying a low-pass digital filter yielded an initial state that was quite similar to that given by the iterative nonlinear normal mode initialization method (state characterized by relations (10.32)). Various improvements to the filter characteristics and a way to take into account the diabatic terms in model integration were proposed by Lynch et al. (1997).

10.4.3 Initialization integrated into data assimilation

Insofar as, during recent years, the development of variational methods of data assimilation uses the model itself as a tool, the problem of initialization may seem outdated, since it is the model that is supposed to ensure the internal coherence of the mass field and the wind field. However, this ideal situation is difficult to achieve in practice; which is why it may be useful, when implementing variational assimilation, to add in the cost function to be minimized a penalty term that is in some sense a '*distance*' between the state to be defined and the balanced state (that is, such that the tendencies of the projections of the atmospheric state considered on the inertia-gravity modes are zero). Gauthier and Thépaut (2001) have shown that in the context of 4D variational assimilation, introducing such a term into the cost function uses did indeed make it possible to obtain a correctly balanced initial state while remaining sufficiently close to the observations.

10.5 Coupled models

10.5.1 Limited area models

The increased spatial resolution of numerical prediction models is very costly since doubling the resolution is generally reflected by a multiplication by a factor of $2^4 = 16$ of the volume of computations to be made. Starting from the principle that forecasts for ranges not exceeding 48 hours over a limited geographical area (a west European country, say) may be made by working on a relatively small domain compared with the sphere, many weather services have developed *limited area models*. The size of the domain is determined by assuming that the perturbation generated at the edges of the domain propagates inwards at the speed of the mean wind. Although this hypothesis omits the effects of gravity waves, which may propagate much faster than the wind speed, experience has shown that the gain from using a greater horizontal resolution exceeds the degradation from perturbations introduced on the lateral boundaries of the working domain.

10.5.2 The classical treatment of lateral boundary conditions

When integrating a limited area model, it is necessary to prescribe the values of the fields on the boundary of the working domain. These values are generally ascertained by interpolating in space and time the forecasts from another model operating over a larger domain at lower resolution. In principle, there is no need to prescribe all of the state variables on the boundary of the domain for the problem to be well-posed and to have a single solution, as shown by Charney (1962), Davies (1973), or Elvius and Sundström (1973). However, given the complexity of such a well-posed formulation, we generally prefer to impose the value of all the state variables on the boundary of the domain and to attenuate the effects of the perturbation thus created locally (which then propagates inwards into the domain) by a suitable damping process.

A first way to damp this noise is to artificially increase the coefficient of the horizontal diffusion terms. The diffusion coefficient keeps a fixed value in the central part of the working domain and then increases progressively as we approach the edges of the domain (Benwell et al., 1971; Burridge, 1975). Another solution is, in a transition zone adjoining the domain boundary, to make a weighted average of the tendencies supplied by the fine mesh prediction model and those supplied by the coupling model, ascribing increasing weight to the coupling model tendencies as we approach the boundary of the fine mesh model (Kesel and Winninghoff, 1972; Perkey and Kreitzberg, 1976).

The method most commonly used for limited area models was proposed by Davies (1976) and involves adding to the prognostic equations an additional term for relaxation towards the forcing values supplied by the coupling model.

The prognostic equation for the generic state variable $\mathbf{X}$, including a term for relaxation towards the forcing values $\mathbf{X}_F$, is written:

$$\frac{\partial \mathbf{X}}{\partial t} = \mathbf{A}(\mathbf{X}) + K_R(\mathbf{X}_F - \mathbf{X}), \qquad (10.37)$$

$\mathbf{A}(\mathbf{X})$ representing the right-hand side of the equation without forcing and K_R the relaxation constant.

Discretization in central differences of the time derivative in the equation without forcing (with $K_R = 0$) yields the value predicted by the model, $\mathbf{X}_M^{t+\Delta t}$:

$$\mathbf{X}_M^{t+\Delta t} = \mathbf{X}^{t-\Delta t} + 2\Delta t \mathbf{A}(\mathbf{X}^t). \qquad (10.38)$$

Discretizing equation (10.37) with implicit treatment of the relaxation term gives:

$$\frac{\mathbf{X}^{t+\Delta t} - \mathbf{X}^{t-\Delta t}}{2\Delta t} = \mathbf{A}(\mathbf{X}^t) + K_R(\mathbf{X}_F^{t+\Delta t} - \mathbf{X}^{t+\Delta t}), \qquad (10.39)$$

where $\mathbf{X}_F^{t+\Delta t}$ denotes the value of forcing at time $t + \Delta t$, interpolated at the grid point in question.

Given expression (10.37), equation (10.39) is written:

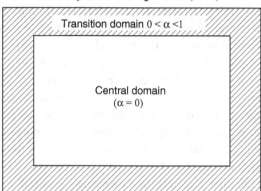

Figure 10.8   The working domain and the values of parameter α.

$$\mathbf{X}^{t+\Delta t} = \frac{1}{1+2\Delta t K_R}\,\mathbf{X}_M^{t+\Delta t} + \frac{2\Delta t K_R}{1+2\Delta t K_R}\,\mathbf{X}_F^{t+\Delta t},$$

or in condensed form:

$$\mathbf{X}^{t+\Delta t} = (1-\alpha)\mathbf{X}_M^{t+\Delta t} + \alpha\mathbf{X}_F^{t+\Delta t}, \text{ with: } \alpha = \frac{2\Delta t K_R}{1+2\Delta t K_R}. \qquad (10.40)$$

The final value $\mathbf{X}^{t+\Delta t}$ is obtained as a linear combination between the value predicted by the model $\mathbf{X}_M^{t+\Delta t}$ and the value given by forcing $\mathbf{X}_F^{t+\Delta t}$. When $\alpha = 1$, the final value is equal to the forcing value; when $\alpha = 0$, the final value is the value supplied by the forecast model. The weighted mean between the value predicted by the model and the forcing is made in a transition zone, taking the form of a rectangular frame of a few grid points, as shown in Figure 10.8: the value of α, which is set to 1 on the boundary, decreases as we move into the domain. The choice of the extent of the transition zone and the law for decrease of parameter α were discussed by Kallberg (1977) and by Lehmann (1993).

This method makes it possible to make a relatively smooth transition between the interior of the domain and its boundary. It should not be overlooked, however, that the intensity of the relaxation force is given by the coefficient K_R while coefficient α involves the product $K_R\Delta t$: this means that for a given value of K_R, the corresponding coefficient α is not independent of the value of the time step.

10.5.3 The principle of nested models

It is fairly easy to show that integrating a coarse mesh model over a large domain, coupled with a fine mesh model over a small domain, requires less computing power than integrating a fine mesh model over a large domain (which, aside from the time constraint, is obviously the best solution). This is why many meteorological services have adopted this *nested model system*. The solution described above can be used for one-way coupling (from large scale to small scale, but two-way coupling can also be contemplated: the values predicted with the fine mesh model are then reinjected into the coarse mesh model (Phillips and Shukla, 1973)). Moreover, there is nothing to stop us multiplying the number of nested models to work with ever finer meshes, especially for forecasting cyclones (Kurihara et al., 1998).

There are also several ways to couple the two models. When the coarse mesh model is integrated before the fine mesh model, the values imposed on the boundary of the fine mesh model correspond in terms of time range and we then speak of *synchronous coupling*; obviously working in this way delays the availability of forecasts from the fine mesh model. Another solution is to begin by integrating the fine mesh model (for which the time to collect the data needed to initialize it may be comparatively short) by using the forecast from a coarse mesh model run earlier (3 to 6 hours) to get the values imposed on its boundaries; such *asynchronous coupling* means that forecasters can be provided with predictions with the best horizontal resolution more quickly.

10.5.4 Using a variable mesh grid

An alternative to coupled models is to use a variable mesh: the central part of the domain is processed with a grid whose mesh size is finer in the centre of the domain and becomes coarser outwards (Coté et al., 1993). It is also possible to use a geometric transformation to move from a regular grid to a distorted grid, with variable mesh spacing, as indicated in Subsection 2.4.4 for the sphere (Courtier and Geleyn, 1988). However, if the time integration scheme is conditionally stable, the CFL condition to be used is that which involves the finest mesh. Accordingly this solution has led to operational applications only for use with time integration schemes allowing large time steps such as the semi-Lagrangian semi-implicit method.

10.6 Post-processing of model output

The results provided by the models cannot be used directly; they are fields defined on the model levels (that follow the orography when we use a vertical coordinate of the normalized pressure type) whereas forecasters work either with charts corresponding to isobaric levels or to constant potential vorticity levels (Malardel, 2005), or alternatively with vertical cross sections using pressure or height as the vertical coordinate. Retrieving meteorological parameters on isobaric or constant height levels therefore requires interpolation in the vertical. Moreover, it is judicious to present the results, which will have to be used to plot charts or supply application programmes, on a grid that is independent of the grid used by the model. In this way, whatever changes in resolution may be made on the models, users may continue to work with the data supplied by this *intermediate grid*. However, care must be taken to perform vertical and horizontal interpolations consistently (i.e. in a determined order), and to reduce the discontinuity arising between neighbouring points when the corresponding data are interpolated in the vertical at different levels; to this end, the technique already used for data assimilation consisting in interpolating only deviations from a more regular background provides a convenient solution.

10.6.1 The problem of topography

The use of a grid for the model and a different grid for presenting the results poses a problem as to the definition of topography. Defining a normalized pressure vertical coordinate at a point requires knowing the surface pressure value and thus the orography height. We have to define a mean orography at each grid point of the model corresponding to the horizontal resolution of the grid; likewise, we define on the intermediate grid a mean orography adapted to the horizontal resolution of this grid. It should be noticed that the heights of these two orographies are generally different (especially in mountain areas) and do not coincide with the true altitude either. These differences must therefore be taken into account carefully when the meteorological parameters have to be interpolated in the vertical by applying, for instance, the technique mentioned in the previous paragraph.

10.6.2 The various interpolations

A great majority of meteorological fields, including those used by forecasters, are plotted on isobaric levels. The move from the model levels to the isobaric levels is usually made by interpolation using either quadratic or cubic polynomials or alternatively spline functions. Determining the value of meteorological parameters on the isobaric levels whose altitude is lower than that of the model's topography, like the value of sea level pressure in mountain areas, raises a specific problem since extrapolation is then necessary. To perform this extrapolation, we use the hydrostatic relation after choosing a temperature gradient (usually the standard gradient of 6.5 °C/km) from the lowest level of the model downwards.

To determine the weather elements such as the wind at 10 m or the temperature at 2 m (heights at which the measurements are made), we must interpolate between the values of the parameters at the lowest level of the model and their surface values ($V_S = 0$, T_S, surface temperature). A clever way of doing this interpolation is to take into account rescaled profiles defined in the boundary layer (Geleyn, 1988). Of course, if we wish to obtain these weather elements on the intermediate grid, a correction is required, above all for temperature, if the two orographies do not coincide.

More generally, some users are especially interested in a detailed description of the lowest layers and even in some instances in vertical cross sections. For this it is then helpful to determine the value of meteorological parameters for given height levels on the intermediate grid; in this case, the interpolation must be done while trying to preserve the structure of the lowest layers as far as can be done.

Determining precipitation on the intermediate grid must also be done with caution so that the quantity of water calculated over a given surface area of the intermediate grid is identical to the quantity of water collected over the equivalent surface area of the model grid.

10.7 Local forecasting

10.7.1 The principle of statistical adaptation

Numerical prediction models provide, for a given time range, the average value of meteorological parameters characterizing the atmosphere at a grid point or, more specifically, for the spatial domain represented by the grid point (that is, the small mesh-sized block on a particular level of the model). The various parameters characterizing a specific location, such as altitude, the nature of the terrain, or the proximity of the sea, may be rather different in the model and in the real world. To obtain locally predicted values that can be compared with the observations made at a meteorological station (temperature and humidity at 2 m, wind at 10 m, precipitation, etc.) it is generally necessary to implement a *local adaptation*.

Such local adaptation of weather forecasts may be done by statistical processing methods for multidimensional data documented very completely by Der Megreditchian (1992, 1993) and Wilks (1995); however, these methods require archives both for the grid point fields (analyses, forecasts) and for the observations of parameters that we want to predict.

The principle of *statistical adaptation* (also termed *statistical interpretation* in the literature) is the following: starting from data records (collated in what we call a *training file*), we look for a statistical relation between a subset of grid point values (the *predictor*) and the parameter to be predicted (the *predictand*). These statistical relations are then applied to the grid point predictions of the day to forecast the values of the local parameters at the observation point. As the values at neighbouring grid points are closely correlated, implementing a technique for compacting the information (by using the principal component method or the canonical analysis method, for example) allows us to reduce the number of variables of the predictor. We then have to select the predictor variables, keeping only those that are most closely correlated with the predictand so as to establish the final statistical relation. For some parameters it may be relevant to split the training file into several subfiles so as to calculate statistical relations with more homogeneous data sets.

When the parameter to be predicted is a quantitative meteorological parameter (like temperature or wind), regression methods are generally used: the value of the predictand is expressed as a function of the selected predictors, the characteristics of this function being obtained from the statistical analysis of the training file. When we are interested in predicting the occurrence of a given phenomenon (presence of fog or ice, precipitation above a given threshold), we use *discrimination* methods: we look to define a discrimination function depending on the selected predictors that can separate the training file into two categories with, on one side, the cases corresponding to the occurrence of the phenomenon and, on the other, the cases corresponding to its non-occurrence; the forecast of the occurrence of the phenomenon is then made from the evaluation of the discrimination function.

The statistical adaptation methods for numerical model predictions may take different forms depending on the predictand concerned, the predictors used, and the characteristics (duration, homogeneity) of the training files: these must be large enough to provide *stable* predictions (of equivalent quality to those obtained with the training file) while containing homogeneous data. We confine ourselves here therefore to succinctly describing the most commonly used methods for local prediction of meteorological parameters.

10.7.2 The perfect prog method

A first method, known as *perfect prog*, consists in determining statistical relations between grid point values of analyses of meteorological fields on the one hand, and observations on the other (Klein et al., 1959). For each local parameter to be forecast, we thus establish a *synchronous relation* (valid at a given time) between the predictor (values of parameters characterizing the analyses) and the predictand (observed value of the local parameter). To obtain a forecast of the local parameter at a given range, this relation is applied by using, instead of the values of the parameters from the analyses, those from the forecasts, obtained as output from the numerical prediction model at the time range in question. The quality of the results is directly related to the skill of the forecast provided by the numerical prediction model and shall be good if the forecast is perfect (identical to the analysis corresponding to the time range in question). Although the method has the drawback of not taking into account systematic errors (bias) of a prediction model, it nonetheless proves very useful for deducing in a simple way the value of local weather parameters from large-scale variables.

10.7.3 The model output statistics method

A second method, known as *model output statistics* or MOS, consists in establishing a statistical relation between the forecast meteorological fields and the observed parameters (Glahn and Lowry, 1972). This can only be done insofar as all of the grid point forecasts of the prediction model are actually archived. Once again we establish a *synchronous relation* (valid at a given time) between the predictor (values of the parameters from the forecasts at a given range) and the predictand (observed value of the local parameter). This method uses relations between values predicted by the model and observed values and takes into account the defects and biases of the model and provides optimal adjustment for each range; however, the relations established with a given model cease to be valid when major modifications are introduced into the model. It is necessary therefore to update the statistical relations when the model is modified, which leads to reducing the size of the training file: it follows that the stability of statistical adaptation may be affected and this cannot be validly applied to the prediction of rarely observed phenomena.

10.7.4 The Kalman filtering correction method

Kalman filtering, the principle of which has already been seen in the context of data assimilation, is also a highly effective tool for statistical adaptation in which statistical relations, as applied, are updated automatically taking into account the data actually observed in the near past (Simonsen, 1991; Persson, 1991). In the case where the parameter to be forecast is obtained by linear regression, for example, recursive Kalman filtering is used to recalculate the mean values and the variances of the regression coefficients each day, taking into account the difference between the last predicted value of the parameter to be forecast and its observed value: the correction made by the filter results from a compromise between that difference and the previous correction (hence it is called recursive filtering). The reactivity of the filter is then adjusted by weighting these two elements: if the weight of the earlier correction dominates, the filter will not be reactivated (the same correction will be made every day) and so not very effective; however, if the weight of the observed error dominates, then the filter will be highly reactive, which may lead to irrelevant corrections in the event of an erroneous local forecast by the model. Filtering must therefore be tuned with caution generally by diminishing its reactivity for the more remote ranges. This method is extremely effective, however, for updating the regression coefficients obtained with the model output statistics method and for taking progressively into account the change in statistical characteristics of the predictor when the numerical model is modified.

10.8 The forecasting process

10.8.1 The various tasks of the forecast suite

What we call the *numerical weather prediction suite* is the sequence of tasks implemented to make a weather forecast from the collection of observations to the dissemination of the forecasts and the way these tasks of the suite are linked together. A forecast suite typically comprises the following tasks:

- observing the various parameters (state variables or derived physical quantities);
- transmitting the data to information processing centres;
- pre-processing data (decoding of messages, data checking, and input into a database);
- determining the initial conditions for the model (objective analysis and initialization);
- forecasting properly using a numerical model;
- determining sensible weather elements at local scale;
- displaying results;
- disseminating forecasts.

Here we look more especially at the way in which the collection and pre-processing of observations, objective analysis, and prediction model are linked together.

10.8.2 The importance of cutoff time

As noted at the beginning of this chapter, the quality of an objective analysis depends both on the quality of the forecast used as the background field or first guess, and on the number and quality of the observations. Meteorological observations arrive continuously in information processing centres with gluts at the main synoptic times. To initialize a model at a given time H_0, we must wait a certain time before launching the analysis process at time H_c. The interval $H_c - H_0$ is called the *cutoff time*, and how it is determined is of capital importance, given the need to provide forecasters with predictions as soon as possible: if it is too short, not all the observations for time H_0 can be collected; if it is too long, it pointlessly extends the lead time available for making the forecasts. It is necessary therefore to strike a compromise: generally the cutoff time is determined once and for all after examining the way in which the forecast degrades when the cutoff time is shortened. It is of the order of 1 to 3 hours depending on the case.

10.8.3 The prediction suite and the assimilation suite

Given the need to provide forecasts as soon as possible, the cutoff time must be relatively short; something in the order of 90 minutes is reasonable if a large enough set of observations is to be collected to make a global forecast. If we stick to this procedure and use the very short range forecasts (3 or 6 hours) of the model as the background to the next objective analysis, all the observations arriving after time H_c are lost for good. This is why most forecasting centres have set up an additional suite specifically for data assimilation: the objective analysis procedure is started with a cutoff time of 6 to 10 hours, which is long enough to take into account the maximum of observations and is followed by a forecast of the model until the time for the next analysis. This procedure provides as a background field for the future analysis a forecast made from an initial state that is as rich in observations as possible.

10.9 Forecast verification

10.9.1 Basic definitions

Evaluating the quality of numerical prediction models is a constant concern both for model developers working to improve models and for forecasters who must ensure the products they use for forecasting are indeed reliable.

First we must fix a geographical domain over which a grid has been defined along with a level. Over this domain, a meteorological field corresponding to a given parameter is represented by a vector $\mathbf{X}\{\mathbf{X}_i, i = 1, ..., N_p\}$, N_p denoting the number of grid points of the domain. We call $\mathbf{X}^f$ the field forecast by the model at a given time and $\mathbf{X}^a$ the field analysed at the forecast time.

We can therefore define the forecast error $\mathbf{E}$:

$$\mathbf{E} = \mathbf{X}^f - \mathbf{X}^a.$$

By mapping this field we can display the geographical distribution of forecast error. It is also possible to condense this information by calculating the *mean error*, $\bar{\varepsilon}$, or *bias* and the *standard deviation* of the error for the forecast, σ, which are written:

$$\bar{\varepsilon} = \frac{1}{N_p} \sum_{i=1}^{N_p} E_i \quad \text{and} \quad \sigma = \sqrt{\frac{1}{N_p} \sum_{i=1}^{N_p} (E_i - \bar{\varepsilon})^2}.$$

We also calculate the *root mean square error* (RMS) δ, which integrates both information on bias and on the standard deviation as it is written:

$$\delta = \sqrt{\frac{1}{N_p} \sum_{i=1}^{N_p} E_i^2}.$$

The RMS error δ and the standard deviation σ are connected by the relation $\delta^2 = \bar{\varepsilon}^2 + \sigma^2$, which shows that these two statistical parameters coincide exactly when the bias $\bar{\varepsilon}$ is zero.

It is also helpful to be able to compare a field $\mathbf{X}$ with a field $\mathbf{Y}$ by calculating the correlation coefficient between these two fields, which is given by the formula:

$$R = \frac{\text{cov}(\mathbf{X}, \mathbf{Y})}{\sigma(\mathbf{X})\sigma(\mathbf{Y})},$$

where $\text{cov}(\mathbf{X}, \mathbf{Y})$ designates the covariance between $\mathbf{X}$ and $\mathbf{Y}$, $\sigma(\mathbf{X})$ and $\sigma(\mathbf{Y})$ the respective standard deviations of $\mathbf{X}$ and $\mathbf{Y}$. The correlation coefficient is 1 when the fields $\mathbf{X}$ and $\mathbf{Y}$ coincide and 0 if the fields $\mathbf{X}$ and $\mathbf{Y}$ are completely uncorrelated (zero covariance).

10.9.2 Some standard verification scores

A fairly detailed inventory of statistical methods used to verify weather forecasts was made by Stanski et al. (1989). We confine ourselves here, then, to providing a few examples of the most common statistical scores and specifying the conditions under which they are used.

The parameters quantifying the forecast error in the absolute are insufficient to provide information about the skill of the forecast: it is useful to compare the forecast error with forecast errors obtained with quite trivial evaluations of forecast fields. An obvious approach is to assume that the prediction field values at a given range are identical to their initial values: this is the *persistence forecast*, noted $\mathbf{X}^p$. Another

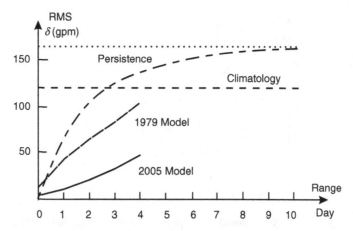

Figure 10.9 Typical plot of the evolution of RMS errors of large-scale operational models used in France.

simple forecast is provided by the climatological value of the field: we then speak of *climatological forecast*, noted $\mathbf{X}^c$ (constant whatever the range). For these trivial forecasts we can calculate the means, standard deviations, or RMS errors $\bar{\varepsilon}^p$, σ^p, δ^p and $\bar{\varepsilon}^c$, σ^c, δ^c. Comparing the model's forecast errors and the errors obtained from these trivial forecasts provides more precise information about the skill of the forecast.

Figure 10.9 gives a typical example of the evolution of RMS errors obtained for a prediction model and for the reference forecasts (persistence, climatic values). The RMS error of the persistence forecast, δ^p, increases with the range and reaches an asymptotic value δ^p_{max} that provides information on the variability of the field concerned. The RMS error of the climatological forecast is constant whatever the range: its value is given by $\delta^c = \delta^p_{max}/\sqrt{2}$.

We observe that the mean RMS error of the forecast for 0 time range is not zero, which seems paradoxical; this nonzero value simply reflects the effect of interpolation, since the statistical parameters are generally calculated on the isobaric levels and on a reference grid (which may be different from the grid used for the model or for the analysis). The RMS error of the prediction grows more or less linearly with the forecast range; the flatter the slope of the RMS error curve, the better the model. In practical terms, the use of a forecast model loses its interest when a trivial forecast is better than the model forecast; so the range D_{max} at which the climatological forecast is better than the forecast of a model gives a fairly clear indication of the model's skill. In 2010, a value of D_{max} of 8 days was obtained with the model of the European Centre for Medium-Range Weather Forecasts (ECMWF), which is generally regarded as the best of the global models in operational use.

For a given range, the RMS error for a model forecast, like that of the persistence forecast, varies notably with the season. In mid-latitudes in particular the variability of meteorological fields is generally lower in summer than in winter and this property is found in the RMS errors. One way of escaping this variability to quantify the quality of the model is to calculate the quality index $Q = 1 - \sigma/\sigma^c$; this index gives a value

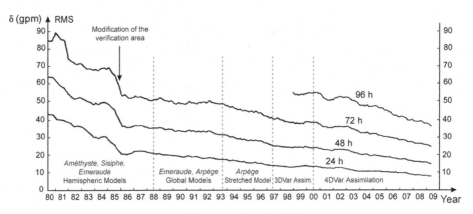

Figure 10.10 RMS error of 500 hPa geopotential height forecast over Europe and the North Atlantic with French large-scale operational models. (Météo-France image)

of 1 for a perfect forecast and a value of 0 for a forecast of equivalent quality to the climatological forecast.

Figure 10.10 shows the evolution of the RMS error of the forecast of the geopotential at 500 hPa, obtained for various ranges with the large-scale operational models used by Météo-France from 1980 until 2009; the verification domain encompasses Western Europe and the North Atlantic and the curves have been smoothed by recalculating moving averages.

Examination of this evolution shows the substantial advance made, since the skill achieved at the end of the period for a four-day forecast is almost as good as for a one-day forecast at the beginning of the period. It is interesting to notice that in terms of scores, progress appears to be much more marked in the first third of the period considered. The regular increase in horizontal resolution has made it possible to better simulate the evolution at the synoptic scale; once a resolution of a hundred or so kilometres was achieved, the improvement rate slowed and the quality gains were made through more detailed processing of physical processes and techniques that better allow for all the observations in the definition of the initial state.

The quality of a short-range forecast (two to three days) is often quantified by calculating the correlation coefficient between the forecast tendencies and the observed tendencies over a given range. If we are interested, however, in medium-range forecasts, it is more worthwhile calculating a correlation coefficient between the forecast *anomalies* (departures from climatological values) and the analysed anomalies.

Figure 10.11 shows the evolution of correlation coefficients for anomalies of the geopotential at 500 hPa with the European Centre's model for various ranges, for the northern hemisphere and for the southern hemisphere.

Because of much poorer observation coverage of the southern hemisphere than the northern hemisphere, it was only from 1996 that the quality of forecasts in the southern hemisphere began to catch up with the quality attained for the northern hemisphere before becoming virtually equivalent: this demonstrates the contribution

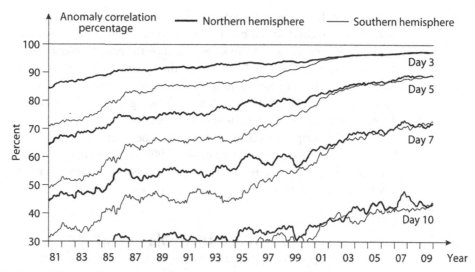

Figure 10.11 Evolution of the anomaly correlation coefficient of geopotential height at 500 hPa for the northern and southern hemispheres with the European Centre model from 1981 to 2010. (ECMWF image)

from the variational assimilation method that has made it possible to effectively take account of the data provided by on-board satellite systems.

Model developers commonly calculate the statistical parameters quantifying the forecast skill by using as a reference the values of the fields analysed at the grid points of the domain concerned. However, in data sparse areas, grid point analyses reflect less the observation of the actual atmosphere than the short-range forecast of the model, used as the background; it is difficult then to accurately compare the performances of two forecast suites that differ in model and in data assimilation.

To overcome this problem and try to better compare various operational forecast systems, we make the statistics at the points where the observations are available. The forecast values then have to be interpolated from the forecast fields at the grid points, if possible with the methods used for defining the observation operators. So as to be able to use the same sample of observation points, the World Meteorological Organization defined for each Region a set of particularly reliable reference stations which are used by the various meteorological services to evaluate the respective performances of the different numerical prediction systems.

10.9.3 The verification of categorized events

The verification methods presented essentially concern large-scale meteorological fields. With the progress of techniques for obtaining a forecast of local weather elements (use of ever finer meshes, statistical interpretation) we have been led to take an interest in the skill of event forecasting: such events may be the occurrence of a given phenomenon (fog, frost, ice, etc.) or the crossing of a critical threshold for a given parameter (variation in temperature, precipitation, etc.).

In the simplest instances we therefore have to compare a binary forecast (event forecast or not forecast) with a binary observation (event observed or not observed). The results fill a double entry table known as a contingency table. From this table we can calculate a range of skill indicators including *probability of detection*, *false alarm rate*, *critical threat score*, or *Rousseau score* (Rousseau, 1980), which has the particularity of giving a zero value for a random draw complying with the frequency of occurrence of the phenomenon. Many commentators have made comparative reviews of the various scores used to appraise yes/no type forecast performances; these include reviews by Woodcock (1976) and by Doswell et al. (1990). It should also be pointed out that skill indicators have also been proposed when the event in question is no longer binary but involves a classification among several categories.

10.10 Ensemble forecasting

10.10.1 The limit of predictability

Despite the constant improvement in weather forecasts made with numerical models, forecasters have had to bow to fact: it is impossible to provide very accurate forecasts beyond a certain range. This *predictability limit*, which is rather variable, is generally an increasing function of the space and time scales of the meteorological phenomena concerned. With current models it is of the order of four to five days for synoptic scales (as shown in Figure 10.12). The nonlinear nature of the evolution equations and the imperfect knowledge of the initial state are the source of this impossibility, even if the model used is perfect (Lorenz, 1969). Like many other dynamic systems, the atmosphere is highly sensitive to initial conditions, in that two similar initial states may lead to divergent outcomes. Accordingly the scientific approach is to forecast for a time range, given our knowledge of the initial state and its errors, the set of possible forecast states, represented by the *probability density function* (PDF). Knowing the

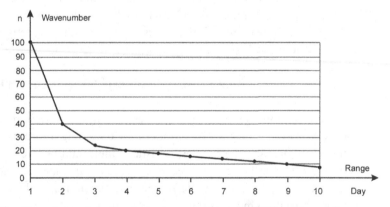

Figure 10.12   Predictability of the various spatial scales expressed as wavenumbers. (ECMWF image)

PDF means we can calculate the ensemble average, its variance, and its higher order statistical moments and make a probabilistic forecast (assigning a probability to each of the forecasts).

The evolution of the PDF is given by the Liouville equation which was introduced in the context of a meteorological model by Gleeson (1966) and Epstein (1969). However, given the number of variables in present-day models, the cost of forecasting the PDF becomes exorbitant (the number of variables to have evolved being equal to the square of the number of variables of the actual model). The *stochastic-dynamic forecast* method proposed by Epstein (1969) and consisting in using a hierarchy of evolution equations for mean values and higher order statistical moments with an appropriate closure hypothesis has run up against the same practical obstacle. This is why we have sought to determine the evolution of the PDF from a sample of the initial PDF and so make what is termed an *ensemble forecast*.

10.10.2 Using an ensemble

The *Monte Carlo method* proposed by Leith (1974) consists in defining a sample of initial situations for making a number of equiprobable forecasts, allowing us first to obtain an evaluation of the forecast PDF from this ensemble and then to calculate the mean values and higher order statistical moments for each of the variables. The mean values then allow us to bring out only the structures that coincide from one forecast to another and to *filter out* the unpredictable elements. This idea has been applied by performing several integrations of the same model from initial states obtained by superimposing on the initial reference state small perturbations distributed randomly, that are compatible with the analysis errors (Hollingsworth, 1980); however, even if care is taken to introduce balanced perturbations only, in keeping with the hydrostatic and geostrophic balance, this method cannot provide sufficient dispersion of the distribution of forecast states.

An original forecasting experience based on the use of an ensemble to evaluate average values was conducted by Rousseau and Chapelet (1985). By simply combining the results of the operational models implemented by seven major meteorological centres (with fairly similar characteristics and performances), the ensemble average of these forecasts showed a gain in predictability compared with each of the individual forecasts. Because it was sparing in its means, this way of making forecasts using the result of a few different deterministic models came to be called *poor man's ensemble forecasting*.

The fundamental problem with ensemble forecasting is to judiciously sample the initial PDF so as to obtain a maximum of solutions that are comparatively distant from each other with a minimum of initial states. The adjoint model operator is used to calculate the disturbances in the initial state that undergo maximum amplification over a given period, in the sense of a suitably chosen norm (Buizza et al., 1993). This *singular vector* technique is used by the European Centre for making some fifty forecasts from which to simulate the forecast PDF. This technique is fairly costly in computing time and is based on an assumption of linear growth of errors, which only holds for the shorter ranges of the forecasts (typically 48 hours) and so cannot be applied for longer ranges. Nonetheless, singular vectors provide by construction an orthogonal basis ensuring a sampling of the PDF in very different directions.

A somewhat different technique has been developed in the United States by the National Center for Environmental Prediction (NCEP); the initial perturbations are generated by the model itself via an iterative selection process termed '*breeding*' of perturbations, proposed by Toth and Kalnay (1993). It is the rapid growth error structure in the recent past that allows us to determine the initial perturbations, whereas the sampling given by singular vectors is based rather on rapid growth errors in the future. However, PDF sampling by this technique does not provide the same qualities of orthogonality as are ensured by sampling obtained by the singular vectors technique. More elaborate techniques have been proposed to overcome this drawback, such as the *ensemble transformation method* (Bishop and Toth, 1999) or *ensemble transform Kalman filter method* (Bishop et al., 2001).

The Canadian Meteorological Centre (CMC) has developed a *perturbed observation* approach (Houtekamer et al., 1996), in which different parallel cycles of forecasts serving as a background field are updated by separate ensembles of perturbed observations providing a set of analyses. This *ensemble Kalman filter* technique (Houtekamer and Mitchell, 1998) has already been mentioned in Subsection 10.3.9 under data assimilation; it is also used during data assimilation to estimate forecast errors depending on the situation and so to improve the overall quality of analyses.

Uncertainty about the initial state is not, though, the only cause of the limitation of predictability. It is also relevant to take into account the uncertainty relating to the imperfection of the model used, termed *model uncertainty*. A first method, called the *stochastic perturbations of physics* (Palmer, 2001), consists in adding a random noise to the physical tendencies; a second method is based on introducing parameterizations of different physical processes from one model to another (Houtekamer and Lefaivre, 1997). Lastly, in the direct line of the ideas developed by Rousseau and Chapelet, the use of several independent models with similar general characteristics in terms of spatial resolution and parameterized physical processes can be used to form multi-model ensembles that are supposed to account for variability resulting from the use of models that all contain imperfections (Krishnamurti et al., 2000; Mylne et al., 2002).

There is no reason, though, why ensemble forecasting models should remain confined to the medium range. The problem of sensitivity to initial conditions and to model imperfections is very general and it is quite logical to try to use these techniques for the short range. Many ensemble prediction experiences using mesoscale models (over a limited area) coupled with global models have been developed. With this type of model, the choice of lateral boundary conditions introduces a new source of variability that ought to be taken into account (Du and Tracton, 1999). Moreover, the use of very sophisticated schemes, which are very sensitive for parameterizing interactions with the ground, also leads us not to ignore the uncertainty that may occur on the various fields of constants defining the characteristics of the soil and of the vegetation (Shapiro and Thorpe, 2004).

10.10.3 Presentation of forecasts

The operational use of ensemble forecasting has led to an effort at synthetic presentation of the results so as to make them usable for forecasters. Among the various products available we can cite the ensemble mean and variance charts, the plots of characteristic iso-values (appositely called *spaghetti plots*), probabilistic charts of parameters exceeding a threshold. The series of local forecasts for the distribution of a given parameter can be visualized in a very eloquent way by using a synthetic '*box-and-whiskers*' plot that indicates, apart from the extreme values of the distribution, the median and the limit values of the first and last quartiles. Figure 10.13 is an example of such a probabilistic forecast for 2 m temperature obtained from a European Centre's ensemble forecast.

This plot shows the simplified distribution of the 50 perturbed 10-day forecasts of the ensemble; the dotted curve is for the unperturbed forecast, while the solid curve gives the values obtained with the high-resolution deterministic model. The information provided by this presentation is useful: the fall in temperature from day 5 onwards coincides with a significant increase in dispersion of the ensemble from day 6 to day 8 that becomes weaker for days 9 and 10.

It is important to notice that probabilistic forecasts can be verified. The *reliability* test consists, for a given forecast, in comparing the forecast probabilities with the observed frequencies. Good reliability is necessary but it is not sufficient, as the climatological forecast gives very good reliability although it is not very informative; a probabilistic forecast must therefore provide a degree of sharpness, that is, it must be able to come down off the fence, to forecast extremes and to separate forecasts into a sufficient number of classes. A detailed analysis of the various methods used to verify

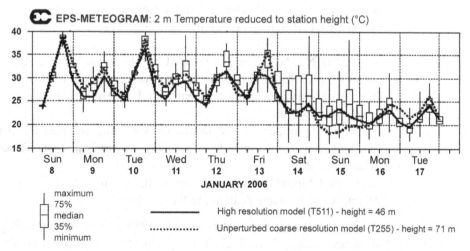

Figure 10.13 Evolution of the distribution of 2m temperature, forecast for the next ten days at Rio Grande (Brazil) for 8–17 January 2006. (From an ECMWF image)

the ensemble forecast and the probabilistic information it provides has been made by
Toth et al. (2003).

10.11 International cooperation

The growing complexity of numerical prediction models and the difficulties encountered in coming up with truly efficient codes on scientific supercomputers have helped
to take things from the stage of individual development work to that of a major scientific project involving specialized teams. Although the early numerical prediction
models were designed, produced, and tested by just one person (whose name they
then took) in general, the development of current models implies cooperation among
many teams and that exceeds the resources of national meteorological services acting
alone.

   Thus we have seen, primarily in Europe, the development of unified models or community models designed for several categories of users in various meteorological services (this road was taken in the 1990s in Germany, Canada, France, the UK, and the
Scandinavian countries). This new form of organization also required the development of genuine model engineering requiring compliance with standards in the writing of software and very strict discipline in updating them. This essential cooperation
extends from the simple exchange of ideas and methods to the production of common
models in various forms in each of the national services participating in the project or
even to the installation of joint centres for the production of weather forecasts.

   The models shared by the various meteorological services and that were operational
in 2010 include:

- the HIRLAM model, the outcome of joint work begun in 1985 by the Scandinavian
  countries, Estonia, Lithuania, Ireland, the Netherlands and Spain;
- the ARPEGE/IFS model developed by Météo-France and the ECMWF, started in
  1987;
- the ALADIN/LACE model developed by Météo-France in conjunction with scientists from central and eastern Europe from 1992, and used by France, Belgium,
  Portugal, and a large number of central and eastern European countries as well as
  North African countries;
- the model developed by the Deutscher Wetterdienst, used by members of the
  COSMO consortium (including Germany, Switzerland, Poland, Romania, Italy,
  Greece, and Russia);
- the Unified Model developed by the MetOffice and operating in Norway, South
  Africa, Australia, New Zealand, India, and Korea.

In Europe, the European Centre set up in 1975 and based in Reading (UK) counts
20 member states in 2011. It is currently equipped with computing facilities capable
of reaching computing speeds of 20 teraflops (that is, 2×10^{13} floating point operations per second). It uses operationally a high-resolution deterministic model and an

ensemble forecasting system using a lower resolution model to make 51 equiprobable forecasts. These high-performance tools mean that it can provide the meteorological services of its member states with very detailed deterministic and probabilistic forecasts for up to 10 days ahead.

There is no question that this development must continue in future years because of the growing complexity of the tools required to simulate the evolution of our planetary environment. Forecasts for the month or the season ahead can only be made by taking into account the evolution of the ocean and its coupling with the atmosphere. Ascertaining the initial states of such models will increasingly call for remote sensing systems which supply data indirectly and require variational assimilation tools that are cumbersome to implement. Lastly, the principle of ensemble forecasting, so successfully used for the medium range, is being extended to all forecasting systems, and the pooling of systems developed by the various meteorological centres can only further improve forecasts.

10.12 Future prospects

10.12.1 Ever more powerful computers

Numerical weather prediction, which began to develop in the 1950s, has become a major discipline the expansion of which is constantly supported by the regular increase in the power of computers. It is accepted that the power of computers on the market doubles about every 18 months, and no dip in this trend has been observed to date. In the future, the probable increase in computing power will therefore continue to improve the performances of operational numerical prediction systems, whether for data assimilation or for numerical modelling.

Concretely this means it will be possible first to increase the spatial resolution of models to take into account ever finer scales and second to treat the various subgrid-scale physical processes in more detail. Computing power will be used to generalize the ensemble prediction systems both in the domain of forecasting at all ranges and for simulating atmospheric flows at all scales. Thus numerical prediction will make it possible not just to imitate better and better the evolution of the actual atmosphere but also to explore the consequences of the various possible evolutions under certain conditions. Lastly, the increase in computing power will allow for better assimilation of the numerous existing or future observations.

10.12.2 Data assimilation

In the field of data assimilation, the systematic use of the various means of remote sensing from space, which has proved its effectiveness for improving the forecasting quality of models, will continue: data from the advanced Atmospheric InfraRed Sounder (AIRS) and the Infrared Atmospheric Sounding Interferometer (IASI) carried by polar orbiting satellites are beginning to be assimilated, as are occultation measurements of the Global Navigation Satellite System (GNSS) using the GNSS Receiver for Atmospheric Sounding (GRAS). In the near future lies the prospect of directly assimilating upper air wind profiles thanks to the Doppler lidar ALADIN (Atmospheric LAser Doppler INstrument) to be carried on board the future polar orbiting satellite ADM-AEOLUS.

For small-scale data assimilation, the use of measurements of the radial wind by networks of Doppler radars and measurements of reflectivity should make it possible to analyse small-scale structures with spatial resolution close to that of the model. In the near future, the improved representation of clouds in models should make it possible to use realistic observation operators and allow, by the variational method, the assimilation of coherent structures appearing on satellite images. Lastly, with the development of methods for locating sensitive zones of the atmosphere (relative to the triggering of severe weather events), the practice of *observation targeting* (that is the possibility of making observations where and when required) is likely to become increasingly frequent. This principle can also be applied to perform a modulation of the *screening process*, which consists in retaining only a relevant subset of observations from among the increasing number of remote sensed data supplied by various instruments.

10.12.3 Prediction models

The development of numerical prediction models obviously depends on the increased spatial resolution that should bring about fairly widespread use of the Euler equations instead of the primitive equations, including for large-scale models; semi-Lagrangian semi-implicit algorithms have now proved effective and make it possible to use a relatively large time step.

In the coming years we can expect to see the use for weather forecasting of global models with a 10 km grid, small-scale models covering domains the size of most European countries with a resolution of 1 km, and models dedicated to studying specific phenomena with a resolution of the order of 100 m. In the area of parameterization, much progress remains to be done on improving the processing of turbulence, convection, interaction among clouds, aerosols, and radiation and to make better allowance for exchanges at the ocean–atmosphere interface and within the ocean mixing layer.

10.12.4 Specialized application models

The output from weather forecasting models is used to initialize a whole range of increasingly complex *application models* for determining certain parameters that are closely dependent on the atmospheric environment. Many application models are already used downstream from weather forecasting models; these include:

- hydrological models for predicting river flow rates;
- models for detecting the formation of hydrometeors (icing zones for aircraft or the formation of thick fog in hazard-prone areas);

- models for the transport of pollutants over various distances;
- atmospheric chemical models for analysing the evolution in the composition of the atmosphere and especially the concentration of gases that are harmful for health;
- models for road temperature change designed to better forecast ice, etc.

10.12.5 Simulation

Meteorological forecasting models are not only used to make predictions from initial data inferred from meteorological observations; they also tend to be increasingly used to investigate the evolution of the atmosphere under a number of particular circumstances that cannot be directly observed or are not amenable to experimentation in the field.

In the domain of small-scale modelling, nonhydrostatic models with grid spacing of the order of 2 km (such as the AROME model developed at Météo-France) are now capable of explicitly simulating deep convection. With even finer meshes they can be used to determine the behaviour of atmospheric circulation in the vicinity of sites of sensitive infrastructures (airports, nuclear power stations, chemical complexes, new agglomerations, etc.). Simulation can be used to determine the harmful effects ahead of time and so modify infrastructure projects.

In the domain of climate, numerical models have been widely used to simulate the evolution of the Earth's atmosphere over the last 150 years and to make predictions for the whole of the twenty-first century, taking into account various scenarios for the evolution of our industrial society (and especially the consequences of the regular increase in the concentration of greenhouse gases). Numerical modelling does not try just to reproduce the real world as best possible, but can also multiply possible experiments by simulating an entire range of possible future evolutions of the atmosphere.

10.12.6 Ensemble forecasting

It is trivial to observe that a weather forecast made with a deterministic model, however sophisticated and accurate it may be, gives just one particular forecast from among a large number of other possible forecasts. The value of ensemble forecasting is that it allows us to better explore the range of possible forecasts and to provide each parameter with its own probability density function, which is what allows us to make probabilistic forecasts. The probabilistic information supplied by ensemble forecasting is currently used as a supplement to deterministic models, since it allows us to appreciate the degree of confidence that can be assigned to the solutions provided by deterministic models. Although they were initially used to provide more complete medium-range information, ensemble forecasting systems are applied to all weather prediction models whatever the scale or the forecast range considered, and also to the various application models, especially nonlinear models. The difficulty with ensemble forecasting lies in the choice of the sample of initial situations and of the boundary conditions, and in the way to generate perturbations that can be worked into the model considered.

It cannot be denied that the possibility of having probabilities for the occurrence of an event is very useful information for any user able to calculate the costs induced by the occurrence of the event and by the protection measures; that information allows users to define an optimization strategy with respect of certain criteria depending on their area of activity. However, the *a priori* assessment of the value of a probabilistic forecast appears very difficult for human judgement and one is entitled to ask what added value forecasting experts may bring to such an automatic forecasting system. Perhaps their knowledge of model systems and experience should be drawn on for selecting initial conditions, choosing certain types of perturbation to be introduced into the model, and defining the relevant application models in the context of such and such a meteorological situation.

10.12.7 Conclusion

As seen in the course of this book, the story of numerical weather prediction is far from over. While there is no denying that the advances in atmospheric modelling have accelerated and forecasting is ever more accurate, as statistics prove, a few instances of poor forecasts of events with sizeable consequences regularly come to trouble the consciences of model makers and sometimes make them doubt the progress made by their discipline. Moreover, the improved quality of weather forecasts at the local scale (of the order of 1 km) and the timing of the corresponding events still remain a major challenge. There is still therefore a huge field of investigation and innovation for fresh generations of model developers.

Appendix A Examples of nonhydrostatic models

A.1 Introduction

The nonhydrostatic Euler equations came into general use in the 1990s, making it possible to deal properly with atmospheric motion on spatial scales of a few kilometres. Advances in numerical methods and faster computers mean nonhydrostatic models can now be used operationally on limited areas. In the medium to long term, then, the nonhydrostatic approach is destined to be applied to all categories of model (including global models) working with grid meshes of less than ten kilometres.

It seemed, therefore, that this book, which is essentially about the primitive equations, needed to be supplemented by two examples of nonhydrostatic models based on the Euler equations. Two models were chosen that have proved their worth for operational forecasting:

- the AROME model developed at Météo-France can be thought of as the nonhydrostatic extension of the baroclinic models described in Chapter 8 and it uses the same type of time integration algorithm (semi-Lagrangian semi-implicit);
- the WRF/ARW model has been developed by US universities and organizations to be made available to various users for research and operational forecasting alike; this model uses a split-explicit time integration algorithm quite unlike those covered elsewhere in this book.

For the sake of consistency, the same notation as used in the other chapters of this book is maintained here, although this entails a few differences with the various papers referred to in this appendix.

A.2 The AROME model

A.2.1 Introduction

The AROME model used by Météo-France for short-term, limited-area forecasting is derived directly from the nonhydrostatic version of the ARPEGE/ALADIN model (Bubnova et al., 1995), for its dynamical part, and, for its physical part, from the mesoscale nonhydrostatic MESO-NH model (Lafore et al., 1998) that allows detailed handling of cloud microphysics and interactions between the Earth's surface and the atmosphere. The nonhydrostatic (or Euler) equations, which are used for the model dynamics and are set out in Subsection 2.2.2 of this book, have as their solutions horizontal elastic waves (also termed sound waves, propagating at speeds close to 340 m/s in the atmosphere), whereas these are filtered out with the primitive equations because of the hydrostatic balance assumption. The implicit processing of the terms responsible for inertia-gravity wave propagation in the primitive equation models can be extended to the terms responsible for the propagation of sound waves in the non-hydrostatic models, so that a comparatively large time step can be used. However, a semi-Lagrangian semi-implicit algorithm of this sort is much less easily implemented when it comes to defining the terms processed implicitly. Of the various algorithms that can be used in the AROME model, we have opted here to describe the semi-Lagrangian semi-implicit two-time-level (SLSI2TL) algorithm, which has been operational at Météo-France since December 2008.

A.2.2 The equations used

The system of equations used is obtained by formulating the Euler equations using the normalized hydrostatic pressure hybrid vertical coordinate, s written, as proposed by Laprise (1992) and described in Subsection 7.5.2. However, to achieve a stable semi-Lagrangian semi-implicit algorithm, first Bubnova et al. (1995) and then Bénard et al. (2010) have shown that new prognostic variables are required to describe the evolution of true pressure p and vertical velocity w.

By calling π hydrostatic pressure defined in Subsection 7.5.1 and by introducing for convenience:

$$\mu = \frac{\partial \pi}{\partial s},$$

these new variables are written:

$$Z_p = \ln(p/\pi)$$

$$\delta_w^\bullet = -g\frac{p}{\mu R_d T}\frac{\partial w}{\partial s} + \mathcal{X}, \quad \text{with} \quad \mathcal{X} = \frac{p}{\mu R T}\nabla\Phi \cdot \frac{\partial \mathbf{V}}{\partial s},$$

R_d and R being the gas constants for dry air and for moist air.

Adopting the notation of Subsection 7.5.3, the set of equations can be written:

$$\frac{d\mathbf{V}}{dt} + \frac{RT}{p}\nabla p + \frac{1}{\mu}\frac{\partial p}{\partial s}\nabla\Phi = \mathbf{F}_H, \tag{A.1}$$

$$\frac{d\delta_w^\bullet}{dt} + g^2\frac{p}{\mu R_d T}\frac{\partial}{\partial s}\left(\frac{1}{\mu}\frac{\partial(p-\pi)}{\partial s}\right) - g\frac{p}{\mu R_d T}\frac{\partial \mathbf{V}}{\partial s}\cdot\nabla w$$
$$-(\delta_w^\bullet - \mathcal{X})(\nabla\cdot\mathbf{V} - D_3) - \frac{d\mathcal{X}}{dt} = -g\frac{p}{\mu R_d T}\frac{\partial F_w}{\partial s}, \tag{A.2}$$

with $D_3 = \nabla \cdot \mathbf{V} + \dfrac{R_d}{R}(\delta_w^* - \mathcal{X}) + \mathcal{X}$,

$$\frac{dT}{dt} + \frac{RT}{C_v} D_3 = \frac{Q}{C_v},\tag{A.3}$$

$$\frac{dZ_p}{dt} + \frac{C_p}{C_v} D_3 + \frac{\dot{\pi}}{\pi} = \frac{Q}{C_v T},\tag{A.4}$$

$$\frac{d\pi_s}{dt} + \int_0^1 \nabla \cdot (\mu V)\, ds = 0.\tag{A.5}$$

In these equations, the constants C_p and C_v represent the specific heats at constant pressure and volume, relative to moist air; $\mathbf{F}_H$, F_w, and Q contain the terms relative to the Coriolis force, the curvature terms, and the forcing terms due to physics for $\mathbf{V}$, w, and T. To these equations should be added equations for the transport of various scalar quantities used in the model physics (hydrometeor mixing ratio, turbulent kinetic energy).

This system is supplemented by the following diagnostic equations:

$$\Phi = \Phi_s + \int_s^1 \frac{\mu RT}{p}\, ds',\tag{A.6}$$

$$\mu \dot{s} = \frac{\partial \pi}{\partial \pi_s} \int_0^1 \nabla \cdot (\mu \mathbf{V})\, ds - \int_0^s \nabla \cdot (\mu \mathbf{V})\, ds',\tag{A.7}$$

$$\dot{\pi} = \nabla \cdot \nabla \pi - \int_0^s \nabla \cdot (\mu \mathbf{V})\, ds',\tag{A.8}$$

$$g\nabla w = g\nabla w_s + \int_s^1 \frac{\mu R_d T}{p}\ \nabla (\,\delta_w^* - \mathcal{X}\,)\, ds' + \int_s^1 (\,\delta_w^* - \mathcal{X}\,)\ \nabla\!\left(\frac{\mu R_d T}{p}\right) ds'.\tag{A.9}$$

Vertical integration is performed from the top of the atmosphere defined by $\pi(s = 0) = 0$ to its base $\pi(s = 1) = \pi_s$, surface hydrostatic pressure. Vertical velocity $\dot{s}$ is zero at the top and bottom of the atmosphere. The variable δ_w^* is defined from $\partial w/\partial s$, the vertical derivative of vertical velocity $w = dz/dt$; the value of w must therefore be defined at the surface. This is done by taking account of the evolution equation of vertical velocity at the surface, which (ignoring the source terms F_w) is written:

$$\left[\frac{g}{\mu}\frac{\partial(p-\pi)}{\partial s}\right]_s = \dot{w}_s,\tag{A.10}$$

with:

$$g w_s = \mathbf{V}_s \cdot \nabla \Phi_s.\tag{A.11}$$

Notice that the occurrence of the term $\partial \mathbf{V}/\partial s$ in the third term of the left-hand side of equation (A.2) implies that the wind is evaluated at the top and base of the atmosphere.

A.2.3 Vertical discretization

The vertical discretization is just like that proposed by Simmons and Burridge (1981) described in Chapter 8 and illustrated in Figure 8.1, being applied here to the equations written with hydrostatic pressure π. The atmosphere is therefore divided into N layers. The variables $\mathbf{V}$, δ_w^*, $\dot{\pi}$, T, and Φ are calculated for the N levels located within these layers (the positions of which will be specified later) whereas the variables $\dot{s}$ and w are calculated at the surfaces separating the layers and at the top and base of the atmosphere.

$$\pi = f(s)\pi_0 + g(s)\pi_s.$$

The functions $f(s)$ and $g(s)$ of the variable s are noted explicitly here to avoid confusion with the Coriolis parameter f and acceleration due to gravity g.

The thickness of the layers is given by:

$$\delta\pi_k = \pi_{\tilde{k}} - \pi_{\widetilde{k-1}},$$

$\pi_{\tilde{k}}$ and $\pi_{\widetilde{k-1}}$ being the hydrostatic pressures at the interfaces bounding layer k.

The vertical integrations that appear in the continuity equation (A.7) and that allow us to calculate vertical velocity $\dot{s}$ on the interlayer surfaces $s_{\tilde{k}}$ and the evolution of surface hydrostatic pressure π_s are computed using discrete sums:

$$\int_0^{s_{\tilde{k}}} \mu X ds' = \sum_{i=1}^{k} \delta\pi_i X_i. \tag{A.12}$$

The integrals allowing us to calculate the geopotential in equations (A.6) and (A.9), and the quantity $\dot{\pi}$ in equation (A.8), at levels s_k of the model are evaluated with the following discrete sums:

$$\int_{s_k}^{1} \frac{\mu}{\pi} X ds' = \sum_{i=k+1}^{N} \delta_i X_i + \alpha_k X_k, \tag{A.13}$$

$$\frac{1}{\pi}\int_0^{s_k} \mu X ds' = \frac{1}{\pi_k} \sum_{i=1}^{k-1} \delta\pi_i X_i + \alpha_k X_k, \tag{A.14}$$

π_k being the pressure at level k, defined by the equation:

$$\pi_k = (\pi_{\tilde{k}} \pi_{\widetilde{k-1}})^{1/2}, \tag{A.15}$$

and the weighting coefficients δ_i and α_k being given by:

$$\delta_k = (\delta\pi_k)/\pi_k \text{ and } \alpha_k = 1 - (\pi_{\widetilde{k-1}}/\pi_{\widetilde{k}})^{1/2}. \tag{A.16}$$

This discretization allows the total energy to be conserved given the boundary conditions adopted at the top and base of the atmosphere under the assumption of suitable lateral boundary conditions (outflow equals inflow).

The nonlinear term for pressure force that appears in the horizontal momentum equation is discretized as follows:

$$\left(RT\frac{\nabla p}{p}\right)_k = RT_k\left(\frac{\nabla\pi}{\pi}\right)_k + RT_k\nabla(Z_p)_k,$$

with:

$$\left(\frac{\nabla\pi}{\pi}\right)_k = \frac{1}{\delta\pi_k}\left[\alpha_k g(s_{\widetilde{k}}) + (\delta_k - \alpha_k)g(s_{\widetilde{k-1}})\right]\nabla\pi_s.$$

This specific formulation is required to ensure the stability of the semi-implicit version of the model and it conserves the total angular momentum of the hydrostatic part of the flow for suitable lateral boundary conditions.

The occurrence of the term $\partial\mathbf{V}/\partial s$ in the third term of the left-hand side of equation (A.2) requires the evaluation of the wind on the interlayer surfaces using a weighted mean of its values taken at the levels of the model. The wind must therefore also be evaluated at the top of the atmosphere and at the surface: at the top, it is taken to be equal to the wind at the top level of the model and at the surface as equal to the wind at the lowest level of the model:

$$\mathbf{V}_{\widetilde{0}} = \mathbf{V}_1 \text{ and } \mathbf{V}_{\widetilde{N}} = \mathbf{V}_N.$$

A.2.4 The choice of the linear part of the model

As explained in Chapters 5, 6, and 8, implicit processing must be applied to a linear part of the model. In the case of the Euler equations, this implicit processing must be applied not just to the terms responsible for the propagation of gravity waves but also to those responsible for the propagation of sound waves. Therefore, the terms of pressure force and of horizontal and vertical divergence have to be linearized.

The chosen basic state is isothermal, dry, at rest, and so consistent with hydrostatic balance; the quantities defining the basic state are marked by an asterisk.

$$T^* = C_1, \qquad \pi_s^* = \pi_{s=1}^* = C_2,$$
$$\Phi_s^* = \Phi_{s=1}^* = 0, \quad \mathbf{V}^* = 0, \quad w^* = 0,$$

C_1 and C_2 being fixed reference values,

from which it is deduced that:

$$D^* = \nabla\cdot\mathbf{V}^* = 0 \text{ and } Z_p^* = 0.$$

By adopting a reference value for surface π_s^* hydrostatic pressure, we can calculate the corresponding values of the reference hydrostatic pressure, $\pi^*(s)$:

$$\pi^* = f(s)\pi_0 + g(s)\pi_s^*,$$

and of the vertical derivative:

$$\mu^* = \frac{\partial \pi^*}{\partial s}.$$

The linear part of the integrals (A.12) to (A.14) is obtained by calculating the discrete sums with new coefficients that are no longer dependent on x, y, and t:

$$\widehat{\delta \pi}_k = \hat{\pi}_{\tilde{k}} - \hat{\pi}_{\widetilde{k-1}}, \text{ with } \hat{\pi}_{\tilde{k}} = \pi^*(s_{\tilde{k}}),$$

$$\hat{\pi}_k = (\hat{\pi}_{\tilde{k}}\hat{\pi}_{\widetilde{k-1}})^{1/2}, \ \hat{\delta}_k = (\widehat{\delta \pi}_{\tilde{k}})/\hat{\pi}_k \text{ and } \hat{\alpha}_k = 1 - (\hat{\pi}_{\widetilde{k-1}}/\hat{\pi}_{\tilde{k}})^{1/2}.$$

The linearization of equations (A.1) to (A.5) provides a set of equations for the departures of the various prognostic variables from the basic state. After calculating the evolution equation for divergence $\delta_w^\bullet$, eliminating T, π_s, and Z, we obtain:

$$\left[-\frac{1}{c^2}\frac{\partial^4}{\partial t^4} + \frac{\partial^2}{\partial t^2}\left(\nabla^2 + \frac{\mathbf{L}^*}{H^2}\right) + N^2\mathbf{T}\nabla^2 \right]\delta_w^\bullet = 0, \tag{A.17}$$

with:

$$c^2 = R_d T^* \frac{C_{pd}}{C_{vd}}, \ H = \frac{R_d T^*}{g}, \text{ and } N^2 = \frac{g^2}{C_{pd}T^*}.$$

c is the velocity of the sound waves, H a characteristic height of the atmosphere, and N the Brunt-Väisälä frequency; R_d, C_{pd}, and C_{vd} represent the gas constant and the specific heats at constant pressure and volume relative to dry air. $\mathbf{L}^*$ is the operator of the discretized second derivative along the vertical and $\mathbf{T} = \mathbf{I} + \mathbf{L}^*\mathbf{Q}$ a tridiagonal matrix ($\mathbf{I}$ is the unit matrix and $\mathbf{Q}$ a diagonal matrix whose elements are $Q_{i,i} = \hat{\delta}_i - 2\hat{\alpha}_i$).

This equation, termed the structure equation, describes in particular the propagation of the various waves (external and internal gravity waves and sound waves) that are solutions to the system obtained after linearization of the discretized model along the vertical.

The structure equation (A.17) is to be compared with the equation obtained with continuous operators (Bubnova et al., 1995):

$$\left[-\frac{1}{c^2}\frac{\partial^4}{\partial t^4} + \frac{\partial^2}{\partial t^2}\left(\nabla^2 + \frac{\mathscr{L}^*}{H^2}\right) + N^2\nabla^2 \right]\delta_w^\bullet = 0, \tag{A.18}$$

with:

$$\mathscr{L}^* = \frac{\pi^*}{\mu^*} \frac{\partial}{\partial s} \left[\frac{\pi^*}{\mu^*} \frac{\partial}{\partial s} + 1 \right],$$

a linear operator bringing out the second derivative along the vertical.

As is clearly detailed in the foregoing reference, the similarity between equation (A.17) corresponding to the discretized linear model and equation (A.18) corresponding to the continuous case is not fortuitous. The choice made for discretization of the integrals (A.12) to (A.14) and the location of the model levels (A.15) was dictated by concern to achieve this result.

A.2.5 Horizontal discretization

The AROME model, being devised for short-term forecasting, uses a rectangular domain on a conformal projection. This rectangular domain is then extended by the addition of an extension zone designed to prolong the field defined on the initial domain so as to give doubly periodic fields on the extended domain (see Section 4.4). In this way, the spectral method can be used to calculate horizontal derivatives. The doubly periodic fields are therefore represented with the help of a truncated series expansion in terms of trigonometric functions with an elliptical truncation. The calculations concerning the linear terms of the model are made in the spectral space, whereas the nonlinear terms must be calculated on an intermediate grid the characteristics of which are chosen depending on the expansion truncation so as to avoid aliasing problems.

A.2.6 Semi-Lagrangian semi-implicit time integration

The time integration algorithm is of the semi-Lagrangian semi-implicit type and consists, as seen in Chapters 6 and 8, in Lagrangian processing of the total derivatives and implicit processing of the linearized terms responsible for fast wave propagation. In the case of the Euler equations, this implicit processing must therefore also be applied to the terms responsible for sound wave propagation. After defining a basic state permitting linearization of these terms, the evolution of a state vector of the model can be written schematically:

$$\frac{d\mathbf{X}}{dt} = \mathcal{N}(\mathbf{X}) + \mathcal{L}(\mathbf{X}) + \mathcal{F}(\mathbf{X}),$$

where $\mathcal{L}$ represents the linear operator, $\mathcal{N}$ the residual nonlinear operator (excluding advection terms), and $\mathcal{F}$ the physical forcings.

This equation is discretized using a two-time-level scheme (Temperton and Staniforth, 1987) and takes the form:

$$\frac{[\mathbf{X}(t+\Delta t)]_G - [\mathbf{X}(t)]_O}{\Delta t} = \left\{ \mathcal{N} \left[\mathbf{X}(t+\Delta t/2) \right] \right\}_I + \mathcal{L} \left(\frac{[\mathbf{X}(t+\Delta t)]_G + [\mathbf{X}(t)]_O}{2} \right) + \left\{ \mathcal{F} [\mathbf{X}(t)] \right\}_O. \quad \text{(A.19)}$$

The subscripts G and O refer respectively to the grid point and origin point of the particle at the time t, whereas subscript I refers to an intermediate point between points O and G. The origin point O is obtained from the iterative algorithm recommended by Robert (1983) and described in Subsection 6.2.5 above. The values at the intermediate point I are calculated by using the extrapolation method proposed by Hortal (2002), which is formulated:

$$\left\{ \mathcal{N}\left[\mathbf{X}(t + \Delta t/2) \right] \right\}_I = \frac{1}{2} \left(\left\{ 2\mathcal{N}\left[\mathbf{X}(t) \right] - \mathcal{N}\left[\mathbf{X}(t - \Delta t) \right] \right\}_O + \left\{ \mathcal{N}\left[\mathbf{X}(t) \right] \right\}_G \right). \qquad \text{(A.20)}$$

The quantities defined at the origin point O in equation (A.19) are evaluated using three-dimensional cubic interpolation, whereas those involved in calculating the residual nonlinear term at the intermediate point I with equation (A.20) are evaluated using three-dimensional linear interpolation.

Linearization done traditionally to apply the semi-implicit method uses an isothermal reference temperature T^* that is higher than the temperatures T encountered in the atmosphere (Simmons et al., 1978). However, as shown by Bénard (2004), this practice applied to the Euler equations with a two-time-level semi-Lagrangian semi-implicit scheme makes it highly unstable. To obtain a stable scheme, it is essential to choose an isothermal reference temperature $T_e^* \ll T^*$ for linearizing the second term of the left-hand side of equation (A.2). Tests have shown that good stability of the model can be ensured with $T_e^* = 100$ K and $T^* = 350$ K.

The expression $d\mathcal{X}/dt$ of in equation (A.2) is normally included in operator $\mathbf{N}$; however, it may be calculated as follows:

$$\left(\frac{d\mathcal{X}}{dt} \right) = \frac{\mathcal{X}_G(t) - \mathcal{X}_O(t - \Delta t)}{\Delta t}.$$

It should be noted that this expression gives only an evaluation of 1st-order accuracy in time. An evaluation with 2nd-order accuracy would involve using an iterative implicit scheme of the ICI (Iterative Centred-Implicit) type (Bénard, 2003).

The Lagrangian formulation also offers an alternative way of calculating the surface boundary condition given by equations (A.10) and (A.11). Thus the surface vertical acceleration $\dot{w}_s$ is given by:

$$\left(\frac{g}{\mu} \frac{\partial(p - \pi)}{\partial s} \right)_{\tilde{N}} = \frac{\left[w_{\tilde{N}}(t + \Delta t) \right]_G - \left[w_{\tilde{N}}(t) \right]_O}{\Delta t},$$

the value of $[w(t + \Delta t)]_G$ being calculated from values obtained using an explicit time step.

A.2.7 Processing horizontal diffusion

With a spectral model, horizontal diffusion, intended to parameterize the effects of turbulent fluxes that are unresolved at grid scale, is processed by selectively damping the spectral coefficients corresponding to the largest wavenumbers. This process is applied to temperature, to the vorticity, and to horizontal wind divergence with a diffusion coefficient corresponding to an e-folding time of two hours for the wave $\lambda = 4\Delta x$, and then increasing with altitude in inverse proportion to mean pressure. For the hydrometeor evolution equations, which are not processed by the spectral method, horizontal advection is computed by the SLHD (Semi-Lagrangian-Horizontal-Diffusion) scheme proposed by Vána et al. (2008). This consists in reinforcing the damping factor of the interpolator used in Lagrangian advection processing. Conversely, no diffusion is introduced into the equation for the evolution of turbulent kinetic energy.

A.2.8 Lateral boundary conditions and coupling

As the AROME model works on a limited area, it must be nested within a larger-scale model operating over a more extensive domain that supplies it with lateral boundary conditions. A transition zone of eight rows of grid points is defined from the boundary towards the interior of the working domain for applying the relaxation method proposed by Davies (1976), described in Subsection 10.5.2, and adapted to the case of a spectral model on a limited area by Radnóti (1995).

A.2.9 Physical parameterizations

A.2.9.1 Introduction

The solutions adopted for handling the various physical processes in the AROME model draw on schemes used in the mesoscale Meso-NH model (Lafore et al., 1998) developed jointly by Météo-France and the Toulouse University's Laboratoire d'Aérologie for research purposes and validated through many measurement campaigns and field experiments.

A.2.9.2 Radiation

To process the solar part of the spectrum, the model uses the radiation scheme of Fouquart and Bonnel (1980) covering six spectral intervals. The optical properties of clouds are derived from work by Morcrette and Fouquart (1986) for liquid water droplet clouds and by Ebert and Curry (1992) for ice clouds. For the infrared part of the spectrum, it uses the RRTM (Rapid Radiative Transfer Model) code proposed by Mlawer et al. (1997). Ozone and four types of aerosols are taken into account based on their climatological distribution. The radiative exchanges, including allowance for clouds calculated in the microphysics scheme and the convection scheme, are not calculated in full for each time step; however, solar flux is changed at each time step to allow for the daytime variation in the Sun's zenith angle.

A.2.9.3 The boundary layer and turbulent diffusion

Allowance for turbulent effects in the planetary boundary layer is based on the use of an equation describing the evolution of a prognostic variable, turbulent kinetic energy, combined with a diagnostic of mixing length. Turbulent kinetic energy is given by the scheme of Cuxart et al. (2000) and the mixing length is obtained from work by Bougeault and Lacarrère (1989). Mixing length is then used to determine the values of exchange coefficients and to calculate turbulent fluxes of momentum, potential temperature, and humidity.

A.2.9.4 Interactions with the Earth's surface

Interactions between the atmosphere and the Earth's surface are taken into account by the SURFEX (SURface EXternalisée) model. It requires a specific type of parameterization depending on the nature of the surface (continental land surface, city, sea, lake):

- over continental land surfaces the ISBA parameterization scheme (Noilhan and Planton, 1989) takes into account the fraction of rainfall intercepted by plant cover and plant transpiration to calculate the temperature and moisture of a surface layer and those same quantities for a deeper layer;
- over urban areas the TEB (Town Energy Balance) model proposed by Masson (2000) simulates the physical processes relating to exchanges between built areas and the atmosphere;
- over ocean surfaces, fluxes are calculated by using the iterative ECUME (Exchange Coefficients from Unified Multi-campaign Estimates) algorithm developed by Belamari and Pirani (2007);
- over lakes, fluxes are obtained by allowing for a roughness length calculated from the formula of Charnock (1955).

Between the surface and the lowest level of the dynamic model a special algorithm known as CANOPY (Masson and Seity, 2009), including a turbulence scheme working on six extra levels, is used for calculating the evolution of horizontal wind, temperature, and turbulent kinetic energy. The following variables are extracted at each time step: temperature and humidity at 2 m and wind at 10 m.

A.2.9.5 Microphysics

The parameterization of cloud microphysics, developed by Pinty and Jabouille (1998), describes in detail interactions among the various water phases in the atmosphere. It introduces, in addition to water vapour, five categories of hydrometeors (liquid water and ice in clouds together with rain, snow, and graupel) characterized by their mixing ratio with dry air. The microphysical processes depending on particle size distribution are parameterized using empirical spectra that provide particle distribution depending on their characteristic diameter. Sedimentation of the various precipitating components is calculated on the basis of a statistical type approach proposed by Bouteloup et al. (2011). The liquid water and ice in suspension provide cloud cover at the grid scale that is used by the radiation scheme in the next time step.

A.2.9.6 Convection

Although the AROME model is designed to work with a horizontal resolution of less than 5 km, allowing it explicitly to handle deep convection, the effects of shallow convection must be taken into account even so. To this effect, we use the EDKF (Eddy-Diffusivity-Kain-Fritsch) scheme developed by Pergaud et al. (2009) from the EDMF (Eddy-Diffusivity/Mass-Flux) scheme proposed by Soares et al. (2004) to parameterize the effects of the development of dry thermals and small cumulus clouds. It is based on the calculation of a single idealized updraft, supposed to represent the mean of all updrafts within a grid, allowing for the processes of entrainment and detrainment of the conservative variables in the upper layers and within the clouds. This scheme also provides a convective cloud cover used by the radiation scheme during the next time step.

A.2.10 Data assimilation and initialization

The initial fields are determined by 3D-Var type intermittent data assimilation (Fischer et al., 2006), in its incremental version (Courtier et al., 1994), the principle of which is briefly set out in Subsections 10.3.5 and 10.3.6 above. This method allows a background field to be corrected by taking into account a wide variety of observed data and background error statistics governing the importance of the changes brought about by these observations to the background field.

In addition to the traditional in-situ observations and measurements provided by the various devices carried on meteorological satellites, the system also assimilates very high spatial resolution data: the delay in propagating the GPS signal received by the European ground station network and the radial velocities deduced from the Doppler signal provided by the weather radar network covering France.

The background error statistics were calculated by the multivariate formulation already implemented for data assimilation with the ALADIN model (Berre, 2000). This relies on a background error statistical model for the vorticity, horizontal wind divergence, temperature, surface pressure, and humidity by scale-dependent regression for calculating cross-covariances. The method of computation is based on ensemble forecasting (Berre et al., 2006): to this effect, 3-hour forecasts of the AROME model from an ensemble of six members for two 14-day periods corresponding to different atmospheric circulation regimes are used. This provides error statistics for the variables analysed that are consistent with the model's spatial resolution (Brousseau et al., 2008).

The intermittent assimilation procedure consists in concatenating assimilation-forecasting cycles of period T. The observations included in a time window from $H - T/2$ to $H + T/2$ around the analysis time H make it possible to initialize the model that makes a forecast for time T, which is used as a background field in analysing the next cycle. As an option, digital filtering (Lynch and Huang, 1992) before starting the forecast can eliminate the propagation of interfering gravity waves at the start-up of the model when the range of the forecast used as background field is very short (typically $T < 3$ hours). At the end of this procedure we obtain the initial values for the two horizontal wind components, temperature, humidity, and surface pressure.

A.2.11 Operational use

The AROME model was brought into operational service in December 2008 for short-range forecasting requirements for France. It works with an E299–255 elliptical truncation and an associated 2.5 km mesh grid. The working domain covers an area of 1250 km × 1 500 km centred on France. This domain is extended to the north and east by adding 12 rows of grid points so fields can be made doubly periodic and the spectral method applied. Vertically the atmosphere is represented by 41 levels, the lowest being at 17 m and the highest at 0.1 hPa. A 60-second time step allows small time scale processes of microphysics to be described and ensures the model's stability. Lateral boundary conditions are interpolated from fields forecast every hour by the ALADIN-FRANCE primitive equations model with a 10 km mesh operating over a domain covering Western Europe and so providing one-way coupling of the AROME model in synchronous mode. This model uses a 3-hour assimilation cycle that assimilates observations at 00.00, 03.00, 06.00, 09.00, 12.00, 15.00, 18.00, and 21.00 UTC and provides analyses for making 3-hour forecasts that act as background fields for the next analysis; in this way all of the data available in the course of a day can be assimilated. Four times a day, for the 00.00, 06.00, 12.00, and 18.00 UTC synoptic hours, about 1 hr 30 min after the synoptic hour, the model is run and provides forecast fields for up to 30 hours that can be exploited by forecasters. This model has been tested pre-operationally and its forecasts compared with those of the ALADIN-France primitive equations model operating with a 10 km mesh. The improvements achieved for predicting surface variables and the model's good behaviour in specific situations (Seity et al., 2011) justified it being put into operational use at Météo-France in 2008.

A.3 The WRF/ARW model

A.3.1 Introduction

The advanced forecasting system of the WRF (Weather Research and Forecasting) group known as ARW (Advanced Research WRF) is a modular tool that fits around a modern forecasting model with its own data assimilation system. It has been developed in the United States by combining the efforts of many scientists from government agencies and various universities. Its nonhydrostatic dynamical core is based on the Euler equations, but it can easily be simplified to accommodate the hydrostatic relation and to operate with primitive equations.

This community model benefits from constant technical improvements and is employed by a wide variety of users worldwide for research, operational forecasting, and teaching. It is documented in a very detailed way in *A Description of the Advanced Research WRF Version 3* prepared by Skamarock et al. (2008), referenced as such in the following paragraphs and available on the website of the system users' community (http://www.mmm.ucar.edu/wrf/users/docs/). Accordingly, the presentation that follows is confined essentially to describing the system of equations used, the time integration algorithm, and the principles of spatial discretization; it also provides a glimpse of the parameterizations that may be used to take into account the physical subgrid-scale processes and other system components: initialization, data assimilation, and nesting mode.

A.3.2 The equations used in the WRF/ARW model

The WRF/ARW model uses the Euler equations written in flux form to work either directly on the sphere (using the geographical latitude-longitude coordinates) or on a conformal projection of the sphere on a plane (using Cartesian coordinates on a polar stereographic, Mercator, or Lambert projection). We have opted here to describe the version applied to a conformal projection defined locally by its map scale factor m.

The vertical coordinate is of the mass type, defined with the hydrostatic pressure of dry air π_d and is written:

$$s = \left[\pi_d - (\pi_d)_t \right] / \mu_d, \text{ with: } \mu_d = (\pi_d)_s - (\pi_d)_t,$$

the subscripts s and t referring to the bottom (surface) and top of the atmosphere, respectively.

The quantity μ_d, which represents the mass of the dry air column between the base and top of the atmosphere, is given by the relation $\partial \pi_d / \partial s = \mu_d$.

To this vertical coordinate is associated the generalized vertical velocity:

$$\dot{s} = \frac{ds}{dt}.$$

The flux form of the equations of momentum and of the thermodynamic equation is obtained by combining the evolution equations for the three components of velocity u, v, w and the potential temperature θ with the continuity equation.

After writing:

$$U = \mu_d u/m, \ V = \mu_d v/m, \ W = \mu_d w/m, \ \dot{S} = \mu_d \dot{s}/m \text{ and } \Theta = \mu_d \theta,$$

we get the equations:

$$\frac{\partial U}{\partial t} + m \left[\frac{\partial (Uu)}{\partial x} + \frac{\partial (Vu)}{\partial y} \right] + \frac{\partial (\dot{S}u)}{\partial s} + \mu_d \alpha \frac{\partial p}{\partial x} + (\alpha / \alpha_d) \frac{\partial p}{\partial s} \frac{\partial \Phi}{\partial x} = F_U, \quad \text{(A.21)}$$

$$\frac{\partial V}{\partial t} + m \left[\frac{\partial (Uv)}{\partial x} + \frac{\partial (Vv)}{\partial y} \right] + \frac{\partial (\dot{S}v)}{\partial s} + \mu_d \alpha \frac{\partial p}{\partial y} + (\alpha / \alpha_d) \frac{\partial p}{\partial s} \frac{\partial \Phi}{\partial y} = F_V, \quad \text{(A.22)}$$

$$\frac{\partial W}{\partial t} + m\left[\frac{\partial(Uw)}{\partial x} + \frac{\partial(Vw)}{\partial y}\right] + \frac{\partial(\dot{S}w)}{\partial s} - \frac{g}{m}\left[(\alpha/\alpha_d)\frac{\partial p}{\partial s} - \mu_d\right] = F_W, \qquad (A.23)$$

$$\frac{\partial \Theta}{\partial t} + m^2\left[\frac{\partial(U\theta)}{\partial x} + \frac{\partial(V\theta)}{\partial y}\right] + m\frac{\partial(\dot{S}\theta)}{\partial s} = F_\Theta, \qquad (A.24)$$

to which we add an evolution equation for $Q_m = \mu_d q_m$ where q_m denotes a scalar (e.g. any of hydrometeor mixing ratios or turbulent kinetic energy):

$$\frac{\partial Q_m}{\partial t} + m^2\left[\frac{\partial(UQ_m)}{\partial x} + \frac{\partial(VQ_m)}{\partial y}\right] + m\frac{\partial(\dot{S}Q_m)}{\partial s} = F_{Q_m}. \qquad (A.25)$$

In these equations, p is the true pressure for moist air, α and α_d represent the densities of moist air and dry air respectively, and Φ is the geopotential.

True pressure p is given by the equation of state for moist air:

$$p = p_0(R_d\theta_m / p_0\alpha_d)^\gamma, \qquad (A.26)$$

with $\theta_m = \theta[1 + (R_v/R_d)q_v]$ a moist potential temperature defined for convenience, and $\gamma = C_p/C_v$ the ratio of the specific heats at constant pressure and volume; R_d and R_v represent perfect gas constants relative to dry air and to water vapour.

The densities α_d and α satisfy the relation:

$$\alpha = \alpha_d(1 + q_v + + q_l + q_i + \ldots),$$

where q_v, q_l, q_i, ... (noted generically q_m) denote the mixing ratio of hydrometeors (water vapour, liquid water, ice, etc.).

The hydrostatic equation is written:

$$\frac{\partial \Phi}{\partial s} = -\alpha_d\mu_d. \qquad (A.27)$$

The definition of vertical velocity w allows us to write the equation for this geopotential:

$$\frac{\partial \Phi}{\partial t} + \frac{1}{\mu_d}\left[m^2\left(U\frac{\partial \Phi}{\partial x} + V\frac{\partial \Phi}{\partial y}\right) + m\dot{S}\frac{\partial \Phi}{\partial s} - mgW\right] = 0. \qquad (A.28)$$

Lastly, the continuity equation, reflecting the conservation of the dry air mass, is written:

$$\frac{\partial \mu_d}{\partial t} + m^2\left(\frac{\partial U}{\partial x} + \frac{\partial V}{\partial y}\right) + m\frac{\partial \dot{S}}{\partial s} = 0. \qquad (A.29)$$

In equations (A.21) to (A.23) the quantities F_U, F_V, and F_w contain the components of the Coriolis force, the curvature terms, and the physical forcings F_{U_ϕ}, F_{V_ϕ}, and F_{w_ϕ} for the three components of momentum, and are written:

$$F_U = + \left(f + u\frac{\partial m}{\partial y} - v\frac{\partial m}{\partial x} \right) V - eW \cos \alpha_r - \frac{uW}{a} + F_{U_\varphi},$$

$$F_V = - \left(f + u\frac{\partial m}{\partial y} - v\frac{\partial m}{\partial x} \right) U + eW \sin \alpha_r - \frac{vW}{a} + F_{V_\varphi}, \qquad (A.30)$$

$$F_w = + e\, (U \cos \alpha_r - V \sin \alpha_r) + \left(\frac{uU + vV}{a} \right) + F_{W_\varphi},$$

α_r being the angle of rotation between the ordinate axis and the meridian, a the radius of the Earth, $f = 2\Omega\sin\varphi$, and $e = 2\Omega\cos\varphi$ (Ω standing for the angular velocity of rotation of the Earth and φ for latitude).

In equations (A.24) and (A.25), F_Θ and F_{Q_m} represent physical forcings for potential temperature and for the other scalar variables q_m processed by the model.

A.3.3 The equations for the perturbations

In order to reduce truncation errors in calculating the pressure gradient term, it is helpful to work with new variables p', Φ', α', and μ'_d (called *perturbation* variables), which are departures from values (functions of z only) that define a reference state in hydrostatic balance:

$$p' = p - \overline{p}(z), \ \Phi' = \Phi - \overline{\Phi}(z), \ \alpha' = \alpha - \overline{\alpha}(z) \text{ and } \mu'_d = \mu_d - \overline{\mu}_d(x,y)$$

As the surfaces $s = $ constant are not horizontal, the reference state values $\overline{p}$, $\overline{\Phi}$, and $\overline{\alpha}$ are therefore functions of x, y, and s whereas $\overline{\mu}_d$ is a function of x and y only.

By using these new variables, the momentum equations are written:

$$\frac{\partial U}{\partial t} + m\left[\frac{\partial(Uu)}{\partial x} + \frac{\partial(Vu)}{\partial y} \right] + \frac{\partial(\dot{S}u)}{\partial s} + \left(\mu_d \alpha \frac{\partial p'}{\partial x} + \mu_d \alpha' \frac{\partial \overline{p}}{\partial x} \right)$$
$$+ (\alpha/\alpha_d)\left(\mu_d \frac{\partial \Phi'}{\partial x} + \frac{\partial p'}{\partial s}\frac{\partial \Phi}{\partial x} - \mu'_d \frac{\partial \Phi}{\partial x} \right) = F_U, \qquad (A.31)$$

$$\frac{\partial V}{\partial t} + m\left[\frac{\partial(Uv)}{\partial x} + \frac{\partial(Vv)}{\partial y} \right] + \frac{\partial(\dot{S}v)}{\partial s} + \left(\mu_d \alpha \frac{\partial p'}{\partial y} + \mu_d \alpha' \frac{\partial \overline{p}}{\partial y} \right)$$
$$+ (\alpha/\alpha_d)\left(\mu_d \frac{\partial \Phi'}{\partial y} + \frac{\partial p'}{\partial s}\frac{\partial \Phi}{\partial y} - \mu'_d \frac{\partial \Phi}{\partial y} \right) = F_V, \qquad (A.32)$$

$$\frac{\partial W}{\partial t} + m\left[\frac{\partial(Uw)}{\partial x} + \frac{\partial(Vw)}{\partial y} \right] + \frac{\partial(\dot{S}w)}{\partial s}$$
$$- \frac{g}{m}(\alpha/\alpha_d)\left[\frac{\partial p'}{\partial s} - \overline{\mu}_d(q_v + q_c + q_r) \right] + \frac{g}{m}\mu'_d = F_W. \qquad (A.33)$$

The continuity equation and the geopotential evolution equation become:

$$\frac{\partial \mu'_d}{\partial t} + m^2 \left(\frac{\partial U}{\partial x} + \frac{\partial V}{\partial y} \right) + m \frac{\partial \dot{S}}{\partial s} = 0, \tag{A.34}$$

$$\frac{\partial \Phi'}{\partial t} + \frac{1}{\mu_d} \left[m^2 \left(U \frac{\partial \Phi}{\partial x} + V \frac{\partial \Phi}{\partial y} \right) + m \dot{S} \frac{\partial \Phi}{\partial s} - mgW \right] = 0. \tag{A.35}$$

The conservation equations (A.24) and (A.25) for potential temperature and for the other scalar variables remain unchanged.

Lastly the hydrostatic equation for perturbations is written:

$$\frac{\partial \Phi'}{\partial s} = -\bar{\mu}_d \alpha'_d - \mu'_d \alpha_d. \tag{A.36}$$

We thus have a system of equations that must be discretized in time and space for the model to be numerically integrated.

A.3.4 Time integration

The WRF/ARW model uses a split-explicit time integration method; this consists of splitting the system of equations into several separate parts which are then integrated with different time steps, adapted to each, so as to ensure the overall stability of the process.

A.3.4.1 The third-order Runge-Kutta scheme

The third-order Runge-Kutta scheme (RK3) allows us to integrate an ordinary differential equation of the type:

$$\frac{dX}{dt} = R(X).$$

The time integration from time t to time $t + \Delta t$ is made in three successive steps yielding intermediate values X^*, X^{**}, and then $X^{t+\Delta t}$:

$$\left. \begin{aligned} X^* &= X^t + \frac{\Delta t}{3} R(X^t), \\ X^{**} &= X^t + \frac{\Delta t}{2} R(X^*), \\ X^{t+\Delta t} &= X^t + \Delta t\, R(X^{**}). \end{aligned} \right\} \tag{A.37}$$

It ensures 3rd-order accuracy for a linear equation and 2nd-order accuracy only for a nonlinear equation. It is applied to partial differential equations (A.24), (A.25), and (A.31) to (A.35).

A.3.4.2 The equations used for small time step integration

The horizontal propagation of fast waves (sound waves, gravity waves) is liable to seriously limit the time step Δt used with the RK3 scheme. Accordingly, at each step of the RK3 method, Wicker and Skamarock (2002) proposed integrating equations bearing on the departures from the values obtained upon completion of the various steps by using a smaller time step $\Delta \tau$ (called the *acoustic* time step). The acoustic integration method specifically for the ARW equations is introduced in detail by Klemp et al. (2007).

These new variables are then written:

$$U'' = U - U^{t^*}, \quad V'' = V - V^{t^*}, \quad W'' = W - W^{t^*},$$

$$S'' = S - S^{t^*}, \quad \Phi'' = \Phi - \Phi^{t^*}, \Theta'' = \Theta - \Theta^{t^*},$$

$$\mu_d'' = \mu_d' - \mu_d'^{t^*}, \alpha_d'' = \alpha_d' - \alpha_d'^{t^*}.$$

The hydrostatic equation becomes:

$$\alpha_d'' = -\frac{1}{\mu_d^{t^*}}\left(\frac{\partial \Phi''}{\partial s} + \alpha_d^{t^*}\mu_d''\right), \tag{A.38}$$

and the equation of state linearized around the values calculated at t^*:

$$p'' = -\frac{c_s^2}{\alpha_d^{t^*}}\left(\frac{\Theta''}{\Theta^{t^*}} - \frac{\alpha_d''}{\alpha_d^{t^*}} - \frac{\mu_d''}{\mu_d^{t^*}}\right), \tag{A.39}$$

with: $c_s^2 = \gamma p^{t^*}\alpha_d^{t^*}$, the square of the speed of sound.

By combining (A.38) and (A.39), the vertical pressure gradient becomes:

$$\frac{\partial p''}{\partial s} = \frac{\partial}{\partial s}\left(C\frac{\partial \Phi''}{\partial s}\right) + \frac{\partial}{\partial s}\left(\frac{c_s^2}{\alpha_d^{t^*}}\frac{\Theta''}{\Theta^{t^*}}\right), \text{ with } C = -\frac{c_s^2}{\mu^{t^*}\alpha_d^{t^{*2}}}. \tag{A.40}$$

After substituting these new variables and equation (A.40) into the prognostic equations (A.24), (A.25), and (A.31) to (A.35), and having replaced the time derivatives by forward explicit differences with a time step $\Delta \tau$, notated:

$$\delta_\tau X = \frac{X^{\tau+\Delta\tau} - X^\tau}{\Delta\tau}, \tag{A.41}$$

we obtain the equations for time integration with the acoustic time step $\Delta \tau$.

$$\delta_\tau U'' + \mu^{t^*}\alpha^{t^*}\frac{\partial p''^\tau}{\partial x} + \left(\mu^{t^*}\frac{\partial \overline{p}}{\partial x}\right)\alpha''^\tau + (\alpha/\alpha_d)\left[\mu^{t^*}\frac{\partial \Phi''^\tau}{\partial x} + \frac{\partial \Phi^{t^*}}{\partial x}\left(\frac{\partial p''}{\partial s} - \mu''\right)^\tau\right] = R_U^{t^*}, \tag{A.42}$$

$$\delta_\tau V'' + \mu^{t^*}\alpha^{t^*}\frac{\partial p''^\tau}{\partial y} + \left(\mu^{t^*}\frac{\partial \overline{p}}{\partial y}\right)\alpha''^\tau + (\alpha/\alpha_d)\left[\mu^{t^*}\frac{\partial \Phi''^\tau}{\partial y} + \frac{\partial \Phi^{t^*}}{\partial y}\left(\frac{\partial p''}{\partial s} - \mu''\right)^\tau\right] = R_V^{t^*}, \tag{A.43}$$

$$\delta_\tau \mu''_d + m^2 \left(\frac{\partial U''}{\partial x} + \frac{\partial V''}{\partial y} \right)^{\tau+\Delta\tau} + m \left(\frac{\partial \dot{S}''^{\tau+\Delta\tau}}{\partial s} \right) = R_\mu^{t^*}, \tag{A.44}$$

$$\delta_\tau \Theta'' + m^2 \left[\frac{\partial(U''\theta^{t^*})}{\partial x} + \frac{\partial(V''\theta^{t^*})}{\partial y} \right]^{\tau+\Delta\tau} + m \left[\frac{\partial(\dot{S}''^{\tau+\Delta\tau}\theta^{t^*})}{\partial s} \right]^{\tau+\Delta\tau} = R_\Theta^{t^*}, \tag{A.45}$$

$$\delta_\tau W'' - \frac{g}{m} \overline{\left[(\alpha/\alpha_d)^{t^*} \frac{\partial}{\partial s} \left(C \frac{\partial\Phi''}{\partial s} \right) + \frac{\partial}{\partial s} \left(\frac{c_s^2}{\alpha^{t^*}} \frac{\Theta''}{\Theta^{t^*}} \right) - \mu''_d \right]}^{\tau} = R_w^{t^*}, \tag{A.46}$$

$$\delta_\tau \Phi'' + \frac{1}{\mu_d^{t^*}} \left[m\dot{S}''^{\tau+\Delta\tau} \frac{\partial\Phi^{t^*}}{\partial s} - \overline{m(gW'')}^{\tau} \right] = R_\Phi^{t^*}. \tag{A.47}$$

Notice that equations (A.46) and (A.47) involve the time-averaging operator:

$$\bar{X}^\tau = \frac{1+\beta}{2} X^{\tau+\Delta\tau} + \frac{1-\beta}{2} X^\tau, \tag{A.48}$$

thereby leading to an implicit system in the vertical. This formulation, by choosing a value for the off-centering parameter β close to 0.1, avoids instability related to the vertical propagation of sound waves.

The right-hand sides of these equations remain fixed for advances with the acoustic time step $\Delta\tau$ for each step of the RK3 method:

$$R_U^{t^*} = -m \left[\frac{\partial(Uu)}{\partial x} + \frac{\partial(Vu)}{\partial y} \right] - \frac{\partial(\dot{S}u)}{\partial s} - \left(\mu_d \alpha \frac{\partial p'}{\partial x} + \mu_d \alpha' \frac{\partial\bar{p}}{\partial x} \right)$$
$$- (\alpha/\alpha_d) \left(\mu_d \frac{\partial\Phi'}{\partial x} - \frac{\partial p'}{\partial s} \frac{\partial\Phi}{\partial x} + \mu_d \frac{\partial\Phi}{\partial x} \right) + F_U, \tag{A.49}$$

$$R_V^{t^*} = -m \left[\frac{\partial(Uv)}{\partial x} + \frac{\partial(Vv)}{\partial y} \right] - \frac{\partial(\dot{S}v)}{\partial s} - \left(\mu_d \alpha \frac{\partial p'}{\partial y} + \mu_d \alpha' \frac{\partial\bar{p}}{\partial y} \right)$$
$$- (\alpha/\alpha_d) \left(\mu_d \frac{\partial\Phi'}{\partial y} - \frac{\partial p'}{\partial s} \frac{\partial\Phi}{\partial y} + \mu_d \frac{\partial\Phi}{\partial y} \right) + F_V, \tag{A.50}$$

$$R_\mu^{t^*} = -m^2 \left(\frac{\partial U}{\partial x} + \frac{\partial V}{\partial y} \right) - m \frac{\partial\dot{S}}{\partial s}, \tag{A.51}$$

$$R_\Theta^{t^*} = -m^2 \left[\frac{\partial(U\theta)}{\partial x} + \frac{\partial(V\theta)}{\partial y} \right] - m \frac{\partial(\dot{S}\theta)}{\partial s} + F_\Theta, \tag{A.52}$$

$$R_w^{t*} = -m^2 \left[\frac{\partial(Uw)}{\partial x} + \frac{\partial(Vw)}{\partial y} \right] - \frac{\partial(\dot{S}w)}{\partial s}$$

$$+ \frac{g}{m}(\alpha/\alpha_d)\left[\frac{\partial p'}{\partial s} - \bar{\mu}_d(q_v + q_c + q_r) \right] - \frac{g}{m}\mu_d' + F_W, \tag{A.53}$$

$$R_\Phi^{t*} = -\frac{1}{\mu_d}\left[m^2\left(U\frac{\partial \Phi}{\partial x} + V\frac{\partial \Phi}{\partial y} \right) - m\dot{S}\frac{\partial \Phi}{\partial s} - mgW \right], \tag{A.54}$$

all of the variables figuring in expressions (A.49) to (A.54) being evaluated at times t, t^* or t^{**} depending on the relevant step of the RK3 scheme defined in (A.37).

A.3.4.3 The successive steps of time integration

Integration with the various acoustic time steps is achieved as follows: beginning with the variables defined for acoustic time steps at time τ, equations (A.42) and (A.43) allow us to obtain $U''^{\,\tau+\Delta\tau}$ and $V''^{\,\tau+\Delta\tau}$. Variables $\mu''^{\,\tau+\Delta\tau}$ and $S''^{\,\tau+\Delta\tau}$ are both calculated from equation (A.44); we begin by integrating equation (A.44) vertically from the surface to the top of the atmosphere, allowing us to eliminate the term $\partial S''/\partial s$:

$$\delta_\tau \mu_d'' = m^2 \int_1^0 \left(\frac{\partial U''}{\partial x} + \frac{\partial V''}{\partial y} \right)^{\tau+\Delta\tau} ds. \tag{A.55}$$

After using equation (A.55) to compute $\mu''^{\,\tau+\Delta\tau}$ we get $S''^{\,\tau+\Delta\tau}$ from equation (A.44) vertically integrated from the surface (where $\dot{S}'' = 0$) up to the desired level. Equation (A.45) allows us to calculate $\Theta''^{\,\tau+\Delta\tau}$. Equations (A.46) and (A.47) combine to form an implicit equation along the vertical for the variable $W''^{\,\tau+\Delta\tau}$, which is solved by taking as boundary conditions $\dot{S} = \dot{S}'' = 0$ at the surface ($z = h(x,y)$, height of topography) and $p' = 0$ at the top of the atmosphere. We then get $\Phi''^{\,\tau+\Delta\tau}$ from (A.47) and finally $\alpha''^{\,\tau+\Delta\tau}$ and $p''^{\,\tau+\Delta\tau}$ from (A.38) and (A.39).

The values of the time step Δt and of the time step $\Delta\tau$ are set to ensure the model is stable for slow waves and for fast waves, respectively. In practice, we must give ourselves the even number n_s allowing us to set the number of time steps n to be made with time step $\Delta\tau$ during the successive steps of the RK3 scheme described in (A.37) in the following way:

$$\begin{cases} \text{step 1}: n = 1, & \Delta\tau = \Delta t/3, \\ \text{step 2}: n = n_s/2, & \Delta\tau = \Delta t/n_s, \\ \text{step 3}: n = n_s, & \Delta\tau = \Delta t/n_s. \end{cases}$$

A.3.4.4 The transition to primitive equations

The transition to hydrostatic primitive equations is made by replacing the solution of the prognostic equations (A.46) and (A.47) giving $W''^{\,\tau+\Delta\tau}$ and then $\Phi''^{\,\tau+\Delta\tau}$

by a diagnostic of hydrostatic pressure and using the definition of the vertical coordinate:

$$\frac{\partial \pi}{\partial s} = \frac{\alpha_d}{\alpha}\mu_d = (1 + q_v + q_l + q_i + \dots)\mu_d.$$

The inverse density α_d is obtained using the equation of state (A.26), since the value of θ is known, and the geopotential is calculated with the hydrostatic equation (A.36).

A.3.5 Spatial discretization

A.3.5.1 The grids used and the arrangement of variables

The vertical arrangement of variables follows Lorenz's arrangement (see Subsection 8.3.8). The atmosphere is therefore subdivided into N layers, within which the N levels of the model are defined. The variables U, V, Θ, μ_d, p, and α are defined on the levels of the model, whereas the variables Φ, W, and $\dot{S}$ are defined on the interlayer boundaries, as shown in Figure A.1; notice therefore that the vertical velocity values W and $\dot{S}$ must be defined at the top and base of the working domain.

Horizontal discretization is achieved by using an Arakawa C grid (see Subsection 8.4.2) on which the horizontal components of the wind U and V are calculated at points which differ from the *mass points* used for calculating pressure p, potential temperature

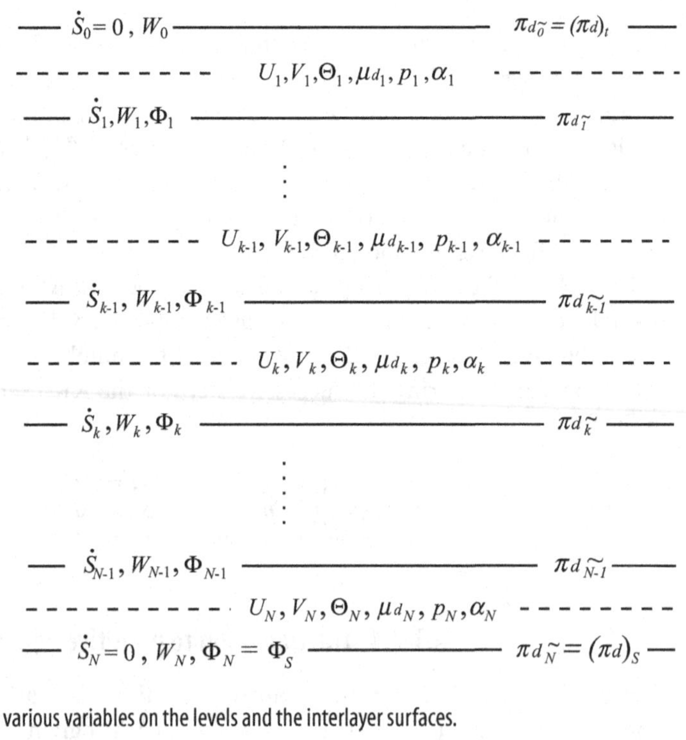

Figure A.1   Location of the various variables on the levels and the interlayer surfaces.

Θ, mass of the atmospheric column μ_d, and inverse density a_d. This requires a number of horizontal interpolations so as to obtain groups of variables assigned to a correct location for calculating the derivative using simple finite differences.

The discretization of equations (A.42) to (A.47) and of the second members (A.49) to (A.54) is documented very completely in the technical note by Skamarock et al. (2008). We therefore confine ourselves here to presenting the general principles behind it.

A.3.5.2 Discretization of non-advective terms

Given the use of a C grid, the variables U and V must be redefined by using the horizontal averaging operator (3.4) introduced in Subsection 3.2.2, in the x- and y-directions, respectively:

$$U = \overline{\mu_d^x u} / \overline{m}^x \text{ and } V = \overline{\mu_d^y v} / \overline{m}^y.$$

In the equations relative to the wind components (A.42), (A.43), and in the expression of the corresponding complementary terms (A.49) and (A.50), all of the terms where variables of type μ_d or α occur, evaluated at the mass points, are interpolated horizontally using the averaging operator (3.4) along x or along y to obtain the values at the points where U and V are calculated. Likewise, in equation (A.45), the averaging operator in the x-direction and in the y-direction is applied to the potential temperature so as to evaluate it at the same points as U and V, respectively, and then to calculate the spatial derivatives at the mass points.

The horizontal derivatives that occur in equations (A.42), (A.43), (A.44), (A.45), and in the complementary terms (A.49) and (A.50) can then be calculated easily with central differences by using the variables that are Δx away on the C grid. Variables of the Φ type, evaluated at the interlayer boundaries, are interpolated vertically to bring them to the model levels before calculating the horizontal derivatives. By contrast, equation (A.46) and the complementary term (A.53) reflect the evolution of vertical velocity calculated at the interfaces between the layers of the model; accordingly the terms involving values expressed on the model levels are vertically interpolated to bring them to the interlayer boundaries.

The vertical derivatives are calculated by using central differences from the values taken either between the base and top of the layer, for calculating a derivative evaluated on a level of the model, or between the lower and upper levels of adjacent layers for a derivative evaluated on a boundary between two layers. Where quantities are not given in the right place, vertical interpolation means they can be evaluated in the right place.

Using the numbering conventions for the levels and interlayer boundaries given in Chapter 8, vertical interpolation of a quantity X from the model levels s_k and s_{k+1} to the surface $s_{\tilde{k}}$ between the two corresponding adjacent layers is written:

$$\overline{X}^s(s_{\tilde{k}}) = \frac{X_{k+1}\Delta s_k + X_k \Delta s_{k+1}}{2(s_{k+1} - s_k)}, \text{ with: } \Delta s_k = s_{\tilde{k}} - s_{\widetilde{k-1}}.$$

The levels s_k are chosen so that vertical interpolation from the surfaces bounding a layer $s_{\tilde{k}}$ and $s_{\widetilde{k-1}}$, to the level s_k located within the layer, comes down to the arithmetic mean:

$$\overline{X}^s(s_k) = (X_{\tilde{k}} + X_{\widetilde{k-1}})/2.$$

A.3.5.3 Discretization of advection terms

Given the flux form adopted for writing equations, the advection terms correspond to the flux divergence terms that appear in the right-hand sides of (A.49) to (A.53) and in equations (A.24) and (A.25).

The advection terms appearing in the right-hand side of the continuity conservation equation (A.51) are calculated with the help of 2nd-order accuracy central differences and are written using notation (3.5) introduced in Subsection 3.2.2:

$$(R_\mu^{t^*})_k = -m^2 \left[(U)_x + (V)_y \right] - m \frac{\dot{S}_{\tilde{k}} - \dot{S}_{\widetilde{k-1}}}{\Delta s_k}. \tag{A.56}$$

The advection terms appearing in the right-hand sides of the prognostic equations for the components of momentum U, V, W, potential temperature Θ, geopotential Φ, and for the other scalar variables Q_m, are discretized using finite differences with an order of accuracy ranging from 2 to 6. For a scalar quantity Q_m, for example, we write:

$$\delta_\tau(Q_m) = -m^2 \left[(U\tilde{Q}_m^x)_x + (V\tilde{Q}_m^y)_y \right] - m \frac{\dot{S}_{\tilde{k}}\overline{Q}_m^s - \dot{S}_{\widetilde{k-1}}\overline{Q}_m^s}{\Delta s_k},, \tag{A.57}$$

where the notations $\tilde{Q}_m^x$ and $\tilde{Q}_m^y$ correspond to horizontal average in the direction of x and y, on level k, the expression of which depends on the order of accuracy of the desired discretization scheme. For 2nd-order accuracy, we use the arithmetic mean:

$$\tilde{Q}_m^x = \overline{Q}_m^x = \left[Q_m(x + \Delta x/2) + Q_m(x - \Delta x/2) \right]/2. \tag{A.58}$$

The expression $\tilde{Q}_m^x$ for the average may be replaced by more complex formulas involving more grid points to obtain accuracy of up to 6th order, as is set out in detail in the technical note by Skamarock et al. (2008).

Some variables like the mixing ratios of the various components of atmospheric water cannot have negative values. However, the discretization given by formulas (A.57) and (A.58) provides no safeguards at all against the occurrence of negative values. To avoid this drawback, a correction may be applied upon completion of the final step in the Runge-Kutta scheme to avoid this non-physical effect (Skamarock and Weisman, 2008; Skamarock, 2006).

A.3.5.4 The choice of time step

Here we examine what time step values can ensure stability of the time integration scheme adopted by taking, for example, a grid mesh $\Delta x = 10$ km. A study of the RK3 scheme on the one-dimensional wave equation (Wicker and Skamarock, 2002) has shown that it remains stable for a Courant number ranging from 1.08 (advection of 6th-order accuracy) to 1.73 (advection of 2nd-order accuracy); furthermore, for applications in three spatial dimensions this value must be divided by the factor $\sqrt{3}$. Allowing for a maximum flow velocity of the order of 100 ms^{-1}, we obtain in the case of a 2nd-order accuracy advection scheme, a time step $\Delta t = 100$ s. The scheme used for acoustic integration remains stable for a Courant number of less than $1/\sqrt{2}$. The characteristic velocity to be taken into account being that of sound in air, which is approximately 340 ms^{-1}, we get a time step $\Delta \tau = 20$ s, that is, $\Delta t/5$. To run the model with a $\Delta x = 10$ km mesh, we can adopt, to allow a good safety margin, $\Delta t = 60$ s and $\Delta \tau = 15$ s.

A.3.5.5 Processing horizontal and vertical diffusion

The WRF/ARW model can take into account the effects of turbulent exchanges that are not resolved at the grid scale by introducing horizontal and vertical diffusion terms expressed by exchange coefficients K_h and K_v. These diffusion terms may be calculated either in the coordinate system specific to the model (the horizontal derivations being taken at constant s), or in the physical space (the horizontal derivatives being taken at constant z).

When the diffusion terms are evaluated in the model's coordinate system, the tendency relative to the diffusion of a variable Q is formulated:

$$\left[\frac{\partial}{\partial t}\left(\mu_d Q\right)\right]_{\text{diff}} = \mu_d m\left[\frac{\partial}{\partial x}\left(mK_h\frac{\partial Q}{\partial x}\right) + \mu_d m\frac{\partial}{\partial y}\left(mK_h\frac{\partial Q}{\partial y}\right)\right] + \frac{g^2}{\mu_d \alpha}\frac{\partial}{\partial s}\left(\frac{1}{\alpha}K_v\frac{\partial Q}{\partial s}\right),$$

the precise discretization of which depends on the various variables concerned.

When the diffusion terms are evaluated in physical space, the computation is more complicated: to obtain the diffusion terms relative to the components of momentum, we must calculate the stress tensor components that are expressed locally as a function of the turbulent exchange coefficients and the deformation tensor components. In addition, curvature effects must be taken into account in calculating the horizontal derivatives.

The exchange coefficients K_h and K_v can be determined in four different ways:

• they may take values set by the user,
• K_h may be determined as a function of horizontal deformation by using the 2D scheme of Smagorinsky (1963); in this case, the effects of turbulent vertical exchanges must be taken into account by a separate parameterization scheme for the planetary boundary layer.

- K_h and K_v may take the same value calculated as a function of the deformation tensor components by using the 3D scheme of Smagorinsky (1963).
- K_h and K_v may be calculated as a function of turbulent kinetic energy: this quantity is then a variable of the model whose evolution equation involves, besides the advection terms, terms for shear, buoyancy, and dissipation.

A.3.6 Filters for the time integration scheme

Three additional filters are used to better control the amplitude of the fast waves that may propagate in the model and to obtain a stable scheme.

A.3.6.1 Three-dimensional divergence damping

Damping three-dimensional momentum divergence allows sound waves to be filtered (Skamarock and Klemp, 1992). Damping is implemented by using modified pressure to calculate the pressure gradient terms in the momentum equations (A.42) and (A.43) for advances with time step $\Delta\tau$. The modified pressure is written:

$$p^{*\tau} = p^\tau + \gamma_d (p^\tau - p^{\tau-\Delta\tau}),$$

where γ_d is a numerical coefficient that is independent of the grid size and of the time step which is set to 0.1. Introducing this modified pressure is equivalent to adding a diffusion term to the equation involving three-dimensional momentum divergence.

A.3.6.2 External gravity wave filtering

External gravity waves are filtered by inserting an additional term into the momentum equations (A.42) and (A.43):

$$\delta_\tau U'' = \ldots - \gamma_e \frac{\partial}{\partial x}\Big[\delta_{\tau-\Delta\tau}(\mu''_d)\Big],$$

$$\delta_\tau V'' = \ldots - \gamma_e \frac{\partial}{\partial y}\Big[\delta_{\tau-\Delta\tau}(\mu''_d)\Big],$$

where the quantity $\delta_{\tau-\Delta\tau}(\mu''_d)$ is the horizontal divergence of momentum integrated along the vertical, (A.55), calculated using the values of U and V at time τ, and γ_e a coefficient that is independent of grid size and time step, of the order of 0.01.

A.3.6.3 Off-centering of the semi-implicit scheme

Instabilities associated with the vertical propagation of sound waves can be damped by introducing the off-centred mean operator (A.48) for discretization of equations (A.46) and (A.47), leading to solving an implicit system (Durran and Klemp, 1983;

Dudhia, 1995). A typical value for the parameter β, independent of grid size and time step, is 0.1.

A.3.7 Inclusion of a gravity wave absorbing layer

It may be helpful to introduce an absorbing layer at the top of the atmosphere to absorb vertically propagating gravity waves and prevent them from reflecting at the top of the atmosphere and contaminating the solution of the model inside the working domain.

A.3.7.1 Spatial filtering

The principle is to introduce horizontal and vertical diffusion terms by using enhanced exchange coefficients K_{dh} and K_{dv} defined as follows:

$$K_{dh} = \frac{\Delta x^2}{\Delta t} \gamma_g \cos\left(\frac{\pi}{2} \frac{z_{top} - z}{z_d} \right),$$

$$K_{dv} = \frac{\Delta z^2}{\Delta t} \gamma_g \cos\left(\frac{\pi}{2} \frac{z_{top} - z}{z_d} \right).$$

γ_g is a user-defined damping coefficient z_{top}, the height of the top of the atmosphere and z_d the depth of the damping layer (from the top of the atmosphere down). When diffusion terms with exchange coefficients K_h and K_v have already been introduced, the maximum of (K_h, K_{dh}) and (K_v, K_{dv}) must be adopted as exchange coefficients. The effects of this filter and the values to be given to the coefficient γ_g are discussed by Klemp and Lilly (1978).

A.3.7.2 Implicit Rayleigh damping

Implicit damping of vertical velocity may be obtained by changing the way to solve the implicit system arising from equations (A.46) and (A.47). The procedure proposed by Klemp et al. (2008) consists in introducing into equation (A.47) for determining geopotential $\Phi''^{\tau+\Delta\tau}$ a modified value of vertical velocity $W''^{\tau+\Delta\tau}$, which is written:

$$W''^{\tau+\Delta\tau} = \widetilde{W}''^{\tau+\Delta\tau} - \tau(z)\Delta\tau W''^{\tau+\Delta\tau},$$

where $\widetilde{W}''^{\tau+\Delta\tau}$ is the solution obtained after solving the tridiagonal implicit system.

The variable $\tau(z)$ defines the vertical structure of the damping layer which is also used in the case of traditional Rayleigh damping described in the next paragraph.

$$\tau(z) = \begin{cases} \gamma_r \sin^2\left[\frac{\pi}{2}\left(1 - \frac{z_{top} - z}{z_d} \right) \right] & \text{if } z \geq (z_{top} - z), \\ 0 & \text{otherwise.} \end{cases}$$

The coefficient γ_r sets the damping intensity to which a typical value of $0.2\ \mathrm{s}^{-1}$ can be assigned, as indicated in the detailed study of this filter by Klemp et al. (2008).

A.3.7.3 Traditional Rayleigh damping

Another solution for damping is to introduce into the equations for u, v, w, and θ terms for relaxing back these variables to horizontally homogeneous reference values $\bar{u}$, $\bar{v}$, $\bar{w} = 0$, and $\bar{\theta}$. We therefore write:

$$\frac{\partial u}{\partial t} = -\tau(z)(u - \bar{u}), \quad \frac{\partial v}{\partial t} = -\tau(z)(v - \bar{v}),$$

$$\frac{\partial w}{\partial t} = -\tau(z)w, \quad \frac{\partial \theta}{\partial t} = -\tau(z)(\theta - \bar{\theta}),$$

keeping the previous formula for the relaxation factor $\tau(z)$ which defines the structure of the damping layer. The reference values can then be updated by linear interpolation based on the height of the model levels at each time step. The effects of this filter and the value to ascribe γ_r to are discussed by Klemp and Lilly (1978).

A.3.8 Physics

The WRF/ARW model provides many possibilities for handling physical processes that are not explicitly taken into account by dynamics: radiation, exchanges with the surface and within the planetary boundary layer, microphysics, and convection. Its modular design means it can use various parameterization schemes developed for other models by the various organizations contributing to the WRF/ARW project. Among them we can cite in particular the Eta model implemented by the National Center for Environmental Prediction (NCEP) and the Geophysical Fluid Dynamics Laboratory (GFDL) and the MM5 model developed by the University of Pennsylvania and the National Center for Atmospheric Research (NCAR).

Detailed allowance for certain physical processes requires additional advected variables (particle concentration, mixing ratio of atmospheric water components, turbulent kinetic energy, etc.). The WRF/ARW model allows the introduction of these additional variables that change with the dynamic constraints of the atmosphere (prognostic equation for Q_m) and the relevant physical parameterizations.

The physical parameterizations generally use the model's state variables in a vertical column and return tendencies for the momentum, potential temperature, and the various moisture fields. These tendencies are calculated in the first step of the Runge-Kutta scheme and then remain constant for the next steps. As concerns microphysics, by contrast, the calculations are made after the final step of the Runge-Kutta scheme and provide final values of the variables upon completion of the time step, thus respecting the equilibrium of the various water phases with respect to saturation.

We confine ourselves here to giving a very brief overview of the variety of schemes available. Fuller information about the possibilities of the continually developing WRF/ARW model and the relevant bibliographic references can be found in the technical note (Skamarock et al., 2008).

A.3.8.1 Radiation

The various schemes proposed for calculating both long- and short-wave radiation fluxes process a vertical column that is assumed to be independent of its neighbours; this is justified so long as the layer is thinner than the grid size. The schemes generally allow for the effects of carbon dioxide, ozone, and clouds and differ in the number of spectral bands taken into account (from 2 to 16 for long-wave radiation and from 1 to 19 for short-wave radiation).

A.3.8.2 Surface boundary layer

The parameterization of turbulent surface fluxes uses schemes based on the Monin-Obukhov similarity theory described in Subsection 9.4.2. The fluxes are calculated by taking into account the nature of the surface via a roughness length and the characteristics of the atmosphere in the vicinity of the surface: wind shear, vertical stability of the atmosphere. The schemes differ essentially in the characteristics of the functions associating flux intensity with the vertical stability of the atmosphere.

A.3.8.3 Earth surface–atmosphere interaction

The models of Earth surface–atmosphere interaction (known as land-surface models or LSMs) use turbulent surface fluxes, radiative fluxes, precipitation fluxes, and information specific to the land surface. They also introduce several levels for calculating temperature and moisture within the soil and at the Earth surface–atmosphere interface more specifically. Four schemes are available; they differ in terms of the processes taken into account (vegetation effects, transformation of snow on the ground), in the number and nature of the variables processed, and in the number of levels, ranging from 2 to 6.

A.3.8.4 The planetary boundary layer and turbulent diffusion

Allowing for the effects of turbulence in the planetary boundary layer and in the free atmosphere involves calculating vertical turbulent fluxes that are expressed using an exchange coefficient K_v. Seven schemes can be used: they differ in how they process the vertical profile of the exchange coefficient for the unstable planetary layer, in how they process entrainment, and in how they compute the top of the planetary boundary layer.

A.3.8.5 Microphysics

Microphysics describes the interactions between water vapour, clouds, and precipitation; they can be allowed for varying degrees of detail by introducing additional variables (mixing ratio of the various components of atmospheric water, particle number concentration). The various parameterization schemes of microphysics available differ in the number and nature of the variables taken into account and the nature of the processes involving ice and mixed phases.

A.3.8.6 Convection

It is essential for convection to be parameterized wherever the model's horizontal resolution precludes it from explicitly taking convective motion into account; it may therefore be useful when grid mesh size is between 5 km and 10 km and remains essential beyond that. In addition to a *'convective adjustment'* type scheme, three other parameterization schemes using a *mass flux* formulation are available; they take detrainment into account and differ in the type of closure applied to the equations.

A.3.9 Initial conditions

The WRF/ARW model can be initialized either from test suites for idealized simulation of various meteorological phenomena (baroclinic waves, squall lines, supercells, mountain waves, sea breezes, etc.) or from data corresponding to actual meteorological situations. In this case, a data preprocessing programme known as WPS (WRF Preprocessing System) computes the various constants relating to the type of projection and the chosen grid, reads and decodes the data from GRIB files (the usual format for data exchange within the meteorological community), and interpolates them horizontally on the model grid. Then a *real data processor* programme calculates the values of the reference state in hydrostatic balance, vertically interpolates meteorological parameters on the levels of the model, and determines the values of the departures from the reference state, which are the forecast model's state variables. To eliminate propagation of gravity and sound waves resulting from insufficient balance between the mass field and the wind field in the initial data, variants of the digital filtering initialization procedure (Lynch and Huang, 1994), already mentioned in Subsection 10.4.2, can be applied.

A.3.10 Data assimilation

The WRF-DA System has been developed to assimilate observation data into the WRF/ARW model. It uses a three-dimensional variational formulation (3DVAR) in incremental mode (Barker et al., 2004). Assimilation minimizes a cost function in the model space involving departures from the background field, departures from observations, and the background error covariance matrices, which determine the analysis response to the observations. The background field generally consists in a recent short-term forecast from the WRF/ARW model (typically from 1 to 6 hours). The observations can be read and decoded from BUFR files (very commonly used for exchanging observations in the meteorological community) and are quality controlled before being taken into account by the assimilation system. Determining background error covariance is primordial for the variational method to be effective; it is done either from differences between two forecasts for different background fields valid at the same time (Parrish and Derber, 1992), or from ensemble forecast results (Fisher, 2003).

A.3.11 Boundary conditions and model coupling

When the WRF/ARW model is operating on a limited geographical domain, it is essential to provide lateral boundary conditions, and many options are available for doing this. For idealized simulations, we can apply periodicity conditions, symmetry conditions, or what are called *open* boundary conditions that allow gravity waves to exit the domain without reflection at the boundaries (Klemp and Lilly, 1978; Klemp and Wilhelmson, 1978).

When the model operates over a limited area with actual data, boundary conditions need to be supplied from a model operating over a wider domain (nesting mode). The data of the coarse mesh parent model must be interpolated in space and time to match the model variables at each grid point and at each instant in time. The procedure employed, developed by Davies and Turner (1977), is analogous to that described in Subsection 10.5.2: calling $\mathbf{X}_F$ the coarse mesh parent model value corresponding to variable $\mathbf{X}$ of the fine mesh child model, two additional terms are introduced into the equations, written:

$$\frac{\partial \mathbf{X}}{\partial t} = \ldots - F_1(\mathbf{X} - \mathbf{X}_F) + F_2 \nabla^2(\mathbf{X} - \mathbf{X}_F).$$

The first is a restore term from the model values $\mathbf{X}$ to the values supplied by the parent model $\mathbf{X}_F$ whereas the second is a dissipation term for damping the noise related to large departures $\mathbf{X} - \mathbf{X}_F$. Inclusion of these additional terms is effective over a transition zone of a few grid points around the lateral boundaries of the geographical domain (see Figure 10.8), the values of functions F_1 and F_2 being maximum on the outer boundary of the domain and diminishing as we move inwards into the domain.

The WRF/ARW system is special in that it offers many possibilities for operating nested models. The nesting may be either one-way, with the child model being passive with respect to the parent model, or two-way, in which case the data of the child model are fed back into the parent model. The models may have multiple levels of nesting: the parent model may operate with two child models located within its geographical domain or the child model may be used to act in turn as a parent model for another model for a yet smaller domain; the only constraint is that the child model domains must be fully contained within the parent model domain and not overlap. Lastly, it is possible for the domain of the child model to be moving within that of the parent model so as to be able to track fast-moving, small-scale meteorological phenomena.

A.3.12 Conclusions

The WRF/ARW system offers a huge choice of possibilities for numerical weather prediction applications. Its modular design means it can be used to run either a primitive equation model over a global domain or a limited area, or a nonhydrostatic model for mesoscale applications. Many parameterization schemes are available, ranging from the quite straightforward to the highly sophisticated, to take into account the effects of various physical processes that are not explicitly resolved at the grid scale. Moreover, the preprocessing and observation data assimilation modules mean the system can be used directly for operational applications. The WRF/ARW system can be implemented easily on a wide variety of computing platforms (scalar or vector type, distributed and shared memory), which is why it is used by a large number of operational meteorological centres and research teams the world over.

Further reading

Here is a selection of books on numerical computing, dynamic meteorology, and atmospheric modelling readers might like to consult to learn more about the topics covered in *Fundamentals of Numerical Weather Prediction*.

Vector analysis and numerical methods in scientific computing

Chan, R., Greif, C., and O'Leary, D. (2007). *Milestones in Matrix Computation: The Selected Works of Gene H. Golub with Commentaries*. New York: Oxford University Press.

Durran, D. R. (1998). *Numerical Methods for Wave Equations in Geophysical Fluid Dynamics: With Applications to Geophysics*. New York: Springer Verlag.

Goldstine, H. H. (1977). *A History of Numerical Analysis from the 16th through 19th Century*. New York: Springer Verlag.

Marchuk, G. I. (1974). *Numerical Methods in Weather Prediction*. New York: Academic Press.

Nebeker, F. (1995). *Calculating the Weather: Meteorology in the 20th Century*, Volume 60 (International Geophysics). New York: Academic Press.

Saad, Y. (2003). *Iterative Methods for Sparse Linear Systems*. Cambridge: Cambridge University Press.

Schey, H. M. (2005). *Div, Grad, Curl and All That. An Informal Text on Vector Calculus*, 4th edn. New York: W.W. Norton & Co.

Schuller, F. (2009). *Grid Computing in Meteorology*. Saarbrücken, Germany: Vdm Verlag.

Strang, G., and Fix, G. (2008). *An Analysis of the Finite Element Method*, 2nd edn. Cambridge: Cambridge University Press.

Dynamic meteorology

Cullen, M. J. P. (2006). *A Mathematical Theory of Large-Scale Atmosphere/Ocean Flow*. London: Imperial College Press.

Gill, A. E. (1982). *Atmosphere-Ocean Dynamics*, Volume 30 (International Geophysics). New York: Academic Press.

Holton, J. R. (2004). *An Introduction to Dynamic Meteorology*, Volume 88 (International Geophysics). 4th edn. New York: Academic Press.

Jacobson, M. (1999). *Fundamentals of Atmospheric Modelling*. Cambridge: Cambridge University Press.

Norbury, J., and Roulstone, I. (2002). *Large-Scale Atmosphere-Ocean Dynamics*. Cambridge: Cambridge University Press.

Pedlosky, J. (1992). *Geophysical Fluid Dynamics*, 2nd edn. New York: Springer Verlag.

Pedlosky, J. (2003). *Waves in the Ocean and Atmosphere: Introduction to Wave Dynamics*. New York: Springer Verlag.

Vallis, G. K. (2006). *Atmospheric and Oceanic Fluid Dynamics: Fundamentals and Large-Scale Circulation*. Cambridge: Cambridge University Press.

Data assimilation and numerical weather prediction

Daley, R. (1993). *Atmospheric Data Analysis*. Cambridge: Cambridge University Press.

Evensen, G. (2009). *Data Assimilation: The Ensemble Kalman Filter*. New York: Springer Verlag.

Haltiner, G. J., and Williams, R. T. (1980). *Numerical Prediction and Dynamic Meteorology*, 2nd edn. New York: John Wiley & Sons.

Joliffe, I., and Stephenson, D. B. (2003). *Forecast Verification: A Practitioner's Guide in Atmospheric Science*. New York: John Wiley & Sons.

Kalnay, E. (2002). *Atmospheric Modeling, Data Assimilation and Predictability*. Cambridge: Cambridge University Press.

Krishnamurti, T. N., and Bounoua, L. (1995). *An Introduction to Numerical Weather Prediction Techniques*. Boca Raton, FL: CRC Press.

Krishnamurti, T. N., Bedi, H. S., Hardiker, V. M., and Ramaswamy, L. (2009). *An Introduction to Global Spectral Modelling*, 2nd edn. New York: Springer Verlag.

Lewis, J., Lakshmivarahan, S., and Dhall, S. (2006). *Dynamic Data Assimilation – A Least Squares Approach*. Cambridge: Cambridge University Press.

Lin, C. A., Laprise, R., and Ritchie, H. Eds. (1997). *Atmospheric and Ocean Modelling. The André J. Robert Memorial Volume*, Canadian Meteorological and Oceanographic Society, Ottawa, Canada: NRC Research Press.

Pielke, R. A. (2001). *Mesoscale Meteorological Modeling*, 2nd edn. New York: Academic Press.

Randall, D. A. (2000). *General Circulation Model Development: Past, Present, and Future*. New York, Academic Press.

Warner, T. T. (2010). *Numerical Weather and Climate Prediction*. Cambridge: Cambridge University Press.

Physical processes and parameterizations

Emanuel, K. A. (1994). *Atmospheric Convection*. New York: McGraw-Hill.

Garatt, J. R. (1994). *The Atmospheric Boundary Layer* (Cambridge Atmospheric and Space Science Series). Cambridge: Cambridge University Press.

Goody, R. M., and Yung, Y. L. (1995). *Atmospheric Radiation: Theoretical Basis*, 2nd edn. New York: Oxford University Press.

Houze, R. A. (1994). *Cloud Dynamics*, Volume 53 (International Geophysics). New York: Academic Press.

Liou, K. N. (2002). *An Introduction to Atmospheric Radiation*, Volume 84 (International Geophysics), 2nd edn. New York: Academic Press.

Mason, B. J. (1971). *Physics of Clouds* (Monographs in Meteorology). Oxford: Oxford University Press.

Petty, G. W. (2006). *A First Course in Atmospheric Radiation*, 2nd edn. Madison, WI: Sundog Publishing.

Pruppacher, H. R., and Klett, J. D. (1996). *Microphysics of Clouds and Precipitation*. Norwell, MA: Kluwer Academic Publishers.

Stensrud, D. J. (2009). *Parameterization Schemes: Keys to Understanding Numerical Weather Prediction Models*. Cambridge: Cambridge University Press.

Stull, R. B. (1988). *An Introduction to Boundary Layer Meteorology*. New York: Springer Verlag.

Yau, M. K., and Rogers, R. R. (1989). *Short Course in Cloud Physics* (International Series in Natural Philosophy), 3rd edn. Oxford: Butterworth-Heinemann.

References

Arakawa, A. (1966). Computational design for long-term numerical integration of the equations of fluid motion. Two-dimensional incompressible flow. Part I. *J. Comput. Phys.*, **1**, 119–43.

Arakawa, A. (1972). *Design of the UCLA general circulation model*. Technical Report No. 7, Department of Meteorology, UCLA. Los Angeles: University of California.

Arakawa, A. (2004). The cumulus parameterization problem. Past, present and future. *J. Climate*, **17**, 2493–525.

Arakawa, A., and Moorthi, S. (1988). Baroclinic instability in vertically discrete systems. *J. Atmos. Sci.*, **45**, 1688–708.

Arakawa, A., and Schubert, W. H. (1974). Interaction of cumulus cloud ensemble with the large scale environment. Part I. *J. Atmos. Sci.*, **31**, 674–701.

Ashford, O. M. (1985). *Prophet or Professor? The Life and Work of Lewis Fry Richardson*. Bristol and Boston: Adam Hilger Ltd.

Asselin, R. (1972). Frequency filter for time integrations. *Mon. Wea. Rev.*, **100**, 487–90.

Baer, F. (1972). An alternate scale representation of atmospheric energy spectra. *J. Atmos. Sci.*, **29**, 649–64.

Baer, F., and Tribbia, J. (1977). On complete filtering of gravity modes through nonlinear initialization. *Mon. Wea. Rev.*, **115**, 272–96.

Ballish, B., Cao, X., Kalnay, E., and Kanamitsu, M. (1992). Incremental nonlinear normal mode initialization. *Mon. Wea. Rev.*, **120**, 1723–34.

Barker, D. M., Huang, W., Guo, Y.-R., Bourgeois, A., and Xiao, X. N. (2004). A three-dimensional variational data assimilation system for MM5: implementation and initial results. *Mon. Wea. Rev.*, **132**, 897–914.

Barré de Saint-Venant, A. J. C. (1871). Théorie du mouvement non permanent des eaux avec application aux crues des rivières et à l'introduction des marées dans leur lit. *Comptes Rendus de l'Academie des Sciences*, Paris, **V, 73**, 147–53.

Bechtold, P., Bazile, E., Guichard, F., Mascart, P., and Richard, E. (2001). A mass flux convection scheme for regional and global models. *Quart. J. Roy. Meteorol. Soc.*, **127**, 869–86.

Belamari, S., and Pirani, A. (2007). Validation of the optimal heat and momentum fluxes using the ORCA2-LIM global oceanic model. Marine environment and security for the European area. MERSEA Integrated Project, Deliverable D4.1.3. Plouzané, France: Institut Français de Recherche pour l'Exploitation de la Mer.

Belousov, S. L. (1962). *Tables of Normalized Associated Legendre-Polynomials*. London: Pergamon Press.

Bénard, P., Marki, A., Neytchev, P. N., and Prtenjak, M. T. (2000). Stabilization of nonlinear vertical diffusion schemes in the context of NWP models. *Mon. Wea. Rev.*, **128**, 1937–48.

Bénard, P. (2003). Stability of semi-implicit and iterative centred-implicit time discretisations for various equation systems used in NWP. *Mon. Wea. Rev.*, **131**, 2479–91.

Bénard, P. (2004). On the use of a wider class of linear systems for the design of constant-coefficients semi-implicit time schemes in NWP. *Mon. Wea. Rev.*, **132**, 1319–24.

Bénard, P., Marki, A., Neytchev, P. N., and Prtenjak, M. T. (2000). Stabilization of nonlinear vertical diffusion schemes in the context of NWP models. *Mon. Wea. Rev.*, **128**, 1937–48.

Bénard, P., Laprise, R., Vivoda, J., and Smolikova, P. (2004). Stability of leap-frog constant-coefficients semi-implicit schemes for the fully elastic system of Euler equations. Flat-terrain case. *Mon. Wea. Rev.*, **132**, 1306–18.

Bénard, P., Vivoda, J., Mašek, J., Smolíková, P., Yessad, K., Smith, Ch., Brožková, R., and Geleyn, J. F. (2010). Dynamical kernel of the Aladin-NH spectral limited-area model: revised formulation and sensitivity experiments. *Quart. J. Roy. Meteor. Soc.*, **136**, 155–69.

Benwell, G. R. R., Gadd, A. J., Keers, J. F., Timpson, M. S., and White, P. W. (1971). *The Bushby-Timpson 10-level model on a fine mesh*. Meteorological Office Scientific Paper No. 32. Bracknell, UK: Meteorological Office.

Bergthorsson, P., and Döös, B. (1955). Numerical weather map analysis. *Tellus*, **7**, 329–40.

Berre, L. (2000). Estimation of synoptic and mesoscale forecast error covariances in a limited area model. *Mon. Wea. Rev.*, **128**, 644–67.

Berre, L., Stefanescu, S. E., and Pereira, M. B. (2006). The representation of the analysis effect in three error simulation techniques. *Tellus*, **58A**, 196–209.

Bhumralkar, C. M. (1975). Numerical experiments on the computation of the ground surface temperature in atmospheric general circulation model. *J. Appl. Meteor.*, **14**, 67–100.

Bishop, C. H., and Toth, Z. (1999). Ensemble transformation and adaptive observations. *J. Atmos. Sci.*, **56**, 1748–65.

Bishop, C., Etherton, B., and Majundar, S. (2001). Adaptive sampling with the ensemble transform Kalman filter. Part I: theoretical aspects. *Mon. Wea. Rev.*, **129**, 420–36.

Bjerknes, V. (1904). Das problem von der Wettervorhersage, betrachtet vom Standpunkt der Mechanik und der Physik. *Meteor. Zeitschrift*, **21**, 1–7. [Le problème de la prévision du temps du point de vue de la mécanique et de la physique, French translation by Gondoin, D. (1995). *La Météorologie*, 8e série, **9**, 55–62.]

Blackadar, A. K. (1962). The vertical distribution of wind and turbulent exchange in a neutral atmosphere. *J. Geophys. Res.*, **67**, 3095–102.

Blackadar, A. K. (1976). Modeling nocturnal boundary layer. In: *Proceedings of the Third Symposium on Atmospheric Turbulence, Diffusion and Air Quality*, Raleigh, NC, USA, Oct. 1976. Boston: American Meteorological Society, 46–49.

Boer, G. J., McFarlane, N. A., Laprise, R., Henderson, J. D., and Blanchet, J. P. (1984). The Canadian Climate Centre spectral atmospheric general circulation model. *Atmosphère-Océan*, **22**, 397–429.

Bolin, B. (1955). Numerical forecasting with the barotropic model. *Tellus*, **7**, 27–49.

Bougeault, P. (1983). A non-reflective boundary condition for limited-height hydrostatic models. *Mon. Wea. Rev.*, **111**, 420–29.

Bougeault, P. (1985). A simple parameterization of the large-scale effects of cumulus convection. *Mon. Wea. Rev.,* **113**, 2108–21.

Bougeault, P., and Lacarrère, P. (1989). Parameterization of orography-induced turbulence in a meso-beta scale model. *Mon. Wea. Rev.*, **117**, 1870–88.

Bourke, W. (1972). An efficient one-level primitive-equation spectral model. *Mon. Wea. Rev.*, **100**, 683–89.

Bourke, W. (1974). A multi-level spectral model. Formulation and hemispheric integrations. *Mon. Wea. Rev.*, **102**, 687–701.

Bourke, W., and McGregor, J. L. (1983). A nonlinear vertical mode initialization scheme for a limited area prediction model. *Mon. Wea. Rev.*, **111**, 2285–97.

Bouteloup, Y., Seity, Y., and Bazile, E. (2011). Description of the sedimentation scheme used operationally in all Météo-France NWP models. *Tellus*, **63**, 300–11.

Bouttier, F., and Courtier, P. (1999). Data assimilation concepts and methods. In: *Training course notes of the European Centre for Medium-range Weather Forecasts.* Reading: European Centre for Medium-range Weather Forecasts.

Bratseth, A. M. (1986). Statistical interpolation by means of successive corrections. *Tellus*, **38A**, 439–47.

Bretherton, C. S., McCaa, J. R., and Grenier, H. (2004). A new parameterization for shallow cumulus convection and its application to marine subtropical cloud-topped boundary layers. Part I: description and 1-D results. *Mon. Wea. Rev.*, **132**, 864–82.

Brière, S. (1982). Nonlinear normal mode initialization of a limited area model. *Mon. Wea. Rev.*, **110**, 1166–88.

Brousseau, P., Bouttier, F., Hello, G., Seity, Y., Fischer, C., Berre, L., Montmerle, T., Auger, L., and Malardel, S. (2008). *A prototype convective-scale data assimilation system for operation: the AROME-RUC.* HIRLAM Technical Report No. 68. Norrköping, Sweden: Swedish Meteorological and Hydrological Institute.

Bubnová, R., Hello, G., Bénard, P., and Geleyn, J.-F. (1995). Integration of the fully elastic equations cast in the hydrostatic pressure terrain-following coordinate in the framework of the ARPEGE/ALADIN NWP System. *Mon. Wea. Rev.*, **123**, 515–35.

Buizza, R., Tribbia, J., Molteni, F., and Palmer, T. N. (1993). Computation of optimal unstable structures for a numerical weather prediction model. *Tellus*, **45A**, 388–407.

Burridge, D. M. (1975). A split semi-implicit reformulation of the Bushby-Timpson 10-level model. *Quart. J. Roy. Meteor. Soc.*, **101**, 777–92.

Bushby, F. H. (1987). A history of numerical weather prediction. Short- and medium-range numerical weather prediction. In: *Collection of papers presented at the WMO/ IUGG Symposium*, Tokyo, 4–6 Aug. 1986. Tokyo: Meteorological Society of Japan, 1–10.

Businger, J. A., Wyngard, J. C., Izumi, Y., and Bradley, E. F. (1971). Flux profile relationships in the atmospheric surface layer. *J. Atmos. Sci.*, **28**, 181–89.

Catry, B., Geleyn, J.-F., Bouyssel, F., Cedilnik, J., Brožková, R., Derkova, M., and Mladek, R. (2008). A new sub-grid scale lift formulation in a mountain drag parameterisation scheme. *Meteor. Zeitschrift*, **17**, 193–208.

Catry, B., Geleyn, J.-F., Tudor, M., Bénard, P., and Trojakova, A. (2007). Flux-conservative thermodynamic equations in a mass-weighted framework. *Tellus*, **59A**, 71–79.

Caya, D., and Laprise, R. (1999). A semi-implicit semi-Lagrangian regional climate model: The Canadian RCM. *Mon. Wea. Rev.*, **127**, 341–62.

Charney, J. G. (1948). On the scale of atmospheric motions. *Geofys. Publ.* **17**, 1–17.

Charney, J. G. (1954). Numerical prediction of cyclogenesis. *Proc. Nat. Acad. Sci. U. S.*, **40**, 99–110.

Charney, J. G. (1955). The use of the primitive equations of motion in numerical prediction. *Tellus*, **7**, 22–26.

Charney, J. G. (1962). Integration of the primitive and balance equations. In: *Proceedings of the International Symposium on Numerical Weather Prediction*, Tokyo, 7–13 Nov. 1960. Tokyo: Meteorological Society of Japan, 131–52.

Charney, J. G., Fjörtoft, R., and von Neumann, J. (1950). Numerical integration of the barotropic vorticity equation. *Tellus*, **2**, 237–54.

Charney, J. G., and Phillips, N. A. (1953). Numerical integration of the quasi-geostrophic equations for barotropic and simple baroclinic flows. *J. Atmos. Sci.*, **10**, 71–99.

Charnock, H. (1955). Wind stress in a water surface. *Quart. J. Roy. Meteorol. Soc.*, **81**, 639–40.

Coiffier, J., Ernie, Y., Geleyn, J.-F., Clochard, J., Hoffman, J. and Dupont, F. (1987a). The operational hemispheric model at the French Meteorological Service. In: *Collection of papers presented at the WMO/IUGG Symposium*, Tokyo, 4–6 Aug. 1986. Tokyo: Meteorological Society of Japan, 337–45.

Coiffier, J., Chapelet, P., and Marie, N. (1987b). Study of various quasi-Lagrangian techniques for numerical models. In: *Proceedings of the ECMWF Workshop on Techniques for Horizontal Discretization in Numerical Weather Prediction Models*, Reading, Nov. 1987. Reading: European Centre for Medium-range Weather Forecasts, 19–46.

Cooley, J. W., and Tukey, J. W. (1965). An algorithm for the machine computation of complex Fourier series. *Math. Comp.*, **19**, 297–301.

Côté, J., Roch, M., Staniforth, A., and Fillion, L. (1993). A variable resolution semi-Lagrangian finite-element global model of the shallow water equations. *Mon. Wea. Rev.*, **121**, 231–43.

Côté, J., Gravel, S., Méthot, A., Patoine, A., Roch, M. and Staniforth, A. (1998). The operational CMC-MRB global environmental multiscale (GEM) model. Part I: design considerations and formulation. *Mon. Wea. Rev.*, **126**, 1373–95.

Courant, R., Friedrichs, K. O., and Lewy, H. (1928). Über die partiellen Differenzen-Gleichungen der mathematischen Physik. *Math. Annalen*, **100**, 32–74. [On the partial difference equations of mathematical physics, English translation by Fox, P. (1967), *IBM Journal*, **11**, 215–34.]

Courtier, P. (1997). Variational methods. *J. Meteor. Soc. Japan*, **75**, 211–18.

Courtier, P., and Geleyn, J.-F. (1988). A global numerical weather prediction model with variable resolution: application to the shallow water equations. *Quart. J. Roy. Meteor. Soc.*, **114**, 1321–46.

Courtier, P., Thépaut, J.N., and Hollingsworth, A. (1994). A strategy for operational implementation of 4D-Var using an incremental approach. *Quart. J. Roy. Meteor. Soc.*, **120**, 1367–87.

Craig, G. C. (1996). Dimensional analysis of a convecting atmosphere in equilibrium with external forcing. *Quart. J. Roy. Meteor. Soc.*, **122**, 1963–67.

Cressman, G. P. (1958). Barotropic divergence and very long atmospheric waves. *Mon. Wea. Rev.*, **86**, 293–97.

Cressman, G. P. (1959). An operational objective analysis system. *Mon. Wea. Rev.*, **87**, 367–74.

Cressman, G. P. (1963). *A three-level model suitable for daily numerical forecasting*. NMC Technical Memorandum No. 22. Washington, DC: Weather Bureau, ESSA, US Department of Commerce.

Cressman, G. P. (1996). The origin and rise of numerical weather prediction. In: *Historical Essays on Meteorology, 1919–1995*, edited by Fleming, J. R. Boston: American Meteorological Society, 21–39.

Cullen, M. J. P. (1973). A simple finite element method for meteorological problems. *J. Inst. Math. Applics.*, **11**, 15–31.

Cullen, M. J. P. (2001). Alternative implementations of the semi-Lagrangian semi-implicit schemes in the ECMWF model. *Quart. J. Roy. Meteor. Soc.*, **127**, 2787–802.

Cullen, M. J. P., and James, J. (1994). A comparison of two different vertical grid staggerings. *Proceedings of the Tenth AMS Conference on Numerical Weather Prediction*, Portland, OR, USA, 18–22 July 1994. Boston: American Meteorological Society, 38–40.

Cuxart, J. P., Bougeault, P., and Redelsperger, J. L. (2000). A turbulence scheme allowing for mesoscale and large-eddy simulations. *Quart. J. Roy. Met. Soc.*, **126**, 1–30.

Dady, G. (1969). *Météorologie Dynamique et Prévision Numérique*. Paris: École de la Météorologie, Direction de la Météorologie Nationale.

Daley, R. (1980). On the optimal specification of the initial state for deterministic forecasting. *Mon. Wea. Rev.*, **108**, 1719–35.

Daley, R. (1991). *Atmospheric Data Analysis*. Cambridge: Cambridge University Press.

Davies, H. C. (1973). On the lateral boundary conditions for primitive equations. *J. Atmos. Sci.*, **30**, 147–50.

Davies, H. C. (1976). A lateral boundary formulation for multi-level prediction models. *Quart. J. Roy. Meteor. Soc.*, **102**, 405–18.

Davies, H. C., and Turner, R. E. (1977). Updating prediction models by dynamical relaxation: An examination of the technique. *Quart. J. Roy. Meteor. Soc.*, **103**, 225–45.

Deardorff, J. W. (1972). Numerical investigation of neutral and unstable planetary boundary layers. *J. Atmos. Sci.*, **29**, 91–115.

Deardorff, J. W. (1977). Efficient prediction of ground surface temperature and moisture with inclusion of a layer of vegetation. *J. Geophys. Res.*, **83**, 1889–1903.

De Moor, G. (2006). *Couche Limite Atmosphérique et Turbulence. Les Bases de la Micrométéorologie Dynamique*. Cours et manuels No. 16, École Nationale de la Météorologie. Toulouse: Météo-France.

Déqué, M., and Cariolle, D. (1986). Some destabilizing properties of the Asselin time-filter modelling. *Mon. Wea. Rev.*, **114**, 880–84.

Déqué, M., Dreveton, C., Braun, A., and Cariolle, D. (1994). The ARPEGE/IFS atmosphere model. A contribution to the French community climate modelling. *Climate Dyn.*, **10**, 249–66.

Der Megreditchian, G. (1992). *Le Traitement Statistique des Données Multidimensionnelles. Application à la Météorologie*, volume I. Cours et manuels No. 8. Toulouse: École Nationale de la Météorologie, Météo-France.

Der Megreditchian, G. (1993). *Le Traitement Statistique des Données Multidimensionnelles. Application à la Météorologie*, volume II. Cours et manuels No. 9. Toulouse: École Nationale de la Météorologie, Météo-France.

Dickinson, R. E., and Williamson, D. L. (1972). Free oscillation of a discrete stratified fluid with application to numerical weather prediction. *J. Atmos. Sci.*, **29**, 623–40.

Doswell III, C. A., Davies-Jones, R., and Keller, D. L. (1990). On summary measures of skill in rare event forecasting based on contingency tables. *Weather and Forecasting*, **5**, 576–85.

Du, J., and Tracton, M. S. (1999). Impact of lateral boundary conditions on regional model ensemble prediction. In: *Research Activities in Atmospheric and Oceanic Modeling*, CAS/JSC Working Group on Numerical Experimentation, Report No. 28, WMO Technical Document No. 942. Geneva: World Meteorological Organization, 6.7–6.8.

Dudhia, J. (1995). Reply to comment on "A nonhydrostatic version of the Penn State / NCAR mesoscale model: Validation tests and simulations of an Atlantic cyclone and cold front" by J. Steppeler. *Mon. Wea. Rev.*, **123**, 2571–75.

Durran, R. D., and Klemp, J. B. (1983). A compressible model for the simulation of moist mountain waves. *Mon. Wea. Rev.*, **111**, 2341–61.

Ebert, E., and Curry, J. A. (1992). A parameterization of ice cloud optical properties for climate models. *J. Geophys. Res.*, **97**, 3831–35.

Edelmann, W. (1972). Initial balancing and damping of gravity oscillations for forecast models including the orography. *Beitr. Phys. Atmos.*, **45**, 94–120.

Eliassen, A. (1949). The quasi-static equations of motion with pressure as independent variable. *Geofys. Publ.*, **17**, 3–44.

Eliassen, A. (1956). *A procedure for numerical integration of the primitive equations of the two-parameter model of the atmosphere*. Scientific Report No. 4, Department of Meteorology, UCLA. Los Angeles: University of California.

Eliassen, A., and Palm, E. (1961). On the transfer of energy in stationary mountain waves. *Geofys. Publ.*, **22**, 1–23.

Eliassen, E., Machenhauer, B., and Rasmussen, E. (1970). *On a numerical method for integration of the hydrodynamical equations with a spectral representation of the horizontal fields*. Institut für Teoretisk Meteorologi Köbenhavns, Report No. 2. Copenhagen: Institute for Theoretical Meteorology.

Elvius, T., and Sundström, A. (1973). Computationally efficient schemes and boundary conditions for a fine mesh barotropic model based on the shallow water equations. *Tellus*, **25**, 132–56.

Emanuel, K. A. (1994). *Atmospheric Convection*. New York: Oxford University Press.

Epstein, E. S. (1969). Stochastic dynamic predictions. *Tellus*, **21**, 739–59.

Evensen, G. (1994). Sequential data assimilation with a nonlinear quasi-geostrophic model using Monte-Carlo methods to forecast error statistics. *J. Geophy. Res.*, **99**, 10, 143–62.

Eymet, V., Dufresne, J. L., Ricchiazzi, P., Fournier, R., and Blanco, S. (2004). Long-wave radiative analysis of cloudy scattering atmospheres using a net exchange formulation. *Atmos. Res.*, **72**, 239–61.

Fischer, C., Montmerle, T., Auger, L., and Lacroix, B. (2006). L'assimilation opérationnelle de données régionales à Météo-France. *La Météorologie, 8e série*, 54, 43–54.

Fisher, M. (2003). Background error covariance modelling. In: *Seminar on Recent Development in Data Assimilation for Atmosphere and Ocean*, Reading, 8–12 Sept. 2003. Reading: European Centre for Medium-range Weather Forecasts, 45–63.

Fjörtoft, R. (1952). On a numerical method of integrating the barotropic vorticity equation. *Tellus*, **4**, 179–94.

Fouquart, Y. and Bonnel, B. (1980). Computations of solar heating of the Earth's atmosphere. A new parameterization. *Beitr. Phys. Atmos.*, **53**, 35–62.

Gandin, L. S. (1963). *Objective Analysis of Meteorological Fields*. Leningrad: Gidromet. Isdaty. Jerusalem: Israel Program for Scientific Translations (1965).

Gauthier, P., and Thépaut, J.-N. (2001). Impact of the digital filter as a weak constraint in the preoperational 4DVAR assimilation system of Météo-France. *Mon. Wea. Rev.*, **129**, 2089–102.

Geleyn, J.-F. (1987). Use of a modified Richardson number for parameterizing the effect of shallow convection. In: *Collection of papers presented at the WMO/IUGG Symposium*, Tokyo, 4–6 Aug. 1986. Tokyo: Meteorological Society of Japan, 141–49.

Geleyn, J.-F. (1988). Interpolation of wind, temperature, humidity values from model levels to the height of measurement. *Tellus*, **40A**, 347–51.

Geleyn J.-F., and Hollingsworth, A. (1979). An economical analytical method for the computation of the interaction between scattering and line absorption of radiation. *Beitr. Phys. Atmos.*, **52**, 1–16.

Geleyn, J.-F., Fournier, R., Hello, G., and Pristov, N. (2005). A new "bracketing" technique for a flexible and economical computation of thermal radiative fluxes, on the basis of the Net Exchange Rate (NER) formalism, In: *WGNE Blue Book – Research Activities in Atmospheric and Ocean Modeling*, 2005 edn. Geneva: World Meteorological Organization, 4–07.

Geleyn, J.-F., Catry, B., Bouteloup, Y., and Brožková, R. (2008). A statistical approach for sedimentation inside a micro-physical precipitation scheme. *Tellus*, **60A**, 649–62.

Gerard, L., Piriou, J.-M., Brožková, R., Geleyn, J.-F. and Banciu, D. (2009). Cloud and precipitation parameterization in a meso-gamma scale operational weather prediction model. *Mon. Wea. Rev.*, **137**, 3960–77.

Gilchrist, B., and Cressman, G. (1954). An experiment in objective analysis. *Tellus*, **8**, 61–75.

Girard, C., and Jarraud, M. (1982). *Short and medium range forecast differences between a spectral and grid point model. An extensive quasi-operational comparison.* ECMWF Technical Report No. 32. Reading: European Centre for Medium-range Weather Forecasts.

Girard, C., and Delage, Y. (1990). Stable schemes for nonlinear vertical diffusion in atmospheric circulation models. *Mon. Wea. Rev.*, **118**, 737–45.

Glahn, H. R., and Lowry, D. A. (1972). The use of model output statistics (MOS) in objective weather forecasting. *J. Appl. Meteor.*, **11**, 1203–11.

Gleeson, T. A. (1966). A causal relation for probabilities in synoptic meteorology. *J. Appl. Meteor.*, **5**, 365–68.

Grenier, H., and Bretherton, C. (2001). A moist PBL parameterization for large scale models and its application to subtropical cloud-topped marine boundary layers. *Mon. Wea. Rev.*, **129**, 357–77.

Grotjahn, R., and O' Brien, J. J. (1976). Some inaccuracies in finite differencing hyperbolic equations. *Mon. Wea. Rev.*, **104**, 180–94.

Guidard, V., Fischer, C., Nuret, M., and Džiedžic, A. (2006). Evaluation of the ALADIN 3D-Var with observations of the MAP campaign. *Meteorol. Atmos. Phys.*, **92**, 161–73.

Halstead, M. H. (1954). The fluxes of momentum, heat, and water vapour in micrometeorology. *Johns Hopkins University Publication in Climatology*, **7**, no. 2, 326.

Herzog, H.-J. (1995). Testing a radiative upper boundary condition in a nonlinear model with hybrid vertical coordinate. *Meteorol. Atmos. Phys.*, **55**, 185–204.

Hildebrand, F. B. (1956). *Introduction to Numerical Analysis*, 2nd edn. (1974). New York: McGraw-Hill (1974). Republication (1987), New York: Dover (1987).

Hinkelmann, K. H. (1951). Der Mechanismus der meteorologisches Lärm. *Tellus*, **3**, 285–96. [The mechanism of meteorological noise. English translation by Rasch, P., Kiehl, J. and Wyman, W. J. (1983), NCAR Technical Note No. 203. Boulder, CO, USA: National Center for Atmospheric Research.]

Hinkelmann, K. H. (1959). Ein numerisches Experiment mit den primitiven Gleichungen. In: *The Atmosphere and Sea in Motion*, Rossby Memorial Volume, edited by Bolin, B. New York: Rockefeller Institute Press and Oxford University Press, 486–500.

Hinkelmann, K. H. (1969). Primitive equations. Lectures in numerical short-range weather prediction. In: *Regional Training Seminar*, Moscow, 17 Nov.–14 Dec. 1965. WMO Publication No. 297. Leningrad: Hydrometeoizdat, 306–75.

Hollingsworth, A. (1980). An experiment in Monte Carlo forecasting procedure. In: *Proceedings of the ECMWF Workshop on Stochastic Dynamic Forecasting*, Reading, Oct. 1979. Reading: European Centre for Medium-range Weather Forecasts, 65–85.

Hollingsworth, A. (1995). *A spurious mode in the 'Lorenz' arrangement of φ and T which does not exist in the 'Charney-Phillips' arrangement*. Technical Memorandum No. 211, ECMWF, Reading: European Centre for Medium-range Weather Forecasts.

Hollmann, G. (1959). Transformation der Grundgleichungen der dynamischen Meteorologie in Koordinaten der stereographischen Projektion zum Zwecke der numerischen Vorhersage. *Beitr. Phys. Atmos.*, **31**, 162–76.

Hortal, M. (2002). The development and testing of a new two-time-level semi-Lagrangian scheme (SETTLS) in the ECMWF forecast model. *Quart. J. Roy. Meteor. Soc.*, **128**, 1671–87.

Hortal, M., and Simmons, A. (1991). Use of reduced Gaussian grids in spectral models. *Mon. Wea. Rev.*, **119**, 1057–74.

Hoskins, B. J., and Simmons, A. J. (1975). A multi-layer spectral model and the semi-implicit method. *Quart. J. Roy. Meteor. Soc.*, **101**, 1231–50.

Houtekamer, P. L., Lefaivre, L., Derome, J., Ritchie, H., and Mitchell, H. L. (1996). A system simulation approach to Ensemble Prediction. *Mon. Wea. Rev.*, **124**, 1225–42.

Houtekamer, P. L., and Lefaivre, L. (1997). Using ensemble forecasts for model validation. *Mon. Wea. Rev.*, **125**, 2416–26.

Houtekamer, P. L., and Mitchell, H. L. (1998). Data assimilation using an ensemble Kalman filter technique. *Mon. Wea. Rev.*, **126**, 796–811.

Hoyer, J. M. (1987). The ECMWF spectral limited-area model. In: *Proceedings of the ECMWF Workshop on Techniques for Horizontal Discretization in Numerical Weather Prediction Models*, Reading, Nov. 1987. Reading: European Centre for Medium-range Weather Forecasts, 343–59.

Imbard, M., Juvanon du Vachat, R., Joly, A., Durand, Y., Craplet, A., Geleyn, J.-F., Audoin, J.-M., Marie, N., and Pairin, J.-M. (1987). The PERIDOT fine-mesh numerical weather prediction system description, evaluation and experiments. In: *Collection of papers presented at the WMO/IUGG Symposium*. Tokyo, 4–6 Aug. 1986. Tokyo: Meteorological Society of Japan, 455–65.

Joly, A. (1992). *ARPEGE/ALADIN: adiabatic model equations and algorithm*. Météo-France Internal Technical Note. Toulouse: Météo-France.

Juvanon du Vachat, R. (1986). A general formulation of normal modes for limited-area models. Applications to initialization. *Mon. Wea. Rev.*, **114**, 2478–87.

Juvanon du Vachat, R. (1988). Non-normal mode initialization: formulation and application to inclusion of the β-terms in the linearization. *Mon. Wea. Rev.*, **116**, 2013–24.

Kaas, E. (1987). *The construction of and tests with a multi-level semi-Lagrangian and semi-implicit limited area model*. Unpublished Ph.D. thesis, Geophysics Institute, Copenhagen University.

Kain, J. S., and Fritsch, J. M. (1990). A one-dimensional entraining/detraining plume model and its application to convective parameterization. *J. Atmos. Sci.*, **47**, 2784–802.

Kallberg, P. (1977). *Test of a lateral boundary relaxation scheme in a barotropic model*. ECMWF Research Department Internal Report No. 3. Reading: European Centre for Medium-range Weather Forecasts.

Katayama, A. (1974). *A simple scheme for computing radiative transfer in the troposphere*. Technical Report No. 6. Department of Meteorology, UCLA. Los Angeles: University of California.

Kershaw, R., and Gregory, D. (1997). Parameterization of momentum transport by the convection. Part I: Theory and cloud modeling results. *Quart. J. Roy. Meteor. Soc.*, **123**, 1135–51.

Kesel, P. G., and Winninghoff, F. G. (1972). The Fleet Numerical Weather Central operational primitive-equation model. *Mon. Wea. Rev.*, **100**, 360–73.

Kessler, E. (1969). On the distribution and continuity of water substance in atmospheric circulation. *Meteorological Monograph*, **32**. Boston, MA: American Meteorological Society.

Klein, W. H., Lewis, B. M., and Enger I. (1959). Objective prediction of five-day mean temperature during winter. *J. Appl. Meteor.*, **16**, 672–82.

Klemp, J. B., and Lilly, D. K. (1978). Numerical simulation of hydrostatic mountain waves. *J. Atmos. Sci.*, **35**, 78–107.

Klemp, J. B., and Wilhelmson, R. B. (1978). The simulation of three-dimensional convective storm dynamics. *J. Atmos. Sci.,* **35**, 1070–96.

Klemp, J. B., and Durran, D. R. (1983). An upper boundary condition permitting internal gravity wave radiation in numerical mesoscale models. *Mon. Wea. Rev.*, **111**, 430–44.

Klemp, J. B., Skamarock, W. C., and Dudhia, J. (2007). Conservative split-explicit time integration methods for the compressible nonhydrostatic equations. *Mon. Wea. Rev.*, **135**, 2897–913.

Klemp, J. B., Dudhia, J., and Hassiotis, A. (2008). An upper gravity wave absorbing layer for NWP applications. *Mon. Wea. Rev.*, **136**, 3987–4004.

Krishnamurti, T. N. (1962). Numerical integration of primitive equations by a quasi-Lagrangian advective scheme. *J. Appl. Meteor.*, **1**, 508–21.

Krishnamurti, T. N., Kishtawal, C. M., Zhang, Z., LaRow, T. E., Bachiochi, D. R., Williford, C. E., Gadgil, S., and Surendran, S. (2000). Improving tropical precipitation forecasts from a multi analysis superensemble. *J. Climate*, **13**, 4217–27.

Krylov, V. I. (1962). *Approximate Calculation of Integrals*. New York: MacMillan. Republication. New York: Dover (2006).

Kuo, H. L. (1965). A theory of parameterization of cumulus convection. *J. Atmos. Sci.*, **22**, 40–43.

Kuo, H. L. (1974). Further studies of the parameterization of the influence of cumulus convection on large scale flows. *J. Atmos. Sci.*, **31**, 1232–40.

Kurihara, Y. (1965). On the use of implicit and iterative methods for the time integration of the wave equation. *Mon. Wea. Rev.*, **93**, 39–46.

Kurihara, Y., Tuleya, R. E., and Bender, M. A. (1998). The GFDL hurricane prediction system and its performance in the 1995 hurricane season. *Mon. Wea. Rev.*, **126**, 1306–22.

Kwizak, M., and Robert, A. (1971). A semi-implicit scheme for grid point atmospheric models of the primitive equations. *Mon. Wea. Rev.*, **99**, 32–36.

Lafore, J.-P., Stein, J., Asencio, N., Bougeault, P., Ducrocq, V., Duron, J., Fischer, C., Hereil, P., Mascart, P., Pinty, J.-P., Redelsperger, J.-L., Richard, E., and Vila-Guerau de Arellano, J. (1998). *The Meso-NH atmospheric simulation system. Part I: adiabatic formulation and control simulations. Annales Geophysicae*, **16**, 90–109.

Lanczos, C. (1956). *Applied Analysis*. Englewood Cliffs, NJ, USA: Prentice Hall. Republication (1988). New York: Dover.

Laprise, R. (1992). The Euler equations of motion with hydrostatic pressure as independent variable. *Mon. Wea. Rev.*, **120**, 197–207.

Le Dimet, F., and Talagrand, O. (1986). Variational algorithms for analysis and assimilation of meteorological observations. Theoretical aspects. *Tellus*, **38A**, 97–110.

Lehmann, R. (1993). On the choice of relaxation coefficients for Davies lateral boundary scheme for regional weather prediction models. *Meteor. Phys. Atmos.*, **52**, 1–14.

Leith, C. E. (1974). Theoretical skill of Monte Carlo forecasts. *Mon. Wea. Rev.*, **102**, 409–18.

Leith, C. E. (1980). Non linear normal mode initialization and quasi-geostrophic theory. *J. Atmos. Sci.*, **37**, 958–68.

Lepas, J. (1963). Prévision barotrope globale au niveau de pression 500 mb. *Journal de mécanique et de physique de l'atmosphère*, **II-19**, 97–104.

Leslie, L. M., and Purser, R. J. (1992). A comparative study of the performance of various vertical discretization schemes. *Meteor. Atmos. Phys.*, **50**, 61–73.

Lewis, J. M., and Derber, J. C. (1985). The use of adjoint equations to solve a variational adjustment problem with advective constraints. *Tellus*, **37A**, 309–22.

Lindzen, R. S. (1981). Turbulence and stress owing to gravity wave and tidal breakdown. *J. Geophys. Res.*, **86**, 9707–14.

Lopez, P. (2002). Implementation and validation of a new prognostic large-scale cloud and precipitation scheme for climate and data-assimilation purposes. *Quart. J. Roy. Meteor. Soc.*, **128**, 229–57.

Lorenc, A. (1981). A global three-dimensional multivariate statistical interpolation scheme. *Mon. Wea. Rev.*, **109**, 701–21.

Lorenz, E. N. (1960). Energy and numerical weather prediction. *Tellus,* **12**, 364–73.

Lorenz, E. N. (1967). *The nature and theory of the general circulation of the atmosphere.* Technical Paper 115, WMO Publication No. 218. Geneva: World Meteorological Organisation.

Lorenz, E. N. (1969). The predictability of a flow which possesses many scales of motion. *Tellus*, **21**, 289–307.

Lott, F. (1999). Alleviation of stationary biases in a GCM through a mountain drag parameterization scheme and a simple representation of mountain lift forces. *Mon. Wea. Rev.*, **127**, 788–801.

Lott, F., and Miller, M. (1997). A new sub-grid scale orographic drag parameterization: its formulation and testing. *Quart. J. Roy. Meteor. Soc.*, **123**, 101–28.

Louis, J.-F. (1979). A parametric model of vertical eddy fluxes. *Boundary Layer Meteorology*, **17**, 187–202.

Lynch, P. (1985). Initialization using Laplace transforms. *Quart. J. Roy. Meteor. Soc.*, **111**, 243–58.

Lynch, P. (1994). Richardson's Marvellous Forecast. In: *The Life Cycles of Extratropical Cyclones*, edited by Shapiro, M. A. and Grønås, S. Boston: American Meteorological Society, 61–73.

Lynch, P., and Huang, X.-Y. (1992). Initialization of the HIRLAM model using a digital filter. *Mon. Wea. Rev.*, **120**, 1019–34.

Lynch, P., and Huang, X.-Y. (1994). Diabatic initialization using recursive filters. *Tellus*, **46A**, 583–97.

Lynch, P., Giard, D., and Ivanovici, V. (1997). Improving the efficiency of a digital filtering scheme. *Mon. Wea. Rev.*, **125**, 1976–82.

Lynch, P. (2006). *The Emergence of Numerical Weather Prediction: Richardson's Dream*. Cambridge: Cambridge University Press.

Machenhauer, B. (1977). On the dynamics of gravity oscillations in a shallow water model with applications to normal mode initialization. *Beitr. Phys. Atmos.*, **50**, 253–71.

Machenhauer, B. (1979). The spectral method. In: *Numerical Methods used in Atmospheric Models*, Vol. II, GARP Publication Series No. 17, ICSU/WMO, Geneva: World Meteorological Organization, 121–75.

Machenhauer, B., and Daley, R. (1972): *A baroclinic primitive equation model with a spectral representation in three dimensions*. Institut für Teoretisk Meteorologi, Technical Report No. 4. Copenhagen: Institute for Theoretical Meteorology.

Machenhauer, B., and Haugen, J. E. (1987). Test of a spectral limited area shallow water model with time dependent lateral boundary conditions and combined normal mode semi-Lagrangian time integration schemes. In: *Proceedings of the ECMWF Workshop on Techniques for Horizontal Discretization in Numerical Prediction Models*, Reading, Nov. 1987. Reading: European Centre for Medium-range Weather Forecasts, 193–200.

Malardel, S. (2005). *Fondamentaux de Météorologie: A l'École du Temps*. Chapitre 14. Toulouse, France: Cépaduès éditions, 305–56.

Malkmus, W. (1967). Random Lorentz band model with exponential-tailed S^{-1} intensity distribution function. *J. Opt. Soc. Amer.*, **57**, 323–29.

Manabe, S., and Strickler, R. F. (1964). On the thermal equilibrium of the atmosphere with a convective adjustment. *J. Atmos. Sci.*, **21**, 361–85.

Marshall, J. S., and Palmer W. M. K. (1948). The distribution of raindrops with size. *J. Meteor.*, **5**, 165–66.

Mašek, J. (2006). *Influence of cloud geometry on saturation effects*. CHMI ALARO Working Paper. Prague: Czech Hydrometeorological Institute.

Masson, V. (2000). A physically-based scheme for the urban energy budget in atmospheric models. *Bound. Layer Meteor.*, **94**, 357–97.

Masson, V., and Seity, Y. (2009). Including atmospheric layers in vegetation and urban offline surface schemes. *J. Appl. Met. Clim.*, **48**, 1377–97.

Mesinger, F., and Arakawa, A. (1976). *Numerical Methods used in Atmospheric Models*. Vol. I, GARP Publications Series No. 17, ICSU/WMO. Geneva: World Meteorological Organization.

Mlawer, E. J., Taubman, S. J., Brown, P. D., Iacono, M. J., and Clough, S. A. (1997). Radiative transfer for inhomogeneous atmospheres: RRTM, a validated correlated k-model for the longwave. *J. Geophys. Res.*, **102**, 16,663–82.

Monin, A. S., and Obukhov, A. M. (1954). Basic laws of turbulent mixing in the ground layer of the atmosphere. *Trans. Geophys. Inst. Akad. Nauk. USSR*, **151**, 163–87.

Morcrette, J. (1991). Radiation and cloud radiative properties in the ECMWF forecasting system. *J. Geophys. Res.*, **96D**, 9121–32.

Morcrette, J. J., and Fouquart, Y. (1986). The overlapping of cloud layers in shortwave radiation parameterizations. *J. Atmos. Sci.*, **43**, 321–28.

Müller, R. (1989). A note on the relation between the traditional approximation and the metric of the primitive equations. *Tellus*, **41A**, 175–78.

Mylne, K. R., Evans, R. E., and Clark, R. T. (2002). Multi-model multi-analysis ensembles in quasi-operational medium-range forecasting. *Quart. J. Roy. Meteor. Soc.*, **128**, 361–84.

Noilhan, J., and Planton, S. (1989). A simple parameterization of land surface processes for meteorological models. *Mon. Wea. Rev.*, **117**, 536–49.

Noilhan, J., and Mahfouf, J.-F. (1996). The ISBA land surface parameterization scheme. *Global and Planet. Change*, **13**, 145–59.

Orlanski, I. (1975). A rational subdivision of scales for atmospheric processes. *Bull. Am. Meteor. Soc.*, **56**, 527–30.

Orlanski, I. (1975). A simple boundary condition for unbounded hyperbolic flows. *J. Comput. Phys.*, **21**, 251–69.

Orszag, S. A. (1970). Transform method for the calculation of vector-coupled sums. Application to the spectral form of the vorticity equation. *J. Atmos. Sci.*, **27**, 890–95.

Pailleux, J., Geleyn, J.-F., and Legrand, E. (2000). La prévision numérique du temps avec les modèles Arpège et Aladin. Bilan et perspectives. *La Météorologie*, 8e série, **30**, 32–60.

Palmer, T. N. (2001). A nonlinear dynamical perspective on model error. A proposal for non-local stochastic-dynamic parameterization in weather and climate prediction models. *Quart. J. Roy. Meteor. Soc.*, **127**, 279–304.

Palmer, T. N., Shutts, G. J., and Swinbank, R. (1986). Alleviation of systematic westerly bias in general circulation and numerical weather prediction models through an orographic gravity wave drag parameterization. *Quart. J. Roy. Meteor. Soc.*, **112**, 2056–66.

Parrish, D. F., and Derber, J. C. (1992). The National Meteorological Center's Spectral Statistical Interpolation analysis system. *Mon. Wea. Rev.*, **120**, 1747–63.

Pergaud, J., Masson, V., Malardel, S., and Couvreux, F. (2009). A parameterization of dry thermals and shallow cumuli for mesoscale numerical weather prediction. *Bound. Layer Meteor.*, **132**, 83–106.

Perkey, D. J., and Kreitzberg, W. (1976). A time-dependent lateral boundary scheme for limited area primitive equation models. *Mon. Wea. Rev.*, **104**, 744–55.

Persson, A. O. (1991). Kalman filtering. A new approach to adaptive statistical interpretation of numerical meteorological forecasts. In: *Lectures presented at the WMO Training Workshop on the Interpretation of NWP Products in Terms of Local Weather Phenomena and their Verification*, Wageningen, 29 July–9 Aug. 1991, PSMP Report Series No. 34, WMO Technical Document No. 421, Geneva: World Meteorological Organization, XX.27–32.

Phillips, N. A. (1957). A coordinate system having some special advantages for numerical forecasting. *J. Meteor.*, **14**, 184–95.

Phillips, N. A. (1959). An example of non-linear computational instability. In: *The Atmosphere and the Sea in Motion*, Rossby Memorial Volume, edited by Bolin, B. New York: Rockefeller Institute Press and Oxford University Press, 501–04.

Phillips, N. A. (1966). The equations of motion for a shallow rotating atmosphere and the 'traditional approximation'. *J. Atmos. Sci.*, **25**, 1154–55.

Phillips, N. A., and Shukla, J. (1973). On the strategy of combining coarse and fine grid meshes in numerical weather prediction. *J. Appl. Met.*, **12**, 736–70.

Phillips, O. M. (1960). The dynamics of unsteady gravity waves of finite amplitude. Part I. *J. Fluid Mech.*, **9**, 193–217.

Pinty, J.-P., and Jabouille, P. (1998). A mixed-phase cloud parameterization for use in mesoscale non-hydrostatic model: simulations of a squall line and of orographic precipitations. In: *Proceedings of the AMS Conference on Cloud Physics*, Everett, WA, USA, Aug. 1999. Boston: American Meteorological Society, 217–20.

Piriou, J.-M. (2005). *Représentation de la convection dans les modèles globaux et régionaux: concepts, équations, tudes de cas*. Unpublished thesis, Université Paul Sabatier, Toulouse.

Piriou, J.-M., Redelsperger, J.-L., Geleyn, J.-F., Lafore, J.-P., and Guichard, F. (2007). An approach of convective parameterization, with memory, in separating microphysics and transport in grid-scale equations. *J. Atmos. Sci.*, **64**, 4127–39.

Plant, R. S. (2010). A review of the theoretical basis for bulk mass flux convective parameterization. *Atmos. Chem. Phys.*, **10**, 3529–44.

Platzman, G. W. (1979). The ENIAC computation of 1950. Gateway to numerical weather prediction. *Bull. Amer. Meteor. Soc.*, **60**, 302–12.

Pône, R. (1993). Les débuts de l'informatique à la division 'Prévision' de la Météorologie Nationale. *La Météorologie*, 8ᵉ série, **3**, 36–43.

Prandtl, L. (1932). Meteorologische Anwendungen der Stromungslehre. *Beitr. Phys. Atmos.*, **19**, 188–202.

Pudykiewicz, J., Benoit, R., and Staniforth, A. (1985). Preliminary results from a partial LRTAP model based on an existing meteorological forecast model. *Atmosphère-Océan*, **23**, 267–303.

Rabier, F., Mahfouf, J.-F., and Klinker, E. (2000). Une nouvelle technique d'assimilation des données d'observation au CEPMMT: l'assimilation variationnelle quadridimensionnelle. *La Météorologie*, 8ᵉ série, **30**, 87–101.

Radnóti, G. (1995). Comments on 'A spectral limited-area formulation with time-dependent boundary conditions for the shallow-water equations'. *Mon. Wea. Rev.*, **123**, 3122–23.

Rasch, P. J. (1985). Developments in normal mode initialization. Part II : a new method and its comparison with currently used schemes. *Mon. Wea. Rev.*, **113**, 1753–70.

Richardson, L. F. (1922). *Weather Prediction by Numerical Process*. Cambridge: Cambridge University Press. Republication (1965), New York: Dover. 2nd edn. (2007), Cambridge: Cambridge University Press.

Richtmyer, R. T., and Morton, W. (1967). *Difference Methods for Initial-value Problems*, 2nd edn. New York: John Wiley & Sons. Reprint of the 2nd edn. (1994), Malabar, FL, USA: Krieger R. E. Publ. Co. Inc.

Ritchie, H. (1986). Eliminating the interpolation associated with the semi-Lagrangian scheme. *Mon. Wea. Rev.,* **114**, 135–46.

Ritchie, H. (1988). Application of the semi-Lagrangian method to a spectral model of the shallow water equations. *Mon. Wea. Rev.*, **116**, 1587–1598.

Ritchie, H., and Tanguay, M. (1996). A comparison of spatially Eulerian and Semi-Lagrangian treatment of mountains. *Mon. Wea. Rev.*, **124**, 167–81.

Ritter, B., and Geleyn, J.-F. (1992). A comprehensive radiation scheme for numerical weather prediction models with potential applications in climate simulations. *Mon. Wea. Rev.*, **120**, 303–25.

Rivest, C., Staniforth, A., and Robert, A. (1994). Spurious resonant response of semi-Lagrangian discretization to orographic forcing: Diagnosis and solution. *Mon. Wea. Rev.,* **122**, 366–76.

Robe, F. R., and Emanuel, K. A. (1996). Moist convective scaling. Some inferences from three-dimensional cloud ensemble simulations. *J. Atmos. Sci.*, **53**, 3265–75.

Robert, A. J. (1966). The integration of a low-order spectral form of the primitive equations. *J. Meteor. Soc. Japan*, **44**, 237–45.

Robert, A. J. (1969). The integration of a spectral model of the atmosphere by the implicit method. In: *Proceedings of the WMO/IUGG Symposium on Numerical Weather Prediction*, Tokyo, Nov. 26–Dec. 4, 1968, Tokyo: Meteorological Society of Japan, VII.19–24.

Robert, A. J. (1981). A stable numerical integration scheme for the primitive meteorological equations. *Atmosphère-Océan*, **19**, 35–46.

Robert, A. J. (1983). The design of efficient time integration schemes for the primitive equations. In: *Proceedings of the ECMWF Seminar on Numerical Methods for Weather Prediction*, Vol. 2, Reading, Sept. 83. Reading: European Centre for Medium-range Weather Forecasts, 193–200.

Rochas, M. (1990). *ARPEGE Documentation*, Part 2, Chapter 6. Toulouse: Météo-France.

Rochas, M., and Courtier, P. (1992). *La méthode spectrale en météorologie*. Note de travail ARPEGE No. 30. Toulouse: Météo-France.

Rochas, M., and Javelle, J.-P. (1993). *La Météorologie – La prévision numérique du temps et du climat*. Paris: Syros, 165–82.

Rodgers, C. D., and Walshaw, C. D. (1966). The computation of infrared cooling rate in planetary atmosphere. *Quart. J. Roy. Meteor. Soc.*, **92**, 67–92.

Rossby, C. G. (1939). Relation between variations in the intensity of the zonal circulation of the atmosphere and the displacements of the semi-permanent centers of action. *Journal of Marine Research (Sears Foundation)*, **2**, 38–55.

Rousseau, D. (1980). A new skill score for the evaluation of yes/no forecasts. In: *Proceedings of the WMO Symposium on Probabilistic and Statistical Methods in Weather Forecasting*, Nice, France, Sept. 1980. Geneva: World Meteorological Orgzanisation, 167–74.

Rousseau, D., and Chapelet, P. (1985). A test of the Monte-Carlo method using the WMO/CAS Intercomparison Project data. In: *Report of the second session of the CAS Working Group on short and medium range weather prediction research*, Belgrade, 26–30 Aug. 1985. PSMP Report Series No. 18, WMO Technical Document No. 421. Geneva: World Meteorological Organization, 53–58.

Rousseau, D., Pham, H. L., and Juvanon du Vachat, R. (1995). Vingt-cinq ans de prévision numérique du temps à échelle fine (1968–1993). De l'adaptation dynamique à maille fine au modèle Péridot. *La Météorologie*, 8e série, Special Issue, April 1995.

Royer, J.-F. (1976). Application d'un modèle d'interpolation statistique itérative au calcul du vent en surface. In: *Comptes rendus des conférences données aux XIVe Journées de l'Hydraulique*, Paris, 1976. Paris: Société Hydrotechnique de France.

Sadourny, R. (1975). The dynamics of finite difference models of the shallow water equations. *J. Atmos. Sci.*, **32**, 680–89.

Sadourny, R., and Maynard, K. (1997). Formulations of lateral diffusion in geophysical fluid dynamics models. In: *Numerical Methods in Atmospheric and Oceanic Modeling*, edited by Lin, C. A., Laprise, R. and Ritchie, H. Ottawa, Canada: NRC Research Press, 547–56.

Sawyer, J. S. (1963). A semi-Lagrangian method of solving the vorticity equation. *Tellus*, **15**, 336–42.

Schmidt, F. (1977). Variable fine mesh in the spectral global models. *Beitr. Phys. Atmos.*, **50**, 211–17.

Scinocca, J. F., and McFarlane, N. A. (2000). The parameterisation of drag induced by stratified flow over anisotropic orography. *Quart. J. Roy. Meteor. Soc.*, **126**, 2353–93.

Seity, Y., Brousseau, P., Malardel, S., Hello, G., Bénard, P., Bouttier, F., Lac, C., and Masson, V. (2011). The AROME-France convective scale operational model. *Mon. Wea. Rev.*, **139**, 976–91.

Shapiro, M. A., and Thorpe, A. J. (2004). *THORPEX International Science Plan, Version 3*. WWRP/THORPEX No. 2, WMO Technical Document No. 1246. Geneva: World Meteorological Organization.

Shuman, F. G. (1957). Numerical methods in weather prediction II. Smoothing and filtering. *Mon. Wea. Rev.*, **85**, 357–61.

Shuman, F. G. (1962). Numerical experiments with the primitive equations. In: *Proceedings of the International Symposium on Numerical Weather Prediction*, Tokyo, Nov. 1960. Tokyo: Meteorological Society of Japan, 85–107.

Shuman, F. G. (1989). History of numerical weather prediction at the National Meteorological Center. *Mon. Wea. Rev.*, **4**, 286–96.

Shuman, F. G., and Hovermale, J. B. (1968). An operational six-layer primitive equation model. *J. Appl. Meteor.*, **7**, 525–47.

Siebesma, A. P., Bretherton, C. S., Brown, A., Chlond, A., Cuxart, J., Duynkerke, P. G., Jiang, H., Khairoutdinov, M., Lewellen, D., Moeng, C. H., Sanchez, E., Stevens, B., and Stevens, D. (2003). A large-eddy simulation intercomparison study of shallow cumulus convection. *J. Atmos. Sci.*, **60**, 1201–19.

Silberman, I. (1954). Planetary waves in the atmosphere. *J. Atmos. Sci.*, **11**, 27–34.

Simmons, A. J., Hoskins, B. J., and Burridge, D. M. (1978). Stability of the semi-implicit method of time integration. *Mon. Wea. Rev.*, **106**, 405–12.

Simmons, A. J., and Burridge, D. M. (1981). An energy and angular momentum conserving vertical finite-difference scheme and hybrid vertical coordinates. *Mon. Wea. Rev.*, **109**, 758–66.

Simonsen, C. (1991). Self adaptive model output statistics based on Kalman filtering. In: *Lectures presented at the WMO Training Workshop on the Interpretation of NWP Products in Terms of Local Weather Phenomena and their Verification*, Wageningen, 29 July–9 Aug. 1991, PSMP Report Series No. 34, WMO Technical Document No. 421. Geneva: World Meteorological Organization, XX.33–37.

Skamarock, W. C., and Weisman, M. L. (2008). The impact of positive-definite moisture transport on NWP precipitation forecasts. *Mon. Wea. Rev.*, **136**, 488–94.

Skamarock, W. C., Klemp, J. B., Dudhia, J., Gill, D. O., Barker, D. M., Duda, M., Huang, X.-Y., Wang, W., and Powers, J. G. (2008). *A description of the Advanced Research WRF* version 3, NCAR Technical Note No. 475, NCAR, Boulder, CO, USA, Internet address: http://www.mmm.ucar.edu/wrf/users/docs/arw_v3.pdf

Smagorinsky, J. (1958). On the numerical integration of the primitive equations of motion for baroclinic flow in a closed region. *Mon. Wea. Rev.*, **86**, 457–66.

Smagorinsky, J. (1962). A primitive equation model including condensation processes. In: *Proceedings of the International Symposium on Numerical Weather Prediction*, Tokyo, 7–13 Nov. 1960. Tokyo: Meteorological Society of Japan, Tokyo, 555.

Smagorinsky, J. (1963). General circulation experiments with primitive equations. 1-The basic experiment. *Mon. Wea. Rev.*, **91**, 99–164.

Smith, W. L. (1968). An improved method for calculating tropospheric temperature and moisture from satellite radiometer measurements. *Mon. Wea. Rev.*, **96**, 387–96.

Soares, P. M. M., Miranda, P. M. A., Siebesma, A. P., and Teixeira, J. (2004). An eddy-diffusivity/mass-flux parameterization for dry and shallow cumulus convection. *Quart. J. Roy. Meteor. Soc.*, **130**, 3055–79.

Staniforth, A. H., and Daley, R. W. (1977). A finite-element formulation for the vertical discretization of sigma-coordinate primitive equation models. *Mon. Wea. Rev.*, **105**, 1108–18.

Staniforth, A. N., and Mitchell, H. L. (1978). A variable-resolution finite-element technique for regional forecasting with the primitive equations. *Mon. Wea. Rev.*, **106**, 439–47.

Stanski, H. R., Wilson, L. J., and Burrows, W. R. (1989). In: *Survey of Common Verification Methods in Meteorology*, WWW Technical Report No. 8, WMO Technical Document No. 358. Geneva: World Meteorological Organization.

Stephens, G. L. (1984). The parameterization of radiation for numerical weather prediction and climate models. *Mon. Wea. Rev.*, **112**, 826–67.

Sundquist, H. (1978). A parameterization scheme for non-convective condensation including precipitation including prediction of cloud water content. *Quart. J. Roy. Meteor. Soc.*, **104**, 677–90.

Talagrand, O. (1997). Assimilation of observations, an introduction. *J. Meteor. Soc. Japan*, **75**, 191–209.

Talagrand, O., and Courtier, P. (1987). Variational assimilation of meteorological observations with the adjoint vorticity equation. I – Theory. *Quart. J. Roy. Meteor. Soc.*, **113**, 1311–28.

Tanguay, M., Robert, A., and Laprise, R. (1990). A semi-implicit fully compressible regional forecast model. *Mon. Wea. Rev.*, **118**, 1970–80.

Tanguay, M., Yakimiw, E., Ritchie, H., and Robert, A. (1992). Advantages of spatial averaging in semi-implicit semi-Lagrangian schemes. *Mon. Wea. Rev.*, **120**, 113–23.

Tatsumi, Y. (1986). A spectral limited-area model with time dependent lateral boundary conditions and its application to a multi-level primitive equation model. *J. Meteor. Soc. Japan*, **64**, 637–63.

Temperton, C. (1988). Implicit normal mode initialization. *Mon. Wea. Rev.*, **116**, 1013–31.

Temperton, C. (1997). Treatment of the Coriolis terms in semi-Lagrangian spectral models. In: *Atmospheric and Ocean Modelling. The André J. Robert Memorial Volume*, Canadian Meteorological and Oceanographic Society, Ottawa, Canada, 293–302.

Temperton, C., and Staniforth, A. (1987). An efficient two-time-level semi-Lagrangian semi-implicit integration scheme. *Quart. J.. Roy. Meteor. Soc.*, **113**, 1025–39.

Teweles, S., and Wobus, H. (1954). Verification of prognostic charts. *Bull. Am. Met. Soc.*, **35**, 286–96.

Tiedtke, M. (1989). A comprehensive mass flux scheme for cumulus parameterization in large-scale models. *Mon. Wea. Rev.*, **117**, 1779–1800.

Toth, Z., and Kalnay, E. (1993). Ensemble forecasting at NMC: the generation of perturbations. *Bull. Am. Met. Soc.*, **74**, 2317–30.

Toth, Z., Talagrand, O., Candille, G., and Zhu, Y. (2003). Probability and ensemble forecasts. Forecast Verification. In: *A Practitioner's Guide in Atmospheric Science*, edited by Jolliffe, I. and Stephenson, D. B. Chichester: John Wiley & Sons, 137–63.

Triplet, J. P., and Roche, G. (1971). *Météorologie Générale*. New edns. (1988, 1996). Trappes, France: Météo-France.

Váňa, F., Bénard, P., Geleyn, J.-F., Simon, A., and Seity, Y. (2008). Semi-Lagrangian advection scheme with controlled damping: an alternative to nonlinear horizontal diffusion in a numerical weather prediction model. *Quart. J. Roy. Meteor. Soc.*, **134**, 523–37.

Van de Hulst, H. C. (1980). *Multiple Light Scattering*. New York: Academic Press.

Wicker, L. J., and Skamarock, W. C. (2002). Time splitting methods for elastic models using forward time schemes. *Mon. Wea. Rev.*, **130**, 2088–97.

Wilks, D. S. (1995). *Statistical Methods in the Atmospheric Sciences*, 2nd edn. (2006). Burlington MA, USA: Academic Press.

Williamson, D. L. (1976). Normal mode initialization procedure applied to forecasts with the global shallow water equations. *Mon. Wea. Rev.*, **104**, 195–206.

Winninghoff, F. J. (1968). *On the adjustment toward a geostrophic balance in a simple primitive equation model with application to the problems of initialization and objective analysis*. Unpublished Ph.D. thesis, University of California, Los Angeles.

Wolff, P. M. (1958). The error in numerical forecasts due to retrogression of ultra-long waves. *Mon. Wea. Rev.*, **86**, 245–50.

Woodcock, F. (1976). The evaluation of yes/no forecasts for scientific and administrative purpose. *Mon. Wea. Rev.*, **104**, 1209–14.

Xu, K. M., and Randall, D. A. (1996). Explicit simulation of cumulus ensembles with the GATE Phase III data. Comparison with observations. *J. Atmos. Sci.*, **53**, 3710–36.

Yanai, M., Esbensen, S., and Chu, J. H. (1973). Determination of bulk properties of tropical cloud clusters from large-scale heat and moisture budgets. *J. Atmos. Sci.*, **30**, 611–27.

Zilitinkevich, S. S., Elperin, T., Kleeorin, N., Rogachevskii, I., Esau, I., Mauritsen, T., and Miles, M. W. (2008). Turbulence energetics in stably stratified geophysical flows: strong and weak mixing regimes. *Quart. J. Roy. Met. Soc.*, **134**, 793–99.

Index

absorbing layer, 309
acoustic time step, 301
adaptation terms, 47
advection terms, 47
aliasing, 71, 72, 78, 110, 113, 187
angular momentum, 17, 152, 160
approximation
  Eddington, 201
  geostrophic, 107
  hydrostatic, 15
  quasi-geostrophic, 2, 6
  thin layer, 15, 17, 18, 25, 152
  traditional, 17
  two flux, 201, 205
asymptotic mixing length, 218

Bjerknes, 2
boundary layer, 213–19, 267, 293
  atmospheric, 213
  planetary, 213, 217, 311
  surface, 213, 214, 311
  system closure, 218
Brunt-Väisälä frequency, 238
buoyancy, 236

CFL condition, 4, 6, 9, 46, 51, 90, 103, 108, 119,
    121, 124, 174, 179, 187, 190, 266
Charney, 2, 15, 23, 106, 107, 166
cloudiness, 208–13
coefficient
  absorption/scattering, 206
  entrainment, 236
  evaporation, 230
  exchange, 217, 219, 223, 307
  Fourier, 71, 128, 186
  horizontal diffusion, 242
  interaction, 61
  molecular diffusion, 228
  normal mode decomposition, 261
  snow melting, 229
  soil thermal conductibility, 221
  spectral, 63, 128, 186
  thermal diffusion, 228
  transfer, 214, 215, 216
computational solution, 91, 92, 95, 98, 102, 103
constant
  soil, 221

solar, 202
Stefan-Boltzmann, 199, 220
von Karman, 214
convection, 193, 211, 219, 231–37, 295, 312
  shallow, 219, 240, 295
  system closure, 236
coordinates
  Cartesian, 32, 33, 74, 167
  curvilinear, 24, 27
  geographic, 25, 127
Coriolis
  acceleration, 17
  force, 213
  parameter, 17, 23, 47, 49, 79, 84, 86, 87, 92, 96,
      98, 108, 116, 119, 125, 134, 241, 262
  terms, 47, 126
curl, 24

damping
  implicit Rayleigh, 309
  traditional Rayleigh, 310
data assimilation, 11, 244, 249, 250–58, 263, 271,
    275, 278, 281, 282, 295, 312
difference
  backward, 40
  central, 40, 42, 44, 45, 82–87, 159, 162, 173, 179,
      186, 223, 264
  forward, 40, 87, 174
diffusion
  horizontal, 241–43, 264, 292, 307
  vertical, 217–19, 223–25
discretization
  horizontal, 291
  spatial, 304
  vertical, 288
divergence, 16, 22, 24, 26, 109, 121, 125, 134, 138,
    158, 169, 175, 183
  discretized, 116, 119
  horizontal, 18
  mean, 107, 108, 109, 139, 179
  term, 23
  vertical, 160, 162, 198, 223, 241

eigenfunction, 57, 61, 67, 74, 260
Eliassen-Palm theory, 238
ENIAC, 2, 3, 12
ensemble forecasting, 245, 276–78, 281, 283

337

Printed in the United States
by Baker & Taylor Publisher Services